THE GRILL BIBLE AND SMOKER COOKBOOK 2021

For Real Pitmasters. Amaze Your Friends with 550 Sweet and Savory Succulent Recipes That Will Make You the MASTER of Smoking Food | INCLUDING DESSERTS

MICHAEL BLACKWOOD

Table of Contents

- PART 1 -

WOOD PELLET SMOKER COOKBOOK

The Wood Pellet Smoker-Grill utilizes wood pellets, which makes temperature and flavor control easier when smoking, grilling, or roasting. The ease of use has made this smoker-grill popular all around the globe.

Each Wood Pellet Smoker-Grill contains a storage hopper. This storage hopper is the place where you add all of the wood pellets. The equipment takes care of the transfer of wood pellets from the storage hopper to the burning area in the correct quantity.

Thus, you get a perfect temperature for the type of cooking approach you are following. The rate of pellet burning increases when you are grilling, and it decreases when you set the smoker-grill at a low temperature. This helps to smoke your food for a long time with a consistent heat.

It only takes about 10–20 minutes to heat and get the smoker-grill ready for cooking. The preheating process usually takes about 15 minutes. This makes cooking efficient and easy for everyone. You can pick any time and work on some delicious recipes.

Internal Design of the Wood Pellet Smoke and Grill

Modern designs of Wood Pellet Smoker-Grills contain electronic functionality. Advanced design allows smoker-grills to manage temperature control and pellets on its own. The wood pellets are transferred to the burning area according to the cooking setting you provide. The one push of a button, you can allow the smoker-grill to take care of the temperature consistency and the flavors of the food inside.

Cooking Options with the Wood Pellet Smoke and Grill

This smoker-grill brings a whole new level of versatility to your cooking. You get more than six ways to cook different kinds of foods. For example, you can bake fish fillets, grill meat, and smoke as well. All in all, the smoker-grill allows you to smoke and grill indirectly and directly along with baking and roasting. This versatile approach to cooking makes this smoker-grill suitable for a variety food options, including chicken, turkey, beef, lamb, pork, and seafood.

Use of Different Wood Pellets In the Wood Pellet Smoke and Grill

Different types of wood pellets, such as apple, cherry, hickory, mesquite, and others, are used to obtain specific flavors in foods. Each type of wood pellet is considered suitable for certain types of foods. Knowing this is critically important so that you can get the best flavors out of your cooking.

1. Apple wood pellets are generally used when the food's main ingredient is pork, chicken, or vegetables.
2. Cherry wood pellets are perfect for baking food, including pork, lamb, chicken, and beef.
3. Hickory wood pellets make pork, beef, vegetables, and even poultry exceptionally delicious.

Along with these three types, there are other wood pellet options, such as alder, maple, mesquite, pecan, and oak. Pork dishes can get the best flavors with almost all kinds of wood pellets except oak and mesquite. Oak, alder, and mesquite types are more effective when you want to cook fish, shrimp, or other kinds of seafood.

The Wood Pellet Smoke and Grill is a durable and cost-effective option for anyone who wants to smoke or grill without worrying all the time. Because of its quality of construction, it works effectively for a long time. You only need to spend a few minutes after cooking to maintain its cleanliness. This keeps the fuel efficiency high and allows for controlled wood pellet burning.

History of the Wood Pellet Smoke and Grill

The very first Wood Pellet Smoker-Grill was introduced in 1985. Joe Traeger was the man behind the concept and the construction of the Wood Pellet Smoker-Grill. After spending a year creating his smoker-grill, he obtained a patent and started production at a commercial level. The smoker-grill looked similar to traditional smokers in terms of its exterior design. There was a drum barrel and a chimney. But the internal components were the true magic. Traeger divided the internal design into three parts. These three parts were the sections where wood pellets had to go in order to get burned.

The storage hopper was the first part, which worked as storage for the wood pellets in the smoker-grill. Then, the next stop for the pellets was the auger, which was a rotating section. This rotation allowed wood pellets to reach the third and final section. This final section was called a firebox or burning box. In this area, a fan allowed the proper distribution of the cultivated heat.

In the early designs, the user had to light the smoker-grill manually. However, the design got updated with time, and now, there are completely automatic Wood Pellet Smoker-Grills available.

The reduction in wood pellet size revolutionized the whole smoking and grilling process. The machine obtained the ability to balance the temperatures on its own for as long as required. This convenience wasn't available with charcoal burning smokers. At the same time, wood pellets also provided more variety based on the flavorful hardwood choices available.

That would not be wrong to say that the BBQ world experienced a revolution with the introduction of the Wood Pellet Smoker-Grill. Cooking got simpler and more comfortable, which gave even newbies a chance to smoke, grill, bake, and roast. The machine was capable of handling the temperature on its own, so the users could be stress-free and safe when cooking. In 2007, after the expiration of Traeger's patent, the Wood Pellet Smoker-Grill market opened for more advanced options. This led to more advancements and automation in the equipment.

Benefits of the Wood Pellet Smoker-Grill

1. Flavorful food

You work on your cooking techniques to get the best flavors possible. However, the techniques alone can't do it all. You need the right kind of equipment to get the correct flavor in your cooked dish. This is why the Wood Pellet Smoker-Grill is considered the best choice in the world of BBQ. The wood pellet flavors, such as cherry, apple, mesquite, and hickory, give different kinds of smoky flavors to the food. This flavor is way better than getting a charcoal aroma or a gaseous aroma when using other types of smokers. The natural flavor in your food enhances the deliciousness.

2. Ease of use

The modern-age designs of this smoker-grill alleviate all stress. You need to click a single button in order to begin the cooking process. The management of wood pellets happens is taken care of by the smoker, so you get the desired fire quality according to the kind of cooking process you want. Hence, barbeque becomes an easy task for everyone.

3. Different smoke temperatures

Since Wood Pellet Smoker-Grills can burn wood pellets in a variety of ways, you need different temperature levels for different processes. The smoke temperature options can range from 180°F to a maximum of 500°F. This wide temperature range makes one machine capable of all kinds of cooking, including smoking, grilling roasting, baking, and searing. You can pick any kind of meat and cook it to your desired level.

4. Temperature consistency

Unlike traditional smokers, Wood Pellet Smoker-Grills offer the consistent temperature you need for grilling or smoking. The wood pellets keep on reaching the burning section as required. This creates and maintains the same temperature during the whole cooking process.

Let's get to the recipes you can cook in your Wood Pellet Smoker-Grill!

WHY A WOOD PELLET SMOKE AND GRILL?

There is nothing more popular in the market nowadays than Pellet Smoker and Grills. And while a very few people claim that the popularity of pellet Smoker and Grill stems from its increase of use and the outstanding marketing for this product, the majority of people agree upon the fact that Pellet Smoker and Grills are acquiring its unrivaled popularity thanks to the effectiveness of this product.

Unlike any traditional grills people could have used in the past, Pellet Smoker and Grills are one of the most versatile, automated and perfect-to use revolutionary grills that one can rely on to get the flavor you dream of tasting. Pellet Smoker and Grills just make the perfect choice and the one and only best solution to cook any type of meat in a healthy way. Not only Pellet Smoker and Grills allow smoking ingredients, but it also allows a slow roasting process, baking a pizza or even perfectly grilling steak. However, this new revolutionary appliance is still not known by many people; so what is a Wood Pellet Smoker and Grill? So how can we use a Pellet Smoker and Grill?

What Is A Pellet Smoker And Grill?

To provide you with a clear answer about the Wood Pellet Smoker and Grill, let us start by defining this grilling appliance. In fact, Pellet Smoker and Grills can be defined as an electric outdoor Smoker and Grill that is only fueled by wood pellets. Wood Pellet is a type of fuel that is characterized by its capsule size and is praised for its ability to enhance more flavors and tastes to the chosen smoked meat. And what is unique and special about wood pellet as a fuel is that it can grill, smoke, roast, braise and even bake according is easily to follow instructions. I equipped with a control board that allows you to automatically maintain your desired temperature for several hours

Why Choose To Use A Wood Pellet Smoker And Grill?

The uniqueness of Pellet Smoker and Grills lies in the combination of the flavor and versatility it offers. Accurate, Pellet Smoker and Grills make an explosive mixture of sublime tastes and incredible deliciousness; it is a great Smoker and Grill appliance that you can use if you want to enjoy the taste of charcoal grill and at the same time you don't want to give up on the traditional taste of ovens. And what is more interesting about pellet Smoker and Grills is that, with a single button, you can grill, roast, bake, braise and smoke, your favorite meat portions. And things can still get better as pellet Smoker and Grills are automatic, so you can just set the temperature of pellet Smoker and Grill and walk away; then when you are back, you will be able to enjoy great flavors you are craving for. But how can we use a Pellet Smoker and Grill?

How to Use a Wood Pellet Smoker and Grill?

Pellet Smoker and Grills function based on an advanced digital technology and many mechanical parts. The pellet Smoker and Grill are then lit while the temperature is usually programmed with the help of a digital control board. Pellet Smoker and Grills work by using algorithm so that it allows calculating the exact number of pellets you should use in order to reach the perfect temperature. Every Wood Pellet Grill is equipped with a rotating auger that allows to automatically feed the fire right from the hopper to the fire in order to maintain the same temperature. And even as the food continues cooking, the wood pellet Smoker and Grill will continue to drop the exact amount of pellets needed to keep the perfect cooking temperature. But what can we cook with a pellet Smoker and Grill.

What Dishes Can We Can On A Wood Pellet Smoker And Grill?

Thanks to its versatile properties to smoke, grill, braise and kitchen oven, Wood Pellet Smoker and Grill can be used to cook endless dishes and recipes. In fact, there is no actual limit of the recipes you can cook like hot dogs, chicken, vegetables, seafood, rabbit, chicken, brisket, turkey and even more. The cooking process is very easy; all you have to do is to pack your favorite wood pellets into the hopper; then program the temperature you desire on the controller; then place the food on the pellet Smoker and Grill. That is it and the pellet will be able to maintain the temperature and keep the pellets burning.

Pellet Smoker and Grills are characterized by being electric and it requires a usual standard outlet of about 110v so that you can power the digital board, fan and auger.

There is a wide variety of types of pellet smoker and grills, like electric pellet smokers, wood fired grills, wood pellet grills and wood pellet smokers; to name a few pellet Smoker and Grill names. But all these names refer to the same outdoor cooker appliance that is only fueled by hardwood pellets. And there are many brands of Wood Pellet Smoker and Grills, like Traeger.

For instance, Traeger is known for being one of the world's most well-known brands of pellet grills. Indeed, Joe Traeger was the person who invented the pellet grill during the mid-1980s and he gave his name to this invention. And when the Traeger patent expired, many other Pellet Smoker and Grills came to life into the market.

Advantages of Wood Pellet Smoker and Grill

Roasting and grilling meat make two cooking methods that play an important role in packing meat with flavorful tastes. And not only grilling makes a healthier cooking technique, but it can also help retain nutrients within the meat portions. And there is no better and useful machine more than the wood pellet Smoker and Grill that you can use to replace the traditional grilling machine known as barbecue. Indeed, Wood Pellet Smoker and Grill have too many advantages. And here are some of the most well-known benefits of using a Pellet Smoker and Grill:

1. Wood Pellet Smoker and Grills are designed to enhance more flavors and

aromas by using as fuel only wood pellets. And by following this method, we can enjoy flavors that we love with a very few ingredients and hardwood pellets.

2. Useful and versatile. Using a wood pellet Smoker and Grill can help you cook a wide variety of dishes and one of the most well-known advantages of this cooking appliance is its versatility. Indeed, wood pellet Smoker and Grills allow cooking different types of ingredients from braised short ribs to chicken drumsticks and wings.
3. Fast and convenient. Have you ever wanted to find a convenient, fast and effortless cooking appliance or technique that can save your time and effort? Indeed, the pellet Smoker and Grill makes a great choice when it comes to saving time and it deserves to give it a try. The idea behind using a wood pellet smoker cooker stems from its great popularity and pellet Smoker and Grills are quickly preheated; thus it can save you so much time.
4. Pellet Smoker and Grill and Temperature regulation. Using a Pellet Smoker and Grill allows you to monitor the temperature of the inner chamber better by regulating it. And while it is very difficult to manage the level of temperature with traditional smokers and grills; pellet Smoker and Grill makes this mission easier. And pellet Smoker and Grills will help keep the temperature under constant check and to retain the delicious flavors of various ingredients.
5. Variety of pellets. Cooking experts with wooden Pellet Smoker and Grills have studied this cooking appliance very well and they even suggest using your favorite flavors. For instance, pellets exist in different flavors like maple, pecan and hickory wood.
6. Evenly-cooked ingredients. Wood Pellet Smoker and Grills offer a better way to get your food ingredients perfectly cooked from the outside to the inside alike. This appliance uses electricity to function and offers fast and even cooking without any difficulties. Besides, wood pellet Smoker and Grills are equipped with a heat diffuser plate that makes the cooking process easier.
7. Easy to clean. Wood pellet Smoker and Grills are equipped with a catch plate that is placed right under the machine and the function of this plate is to catch the drips during the cooking process. Thus, it becomes very easy to clean the wood pellet Smoker and Grill.

TEMPERATURE CONTROL

A pellet flame broil is a kind of barbecue that depends on tube-shaped hardwood sawdust pellets as fuel for the barbecuing. The sawdust is sourced from spots, for example, saw factories and timber yard. The wood pellet resembles a long pill and has a breadth of about ¼ inch. The small size of the pellets empowers them to consume neatly without leaving a great deal of fiery debris. A concoction called lignin will be discharged into the smoke when the wood pellets are copied and add a wood terminated flavor to the meat. Other than that, it doesn't contain some other added substance.

What are the Portions of a Pellet Grill?

The enlistment fan guarantees that the smoke from the hardwood pellet is coursed appropriately in the cook chamber. This ensures the flavor is circulated equally on the meat. The twist drill moves the wood pellets into the flame pot. The twist drill can move delayed for low-temperature cooking, or it can run at a quick rate for high-temperature baking.

The warmth diffuser transmits and scatters the warmth equitably on the cooking surface of the barbecue to guarantee that all zones of the meat are cooked well. The dribble dish is situated over the warmth diffuser, and it gets the oil that tumbles from the flame broil. The capacity container is the place you store the wood pellet energies. Topping the capacity container off to the edge keeps you from always having to return to refill it. The speedster will sparkle red and consume the pellets in smoke while you go to unwind and doing different things.

Points of Interest

You can cook practically any sort of meat on the pellet barbecue.

The flame broil can be preheated inside a period of 10 - 15 minutes.

You can set physically set the temperature on the advanced controller anyplace in the middle of 175° F to 500° F. Some pellet flame broils enables you to change the temperature by a 5 ° F increase.

Many pellet barbecues offer Bluetooth include that enables you to utilize a Bluetooth gadget to screen the cooking. It additionally accompanies a meat test for checking the cooking time.

Different sizes of pellet flame broils are accessible from family to business size units. The drunker the capacities, the more costly it is. The business size unit offers more spaces to flame broil meat as enormous overall hoard for a horde of individuals. A portion of the

leading brand names is Traeger, Yoder Smokers and Memphis Wood Fire Grills.
The wood pellets that fuel the barbecuing is accessible in a wide range of sorts of flavors including cherry, birch, apple, maple, whiskey and hickory. You can blend the pellets add more than one characters to the meat.
A single 20-pound sack of pellets is adequate for flame broiling the sustenance a few times. The flame broil will expend around 2 pounds of pellets consistently. Be that as it may, the real utilization of the pellets will rely upon different factors, for example, temperature and wind. In case you are fire cooking open air and there happens to be a great deal of wind, you should utilize more pellets to give the fuel.

Burdens

It is subject to power so it might be poorly arranged for you to flame broil the meat in a spot that doesn't have any electrical outlet close by. You have to connect it to a standard 110v electrical outlet in the house.
A pellet barbecue can be costly, and the littlest family unit will cost at any rate a couple of hundred dollars.
There will, in general, be secondary smoke when you set a higher temperature. The best temperature to cook is 250 degrees.

Extra Tips

It is savvy to contribute more cash forthright to purchase a quality pellet flame broil that will keep going for a long time. Pellet flame broils produced using 304 or 430 evaluation tempered steel is the best as they are safe against rust. To see whether a barbecue is a high quality, you can check whether it has an active development including equipment, joints and flame broil. Perusing surveys and posing inquiries on the discussion can assist you with making the correct choice.

BARBECUE RECIPES

1. Barbecue Pork Butt

Preparation Time: 10 minutes

Cooking Time: 5 hours

Servings: 8

Ingredients:

- 1 pork butt, boneless
- 2 tablespoons sweet dry rub
- 12 oz. dark beer
- 1 tablespoon olive oil
- 1/2 cup brown sugar
- 4 tablespoons honey
- 1 cup ketchup
- 4 tablespoons yellow mustard
- 2 tablespoons granulated garlic
- 2 tablespoons Worcestershire sauce

Directions:

1. Season the pork butt with the dry rub. Place it on a roasting pan. Add half of the beer to the pan.
2. Set the wood pellet grill to high. Add the pan to the grill. Grill for 30 minutes. Remove the pan from the grill.
3. Lessen the heat to 325 degrees F. In a bowl, mix the remaining ingredients. Pour this mixture on top of the roast. Cook for another 5 hours. Shred the pork. Coat with the barbecue sauce.

Nutrition: Calories: 192 Carbs: 0g Fat: 14g Protein: 20g

2. Cajun Barbecue Chicken

Preparation Time: 15 minutes
Cooking Time: 25 minutes
Servings: 4

Ingredients:

- 2 tablespoons sweet spicy dry rub
- 1/4 teaspoon ground thyme
- 1/2 teaspoon oregano
- 1 tablespoon olive oil
- 1 lb. chicken breast fillet
- 2 cloves garlic clove, minced
- 1/2 cup barbecue sauce
- 1 tablespoon butter
- 1/4 cup beer
- 1 tablespoon Worcestershire sauce
- 1 tablespoon lime juice
- 1 teaspoon hot sauce

Directions:

1. Combine the dry rub, thyme and oregano in a bowl.
2. Coat the chicken breasts with olive oil.
3. Season both sides with the dry rub mixture.
4. Set the wood pellet grill to 350 degrees F.
5. Add the chicken breast to the grill.
6. Grill for 7 to 8 minutes.
7. Let rest for 10 minutes.
8. Combine rest of the ingredients in a saucepan. Bring to a boil.
9. Serve the chicken with the sauce.

Nutrition: Calories: 286 Carbs: 14g Fat: 5g Protein: 38g

3. Barbecue Mustard Ribs

Preparation Time: 10 minutes
Cooking Time: 5 hours
Servings: 4

Ingredients:

- Sauce
- 1/4 cup dark brown sugar
- 1/4 cup cider vinegar
- 1 cup yellow mustard
- 2 tablespoons ketchup
- 1/4 cup honey
- 1 tablespoon hot sauce
- 1 tablespoon Worcestershire sauce
- 1 tablespoon sweet dry rub
- Ribs
- 1 rack ribs
- 1 cup yellow mustard
- 6 tablespoons sweet dry rub
- 2 cups apple juice

Directions:

1. Combine the sauce ingredients in a bowl.

2. Refrigerate while preparing the ribs.
3. Coat the ribs with the remaining mustard.
4. Season both sides with the sweet dry rub.
5. Set the wood pellet grill to 275 degrees F.
6. Grill the ribs for 3 hours.
7. Spray with the apple juice every 30 minutes. .
8. Brush with the sauce mixture.
9. Slice and serve.

Nutrition: Calories: 231 Carbs: 2g Fat: 15g Protein: 17g

4. Barbecue Raspberry Pork Ribs

Preparation Time: 5 minutes
Cooking Time: 3 hours
Servings: 6

Ingredients:

- 4 lb. baby back ribs
- 3 tablespoons raspberry chipotle dry rub
- 1 cup barbecue sauce

Directions:

1. Season the ribs with the dry rub.
2. Cover with foil.
3. Refrigerate for 1 hour.
4. Turn on the wood pellet grill.
5. Preheat your grill to 250 degrees F.
6. Add the ribs to the grill.
7. Cook for 2 hours.
8. Brush with the barbecue sauce.
9. Increase temperature to 300 degrees F.
10. Grill for another 1 hour.

Nutrition: Calories: 114 Carbs: 6g Fat: 3g Protein: 14g

5. Barbecue Steak

Preparation Time: 5 minutes
Cooking Time: 45 minutes
Servings: 4

Ingredients:

- 2 bone-in rib eye steaks
- Steak seasoning
- 1 cup barbecue sauce

Directions:

1. Sprinkle both sides of the steak with the rub.
2. Turn on the wood pellet grill.
3. Set it to 400 degrees F.
4. Grill it for 7 to 8 minutes per side.
5. Transfer the steaks to a cutting board.
6. Let sit for 10 minutes before slicing.
7. Using a pan over medium heat, simmer the barbecue sauce for 5 minutes.
8. Serve the grilled steak with the barbecue sauce.

Nutrition: Calories: 330 Carbs: 47g

Fat: 6g Protein: 20g

6. Barbecue Kebabs

Preparation Time: 1 hour
Cooking Time: 25 minutes
Servings: 6

Ingredients:

- 1 tablespoon olive oil
- 1/8 cup apple cider vinegar
- 1 tablespoon honey
- 2 tablespoons raspberry chipotle rub
- 1 lb. pork loin, sliced into cubes
- 1 red onion, sliced
- 3 green bell peppers, sliced
- 1/2 cup barbecue sauce

Directions:

1. In a mixing bowl, mix the oil, vinegar, honey and rub.
2. Marinate the pork slices in this mixture.
3. Cover it with a foil and marinate in the refrigerator for 1 hour.
4. Thread the pork cubes into skewers alternating with the red onion and green bell pepper.
5. Set the wood pellet grill to 400 degrees F.
6. Grill the kebabs for 15 to 20 minutes, rotating every 5 minutes.
7. Brush with the barbecue sauce.

Nutrition: Calories: 165 Carbs: 3g Fat: 10g Protein: 10g

7. Cheesy Barbecue Chicken

Preparation Time: 10 minutes
Cooking Time: 2 hours
Servings: 4

Ingredients:

- 4 chicken thigh fillets, skin removed
- 1 teaspoon olive oil
- 1/4 cup barbecue sauce
- 1 tablespoon sweet spicy dry rub
- 4 slices cheese

Directions:

1. Preheat your wood pellet grill to 350 degrees F.
2. Brush the chicken with the mixture of olive oil and barbecue sauce.
3. Sprinkle with the rub.
4. Grill the chicken for 30 minutes.
5. Add some cheese on top of the chicken.
6. Grill until the cheese has melted.

Nutrition: Calories: 378 Carbs: 7g Fat: 22g Protein: 38g

8. Lemon Pepper Barbecue Wings

Preparation Time: 10 minutes
Cooking Time: 30 minutes
Servings: 4

Ingredients:

- 1 tablespoon barbecue sauce
- 2 tablespoons lemon zest
- 1/4 cup ground black pepper
- 2 teaspoons ground coriander
- 2 teaspoons garlic powder
- 3 teaspoons dried thyme
- Salt to taste
- 4 lb. chicken wings

Directions:

1. Set your wood pellet grill to smoke.
2. Set it to 400 degrees F.
3. Using a bowl mix all the ingredients except the chicken wings.
4. Soak the chicken wings in half of the mixture.
5. Place the wings on the grill.
6. Grill for 15 minutes per side.
7. Brush the chicken with the remaining sauce.

Nutrition: Calories: 211 Carbs: 14g Fat: 13g Protein: 9g

9. Barbecue Pulled Pork

Preparation Time: 10 minutes
Cooking Time: 8 hours
Servings: 8

Ingredients:

- 1 teaspoon ground coriander
- 2 teaspoons black pepper
- 1 teaspoon cumin
- Salt to taste
- 7 lb. pork shoulder
- 4 cups chicken broth
- 2 cups barbecue sauce

Directions:

1. Set your wood pellet grill to smoke. Set it to 300 degrees F.
2. Combine the coriander, pepper, cumin and salt in a container. Put the mixture on all sides of the pork shoulder. Pour the broth into a roasting pan. Add the pork shoulder. Add the pan to the grill.
3. Cook for 8 hours. Shred the pork. Toss in the barbecue sauce. Heat through in a grill pan and serve.

Nutrition: Calories: 207 Carbs: 2g Fat: 5g Protein: 34g

10. Barbecue Pork Shoulder

Preparation Time: 10 minutes
Cooking Time: 7 hours
Servings: 8

Ingredients:

- 6 lb. pork shoulder
- 2 tablespoons hickory bacon seasoning
- 1 cup apple cider vinegar
- 1 tablespoon sugar

Directions:

1. Turn on your wood pellet grill. Set it to 225 degrees F. Sprinkle all sides of the pork with the seasoning.
2. In a bowl, mix the vinegar and sugar. Mix until the sugar has been dissolved. Inject the vinegar mixture into the pork shoulder.
3. Wrap the pork with foil. Place on top of the grill. Close the lid. Smoke for 6 hours. Uncover the pork slowly.
4. Let rest on a cutting board for 30 minutes.

Nutrition: Calories: 260 Carbs: 3g Fat: 20g Protein: 16g

APPETIZERS AND SIDES RECIPES

11. Mushrooms Stuffed With Crab Meat

Preparation Time: 20 Minutes

Cooking Time: 30 to 45 Minutes

Servings: 4 to 6

Ingredients:

- 6 medium-sized Portobello mushrooms
- Extra virgin olive oil
- ⅓ Grated parmesan cheese cup
- Club Beat Staffing:
- 8 ounces fresh crab meat or canned or imitation crab meat
- 2 tablespoons extra virgin olive oil
- ⅓Chopped celery
- Chopped red peppers
- ½ cup chopped green onion
- ½ cup Italian bread crumbs
- ½Cup mayonnaise
- 8 ounces cream cheese at room temperature
- ½ teaspoon of garlic
- 1 tablespoon dried parsley
- Grated parmesan cheese cup
- 1 teaspoon of Old Bay seasoning
- ¼ teaspoon of kosher salt
- ¼ teaspoon black pepper

Directions:

1. Clean up the mushroom cap with a damp paper towel. Cut off the stem and save it.

2. Remove the brown gills from the bottom of the mushroom cap with a spoon and discard.
3. Prepare crab meat stuffing. If you are fan of using canned crab meat, drain, rinse, and remove shellfish.
4. Heat the pan with olive oil first over medium high heat. Add celery, peppers and green onions and fry for 5 minutes. Set aside for cooling.
5. Gently pour the chilled sautéed vegetables and the remaining ingredients into a large bowl.
6. Cover and refrigerate crab meat stuffing until ready to use.
7. Put the crab mixture in each mushroom cap and make a mound in the center.
8. Sprinkle extra virgin olive oil and sprinkle parmesan cheese on each stuffed mushroom cap. Put the mushrooms in a 10 x 15 inch baking dish.
9. Use the pellets to set the wood pellet smoker and grill to indirect heating and preheat to 375 ° F.
10. Bake for 30-45 minutes until the filling becomes hot (165 degrees Fahrenheit as measured by an instant-read digital thermometer) and the mushrooms begin to release juice.

Nutrition: Calories: 160 Carbs: 14g Fat: 8g Protein: 10g

12. Bacon Wrapped With Asparagus

Preparation Time: 15 Minutes

Cooking Time: 25 to 30 Minutes

Servings: 4 to 6

Ingredients:

- 1 pound fresh thick asparagus (15-20 spears)
- Extra virgin olive oil
- 5 sliced bacon
- 1 teaspoon of Western Love or salted pepper

Directions:

1. Cut off each wooden ends of the asparagus and make them all the same length.
2. Divide the asparagus into a bundle of three spears and split with olive oil. Wrap each others bundle with a piece of bacon, then dust with seasonings or salt pepper for seasoning.
3. Set the wood pellet smoker and grill for indirect cooking and place a Teflon coated fiberglass mat on the grate (to prevent asparagus from sticking to the grate grate). Preheat to 400 degrees Fahrenheit using all types of pellets. The grill can be preheated during asparagus preparation.
4. Bake the asparagus wrapped in bacon for 25-30 minutes until the asparagus is soft and the bacon is cooked and crispy.

Nutrition: Calories: 71 Carbs: 1g Fat: 4g Protein: 6g

13. Brisket Baked Beans

Preparation Time: 20 Minutes
Cooking Time: 1 to 2 hours
Servings: 10

Ingredients:

- 2 tablespoons extra virgin olive oil
- 1 large diced onion
- 1 diced green pepper
- 1 red pepper diced
- 2 to 6 jalapeno peppers diced
- Texas style brisket flat chopped 3 pieces
- 1 baked bean, like Bush's country style baked beans
- 1 pork and beans
- 1 red kidney beans, rinse, drain
- 1 cup barbecue sauce like Sweet Baby Ray's barbecue sauce
- ½ cup stuffed brown sugar
- 3 garlics, chopped
- 2 teaspoons of mustard
- ½ teaspoon kosher salt
- ½ teaspoon black pepper

Directions:

1. Heat the skillet with olive oil over medium heat and add the diced onion, peppers and jalapeno. Sauté the food for about 8-10 minutes until the onion is translucent.
2. In a 4 quart casserole dish, mix chopped brisket, baked beans, pork beans, kidney beans, cooked onions, peppers, barbecue sauce, brown sugar, garlic, mustard, salt and black pepper.
3. Using the selected pellets, configure a wood pellet smoking grill for indirect cooking and preheat to 325 ° F. Cook the beans baked in the brisket for 1.5 to 2 hours until they become bare beans. Rest for 15 minutes before eating.

Nutrition: Calories: 199 Carbs: 35g Fat: 2g Protein: 9g

14. Bacon Cheddar Slider

Preparation Time: 30 Minutes
Cooking Time: 15 Minutes
Servings: 6 To 10

Ingredients:

- 1 pound ground beef (80% lean)
- ½ teaspoon of garlic salt
- ½ teaspoon salt
- ½ teaspoon of garlic
- ½ teaspoon onion
- ½ teaspoon black pepper
- 6 bacon slices, cut in half
- ½ Cup mayonnaise
- 2 teaspoons of creamy wasabi (optional)
- 6 sliced sharp cheddar cheese, cut in half (optional)
- Sliced red onion
- ½ Cup sliced kosher dill pickles
- 12 mini breads sliced horizontally
- Ketchup

Directions:

1. Place ground beef, garlic salt, seasoned salt, garlic powder, onion powder and black hupe pepper in a medium bowl.
2. Divide in 12 equal parts the meat mixture into shape into small thin round patties (about 2 ounces each) and save.

3. Cook the bacon on medium heat over medium heat for 5-8 minutes until crunchy. Set aside.
4. To make the sauce, mix the mayonnaise and horseradish in a small bowl, if used.
5. Set up a wood pellet smoker and grill for direct cooking to use griddle accessories. Look for the manufacturer to see if there is a griddle accessory that works with the particular wooden pellet smoker and grill.
6. Spray a cooking spray on the griddle cooking surface for best non-stick results.
7. Preheat wood pellet smoker and grill to 350 ° F using selected pellets. Griddle surface should be approximately 400 ° F.
8. Grill the putty for 3-4 minutes each until the internal temperature reaches 160 ° F.
9. If necessary, place a sharp cheddar cheese slice on each patty while the patty is on the griddle or after the patty is removed from the griddle. Place a small amount of mayonnaise mixture, a slice of red onion, and a hamburger pate in the lower half of each roll. Pickled slices, bacon and ketchup

Nutrition: Calories: 80 Carbs: 0g Fat: 5g Protein: 0g

15. Apple Wood Smoked Cheese

Preparation Time: 1 Hour 15 Minutes

Cooking Time: 2 Hours

Servings: 6

Ingredients:

- Gouda
- Sharp cheddar
- Very sharp 3 year cheddar
- Monterey Jack
- Pepper jack
- Swiss

Directions:

1. According to the shape of the cheese block, cut the cheese block into an easy-to-handle size (approximately 4 x 4 inch block) to promote smoke penetration.
2. Leave the cheese on the counter for one hour to form a very thin skin or crust, which acts as a heat barrier, but allows smoke to penetrate.
3. Configure the wood pellet smoking grill for indirect heating and install a cold smoke box to prepare for cold smoke. Make sure that the louvers on the smoking box are fully open to allow moisture to escape from the box.
4. Preheat the wood pellet smoker and grill to 180 ° F or use apple pellets and smoke settings, if any, to get a milder smoke flavor.
5. Place the cheese on a Teflon-coated fiberglass non-stick grill mat and let cool for 2 hours.
6. Remove the smoked cheese and cool for 1 hour on the counter using a cooling rack.

7. After labelling the smoked cheese with a vacuum seal, refrigerate for 2 weeks or more, then smoke will permeate and the cheese flavor will become milder.

Nutrition: Calories: 102 Carbs: 0g Fat: 9g Protein: 6g

16. Hickory Smoked Moink Ball Skewer

Preparation Time: 30 Minutes

Cooking Time: 1 Hour 30 Minutes

Servings: 9

Ingredients:

- ½ pound ground beef (80% lean)
- ½ pound pork sausage
- 1 large egg
- ½ cup Italian bread crumbs
- ½ cup chopped red onion
- Grated parmesan cheese cup
- ¼ Cup finely chopped parsley
- ¼ cup whole milk
- 2 pieces of garlic, 1 chopped or crushed garlic
- 1 teaspoon oregano
- ½ teaspoon kosher salt
- ½ teaspoon black pepper
- ¼ cup barbecue sauce like Sweet Baby Ray
- ½ pound bacon cut in half, cut in half

Directions:

1. In a container, mix ground beef, ground pork sausage, eggs, crumbs, onions, parmesan cheese, parsley, milk, garlic,
2. salt, oregano, and pepper. Do not overuse the meat.
3. Form a 1½ ounces meatball about 1.5 inches in diameter and place on a Teflon-coated fiberglass mat.
4. Wrap each meatball in half thin bacon. Stab moink balls on 6 skewers (3 balls per skewer).
5. Set up wood pellet smoker and grill for indirect cooking.
6. Preheat wood pellet smoker and grill to 225 ° F using hickory pellets.
7. Tap the moink ball skewer for 30 minutes.
8. Raise the pit temperature to 350 ° F until the meatball internal temperature reaches 175 ° F and the bacon is crisp (about 40-45 minutes).
9. For the last 5 minutes, brush your moink ball with your favorite barbecue sauce.
10. While still hot, offer moink ball skewers.

Nutrition: Calories: 170 Carbs: 2g Fat: 15g Protein: 7g

17. Garlic Parmesan Wedge

Preparation Time: 15 Minutes

Cooking Time: 30 to 35 Minutes

Servings: 4

Ingredients:

- 3 large russet potatoes
- ¼ cup of extra virgin olive oil
- 1 teaspoon salt
- ¾ teaspoon black hu pepper
- 2 teaspoon garlic powder
- ¾ cup grated parmesan cheese
- 3 tablespoons of fresh coriander or flat leaf parsley (optional)
- ½ cup blue cheese or ranch dressing per serving, for soaking (optional)

Directions:

1. Gently rub the potatoes with cold water using a vegetable brush to dry the potatoes.
2. Cut the potatoes in half vertically and cut them in half.
3. Wipe off any water released when cutting potatoes with a paper towel. Moisture prevents wedges from becoming crunchy.
4. Put the potato wedge, olive oil, salt, pepper and garlic powder in a large bowl and shake lightly by hand to distribute the oil and spices evenly.
5. Place the wedges on a single layer of non-stick grill tray / pan / basket (about 15 x 12 inches).
6. Set the wood pellet r grill for indirect cooking and use all types of wood pellets to preheat to 425 degrees Fahrenheit.
7. Put the grill tray in the preheated smoker and grill, roast the potato wedge for 15 minutes, and turn. Roast the potato wedge for an additional 15-20 minutes until the potatoes are soft inside and crispy golden on the outside.
8. Sprinkle potato wedge with parmesan cheese and add coriander or parsley as needed. If necessary, add blue cheese or ranch dressing for the dip.

Nutrition: Calories: 112 Carbs: 17g Fat: 3g Protein: 3g

18. Roasted Vegetables

Preparation Time: 20 Minutes

Cooking Time: 20 to 40 Minutes

Servings: 4

Ingredients:

- 1 cup cauliflower floret
- 1 cup small mushroom, half
- 1 medium zucchini, sliced in half
- 1 medium yellow squash, sliced in half
- 1 medium-sized red pepper, chopped to 1.5-2 inches
- 1 small red onion, chopped to 1½-2 inch
- 6 ounces small baby carrot
- 6 mid-stem asparagus spears, cut into 1-inch pieces
- 1 cup cherry or grape tomato
- ¼ Extra virgin olive oil with cup roasted garlic flavor
- 2 tablespoons of balsamic vinegar
- 3 garlics, chopped
- 1 teaspoon dry time

- 1 teaspoon dried oregano
- 1 teaspoon of garlic salt
- ½ teaspoon black pepper

Directions:

1. Put cauliflower florets, mushrooms, zucchini, yellow pumpkin, red peppers, red onions, carrots, asparagus and tomatoes in a large bowl.
2. Add olive oil, balsamic vinegar, garlic, thyme, oregano, garlic salt, and black hu add to the vegetables.
3. Gently throw the vegetables by hand until completely covered with olive oil, herbs and spices.
4. Spread the seasoned vegetables evenly on a non-stick grill tray / bread / basket (about 15 x 12 inches).
5. Set the wood pellet smoker and grill for indirect cooking and preheat to 425 degrees Fahrenheit using all types of wood pellets.
6. Transfer the grill tray to a preheated smoker and grill and roast the vegetables for 20-40 minutes or until the vegetables are perfectly cooked. Please put it out immediately.

Nutrition: Calories: 114 Carbs: 17g Fat: 4g Protein: 3g

19. Atomic Buffalo Turds

Preparation Time: 30 to 45 Minutes

Cooking Time: 1.5 Hours to 2 Hours

Servings: 6

Ingredients:

- 10 Medium Jalapeno Pepper
- 8 ounces regular cream cheese at room temperature
- ¾Cup Monterey Jack and Cheddar Cheese Blend Shred (optional)
- 1 teaspoon smoked paprika
- 1 teaspoon garlic powder
- ½ teaspoon cayenne pepper
- Teaspoon red pepper flakes (optional)
- 20 smoky sausages
- 10 sliced bacon, cut in half

Directions:

1. Wear food service gloves when using. Jalapeno peppers are washed vertically and sliced. Carefully remove seeds and veins using a spoon or paring knife and discard. Place Jalapeno on a grilled vegetable tray and set aside.
2. In a small bowl, mix cream cheese, shredded cheese, paprika, garlic powder, cayenne pepper if used, and red pepper flakes if used, until thoroughly mixed.
3. Mix cream cheese with half of the jalapeno pepper.
4. Place the Little Smokiness sausage on half of the filled jalapeno pepper.
5. Wrap half of the thin bacon around half of each jalapeno pepper.
6. Fix the bacon to the sausage with a toothpick so that the pepper does not

pierce. Place the ABT on the grill tray or pan.

7. Set the wood pellet smoker and grill for indirect cooking and preheat to 250 degrees Fahrenheit using hickory pellets or blends.
8. Suck jalapeno peppers at 250 ° F for about 1.5 to 2 hours until the bacon is cooked and crisp.
9. Remove the ABT from the grill and let it rest for 5 minutes before hors d'oeuvres.

Nutrition: Calories: 131 Carbs: 1g Fat: 12g Protein: 5g

20. Twice-Baked Spaghetti Squash

Preparation Time: 15 Minutes

Cooking Time: 45 to 60 Minutes

Servings: 2

Ingredients:

- 1 medium spaghetti squash
- 1 tablespoon extra-virgin olive oil
- 1 teaspoon salt
- ½ teaspoon pepper
- ½ Chop shredded mozzarella cheese, split
- ½ cup of parmesan cheese, split

Directions:

1. Using a beautiful large and sharp knife, carefully cut the pumpkin in half lengthwise. Use a spoon to remove each half of the seeds and pulp.
2. Rub olive oil inside half of the squash and sprinkle with salt and pepper.
3. Set wood pellet smoker and grill for indirect cooking and preheat to 375 degrees Fahrenheit using all types of wood pellets.
4. Place the squash half-side up directly on the hot grill grate.
5. Cook the squash for about 45 minutes until the internal temperature reaches 170 degrees Fahrenheit. When complete, the spaghetti squash softens and is easily pierced with a fork.
6. Transfer the squash to the cutting board and cool for 10 minutes.
4. Raise wood pellet smoker and grill temperature to 425 ° F.
7. Using a fork, scoop the squash back and forth, taking care not to damage the shell, and remove the chunks of meat. Note that the stand looks like spaghetti.
8. Transfer the strand to a large bowl. Add half the mozzarella cheese and parmesan cheese and mix.
9. Fill half the squash shell with the mixture and sprinkle the remaining mozzarella and parmesan cheese on top.
10. Bake the packed spaghetti squash halves at 425 ° F for another 15 minutes or until the cheese is brown.

Nutrition: Calories: 230 Carbs: 9g Fat: 17g Protein: 12g

21. Grilled Corn

Preparation Time: 15 minutes

Cooking Time: 25 minutes

Servings: 6

Ingredients:

- 6 fresh ears corn
- Salt
- Black pepper
- Olive oil
- Vegetable seasoning
- Butter for serving

Directions:

1. Preheat the grill to high with closed lid.
2. Peel the husks. Remove the corn's silk. Rub with black pepper, salt, vegetable seasoning, and oil.
3. Close the husks and grill for 25 minutes. Turn them occasionally.
4. Serve topped with butter and enjoy.

Nutrition: Calories: 70 Protein: 3g Carbs: 18g Fat: 2g

22. Thyme - Rosemary Mash Potatoes

Preparation Time: 20 minutes
Cooking Time: 1 hour
Servings: 6

Ingredients:

- 4 ½ lbs. Potatoes, russet
- Salt
- 1 pint of Heavy cream
- 3 Thyme sprigs + 2 tablespoons for garnish
- 2 Rosemary sprigs
- 6 - 7 Sage leaves
- 6 - 7 Black peppercorns
- Black pepper to taste
- 2 stick Butter, softened
- 2 Garlic cloves, chopped

Directions:

1. Preheat the grill to 350F with closed lid.
2. Peel the russet potatoes. Cut into small pieces and place them in a baking dish. Fill it with water (1 ½ cups). Place on the grill and cook with closed lid for about 1 hour.
3. In the meantime in a saucepan combine the garlic, peppercorns, herbs, and cream. Place on the grate and cook covered for about 15 minutes. Once done, strain to remove the garlic and herbs. Keep warm.
4. Take out the water of the potatoes and place them in a stockpot. Rice them with a fork and pour 2/3 of the mixture. Add 1 stick softened butter and salt.
5. Serve right away.

23. Grilled Broccoli

Preparation Time: 15 minutes

Cooking Time: 10 minutes

Servings: 4 to 6

Ingredients:

- 4 bunches of Broccoli
- 4 tablespoons Olive oil
- Black pepper and salt to taste
- ½ Lemon, the juice
- ½ Lemon cut into wedges

Directions:

1. Preheat the grill to High with closed lid.
2. In a bowl add the broccoli and drizzle with oil. Coat well. Season with salt.
3. Grill for 5 minutes and then flip. Cook for 3 minutes more.
4. Once done transfer on a plate. Squeeze lemon on top and serve with lemon wedges. Enjoy!

Nutrition: Calories: 35g Protein: 2.5g Carbs: 5g Fat: 1g

24. Smoked Coleslaw

Preparation Time: 15 minutes

Cooking Time: 25 minutes

Servings: 8

Ingredients:

- 1 shredded Purple Cabbage
- 1 shredded Green Cabbage
- 2 Scallions, sliced
- 1 cup Carrots, shredded
- Dressing
- 1 tablespoon of Celery Seed
- 1/8 cup of White vinegar
- 1 ½ cups Mayo
- Black pepper and salt to taste

Directions:

1. Preheat the grill to 180F with closed lid.
2. On a tray spread the carrots and cabbage. Place the tray on the grate and smoke for about 25 minutes.
3. Transfer in the fridge to cool.
4. In the meantime make the dressing. In a bowl combine the ingredients. Mix well.
5. Transfer the veggies in a bowl. Drizzle with the sauce and toss
6. Serve sprinkled with scallions.

Nutrition: Calories: 35g Protein: 1g Carbs: 5g Fat: 5g

25. The Best Potato Roast

Preparation Time: 15 minutes

Cooking Time: 35 minutes

Servings: 6

Ingredients:

- 4 Potatoes, large (scrubbed)
- 1 ½ cups gravy (beef or chicken)
- Rib seasoning to taste
- 1 ½ cups Cheddar cheese
- Black pepper and salt to taste
- 2 tablespoons sliced Scallions

Directions:

1. Preheat the grill to high with closed lid.
2. Slice each potato into wedges or fries. Transfer into a bowl and drizzle with oil. Season with Rib seasoning.
3. Spread the wedges/fries on a baking sheet (rimmed). Roast for about 20 minutes. Turn the wedges/fries and cook for 15 minutes more.
4. In the meantime in a saucepan warm the chicken/beef gravy. Cut the cheese into small cubes.
5. Once done cooking place the potatoes on a plate or into a bowl. Distribute the cut cheese and pour hot gravy on top.
6. Serve garnished with scallion. Season with pepper. Enjoy!

Nutrition: Calories: 220 Protein: 3g Carbs: 38g Fat: 15g

26. Corn Salsa

Preparation Time: 10 minutes

Cooking Time: 15 minutes

Servings: 4

Ingredients:

- 4 Ears Corn, large with the husk on
- 4 Tomatoes (Roma) diced and seeded
- 1 teaspoon of Onion powder
- 1 teaspoon of Garlic powder
- 1 Onion, diced
- ½ cup chopped Cilantro
- Black pepper and salt to taste
- 1 lime, the juice
- 1 grille jalapeno, diced

Directions:

1. Preheat the grill to 450F.
2. Place the ears corn on the grate and cook until charred. Remove husk. Cut into kernels.
3. Combine all ingredients, plus the corn and mix well. Refrigerate before serving.

Nutrition: Calories: 120 Protein: 2g Carbs: 4g Fat: 1g

27. Nut Mix on the Grill

Preparation Time: 10 minutes

Cooking Time: 20 minutes

Servings: 8

Ingredients:

- 3 cups Mixed Nuts, salted
- 1 teaspoon Thyme, dried
- 1 ½ tablespoon brown sugar, packed
- 1 tablespoon Olive oil
- ¼ teaspoon of Mustard powder
- ¼ teaspoon Cayenne pepper

Directions:

1. Preheat the grill to 250F with closed lid.
2. In a bowl combine the ingredients and place the nuts on a baking tray lined with parchment paper. Place the try on the grill. Cook 20 minutes.
3. Serve and enjoy.

Nutrition: Calories: 65 Protein: 23g Carbs: 4g Fat: 52g

28. Cinnamon Almonds

Preparation Time: 10 minutes

Cooking Time: 1 hour 30 minutes

Servings: 4

Ingredients:

- 1 egg, the white
- 1lb. Almonds
- ½ cup of Brown Sugar
- ½ cup of Granulated sugar
- 1/8 teaspoon Salt
- 1 tablespoon ground Cinnamon

Directions:

1. Whisk the egg white until frothy. Add the salt, cinnamon, and sugars. Add the almonds and toss to coat.
2. Spread the almonds on a baking dish lined with parchment paper. Make sure they are in a single layer.
3. Preheat the grill to 225F with closed lid.
4. Grill for 1 h and 30 minutes. Stir often.
5. Serve slightly cooled and enjoy.

Nutrition: Calories: 280 Protein: 10g Carbs: 38g Fat: 13g

29. Grilled French Dip

Preparation Time: 15 minutes

Cooking Time: 35 minutes

Servings: 8

Ingredients:

- 3 lbs. onions, thinly sliced (yellow)
- 2 tablespoons oil
- 2 tablespoons of Butter
- Salt to taste
- Black pepper to taste
- 1 teaspoon Thyme, chopped
- 2 teaspoon of Lemon juice
- 1 cup Mayo
- 1 cup of Sour cream

Directions:

1. Preheat the grill to high with closed lid.
2. In a pan combine the oil and butter. Place on the grill to melt. Add 2 teaspoons salt and add the onions.
3. Mix it well and close the lid of the grill. Cook 30 minutes stirring often.
4. Add the thyme. Cook for an additional 3 minutes. Set aside and add black pepper.
5. Once cooled add lemon juice, mayo, and sour cream. Stir to combine. Taste and add more black pepper and salt if needed.
6. Serve with veggies or chips. Enjoy.

Nutrition: Calories: 60 Protein: 4g Carbs: 5g Fat: 6g

30. Roasted Cashews

Preparation Time: 15 minutes

Cooking Time: 12 minutes

Servings: 6

Ingredients:

- ¼ cup Rosemary, chopped
- 2 ½ tablespoons Butter, melted
- 2 cups Cashews, raw
- ½ teaspoon of Cayenne pepper
- 1 teaspoon of salt

Directions:

1. Preheat the grill to 350F with closed lid.
2. In a baking dish layer the nuts. Combine the cayenne, salt rosemary, and butter. Add on top. Toss to combine.
3. Grill for 12 minutes.
4. Serve and enjoy.

Nutrition: Calories: 150 Proteins: 5g Carbs: 7g Fat: 15g

31. Smoked Jerky

Preparation Time: 20 minutes

Cooking Time: 6 hours

Servings: 6

Ingredients:

- 1 Flank Steak (3lb.)
- ½ cup of Brown Sugar
- 1 cup of Bourbon
- ¼ cup Jerky rub
- 2 tablespoons of Worcestershire sauce
- 1 can of Chioplete
- ½ cup Cider Vinegar

Directions:

1. Slice the steak into ¼ inch slices.
2. Combine the remaining ingredients in a bowl. Stir well.
3. Put the steak in an empty bag and add the marinade sauce. Marinade in the fridge overnight.
4. Preheat the grill to 180F with closed lid.
5. Remove the flank from marinade. Place directly on a rack and on the grill.
6. Smoke for 6 hours.
7. Cover them lightly for 1 hour before serving. Store leftovers in the fridge.

Nutrition: Calories: 105 Protein: 14g Carbs: 4g Fat: 3g

32. Bacon BBQ Bites

Preparation Time: 10 minutes

Cooking Time: 25 minutes

Servings: 4

Ingredients:

- 1 tablespoon Fennel, ground
- ½ cup of Brown Sugar
- 1 lb. Slab Bacon, cut into cubes (1 inch)
- 1 tablespoon Black pepper
- Salt

Directions:

1. Take an aluminum foil and then fold in half. Once you do that, then turn the edges so that a rim is made. With a fork make small holes on the bottom. In this way, the excess fat will escape and will make the bites crispy.
2. Preheat the grill to 350F with closed lid.
3. In a bowl combine the black pepper, salt, fennel, and sugar. Stir.
4. Place the pork in the seasoning mixture. Toss to coat. Transfer on the foil.
5. Place the foil on the grill. Bake the mixture for 25 minutes, or until crispy and bubbly.
6. Serve and enjoy.

Nutrition: Calories: 300 Protein: 27g Carbs: 4g Fat: 36g

33. Smoked Guacamole

Preparation Time: 15 minutes

Cooking Time: 30 minutes

Servings: 8

Ingredients:

- ¼ cup chopped Cilantro
- 7 Avocados, peeled and seeded
- ¼ cup chopped Onion, red
- ¼ cup chopped tomato
- 3 ears corn
- 1 teaspoon of Chile Powder
- 1 teaspoon of Cumin
- 2 tablespoons of Lime juice
- 1 tablespoon minced Garlic
- 1 Chile, poblano
- Black pepper and salt to taste

Directions:

1. Preheat the grill to 180F with closed lid.
2. Smoke the avocado for 10 min.
3. Set the avocados aside and increase the temperature of the girl to high.
4. Once heated grill the corn and chili. Roast for 20 minutes.
5. Cut the corn. Set aside. Place the chili in a bowl. Cover it with a wrapper and let it sit for about 10 minutes. Peel the chili and dice. Add it to the kernels.
6. In a bowl mash the avocados, leave few chunks. Add the remaining ingredients and mix.
7. Serve right away because it is best eaten fresh. Enjoy!

Nutrition: Calories: 51 Protein: 1g Carbs: 3g Fat: 4.5g

34. Jalapeno Poppers

Preparation Time: 15 minutes

Cooking Time: 1 hour

Servings: 4

Ingredients:

- 6 Bacon slices halved
- 12 Jalapenos, medium
- 1 cup grated Cheese
- 8 ounces softened Cream cheese
- 2 tablespoons Poultry seasoning

Directions:

1. Preheat the grill to 180F with closed lid.
2. Cut the jalapenos lengthwise. Clean them from the ribs and seeds.
3. Mix the poultry seasoning, grated cheese, and cream fill each jalapeno with the mixture and wrap with 1 half bacon. Place a toothpick to secure it. Place them on a baking sheet and smoke and grill 20 minutes.
4. Increase the temperature of the grill to 375F. Cook for 30 minutes more.

Nutrition: Calories: 60 Protein: 4g Carbs: 2g Fat: 8g

35. Shrimp Cocktail

Preparation Time: 10 minutes

Cooking Time: 15 minutes

Servings: 4

Ingredients:

- 2 lbs. of Shrimp with tails, deveined
- Black pepper and salt
- 1 teaspoon of Old Bay
- 2 tablespoons Oil
- ½ cup of Ketchup
- 1 tablespoon of Lemon Juice
- 2 tablespoons Horseradish, Prep Timeared
- 1 tablespoon of Lemon juice
- For garnish: chopped parsley
- Optional: Hot sauce

Directions:

1. Preheat the grill to 350F with closed lid.
2. Clean the shrimp. Pat dry using paper towels.
3. In a bowl add the shrimp, Old Bay, and oil. Toss to coat. Spread on a baking tray. Place the tray on the grill and let it cook for 7 minutes.
4. In the meantime make the sauce: Combine the lemon juice, horseradish, and ketchup. Season with black pepper and sauce and if you like add hot sauce. Stir.
5. Serve the shrimp with the sauce and enjoy.

Nutrition: Calories: 80 Protein: 8g Carbs: 5g Fat: 1g

36. Perfectly Smoked Artichoke Hearts

Preparation Time: 2 hours and 10 minutes

Servings: 6

Ingredients:

- 12 canned whole artichoke hearts
- 1/4 cup of extra virgin olive oil
- 4 cloves of garlic minced
- 2 Tbsp of fresh parsley finely chopped (leaves)
- 1 Tbsp of fresh lemon juice freshly squeezed
- Salt to taste
- Lemon for garnish

Directions:

1. Start the pellet grill on SMOKE with the lid open until the fire is established. Set the temperature to 350 °F and preheat, lid closed, for 10 to 15 minutes.
2. In a bowl, combine all remaining ingredients and pour over artichokes.
3. Place artichokes on a grill rack and smoke for 2 hours or so.
4. Serve hot with extra olive oil, and lemon halves.

Nutrition: Calories: 105.5; Carbs: 19g; Fat: 0.5g; Fiber: 11g; Protein: 7.7g

37. Finely Smoked Russet Potatoes

Preparation Time: 2 hours and 15 minutes

Servings: 6

Ingredients:

- 8 large Russet potatoes
- 1/2 cup of garlic-infused olive oil
- Kosher salt and black pepper to taste

Directions:

1. Start the pellet grill on SMOKE with the lid open until the fire is established. Set the temperature to 225 °F and preheat, lid closed, for 10 to 15 minutes.
2. Rinse and dry your potatoes; pierce with a fork on all sides.
3. Drizzle with garlic-infused olive oil and rub generously all your potatoes with the salt and pepper.
4. Place the potatoes on the pellet smoker and close the lid.
5. Smoke potatoes for about 2 hours.
6. Serve hot with your favorite dressing.

Nutrition: Calories: 384; Carbs: 48g; Fat: 18.2g; Fiber: 3.7g; Protein: 6g

38. Simple Smoked Green Cabbage (Pellet)

Preparation Time: 55 minutes
Servings: 4

Ingredients:

- 1 medium head of green cabbage
- 1/2 cup of olive oil
- salt and ground white pepper to taste

Directions:

1. Start the pellet grill on SMOKE with the lid open until the fire is established. Set the temperature to 250 °F and preheat, lid closed, for 10 to 15 minutes.
2. Clean and rinse cabbage under running water.
3. Cut the stem and then cut it in half, then each half in 2 to 3 pieces.
4. Season generously cabbage with the salt and white ground pepper; drizzle with olive oil.
5. Arrange the cabbage peace on their side on a smoker tray and cover.
6. Smoke the cabbage for 20 minutes per side.
7. Remove cabbage and let rest for 5 minutes.
8. Serve immediately.

Nutrition: Calories: 57; Carbs: 13g; Fat: 0.2g; Fiber: 6g; Protein: 3g

39. Smoked Asparagus with Parsley and Garlic

Preparation Time: 1 hour and 10 minutes

Servings: 3

Ingredients:

- 1 bunch of fresh asparagus, cleaned
- 1 Tbs of finely chopped parsley
- 1 Tbs of minced garlic
- 1/2 cup of olive oil
- Salt and ground black pepper to taste

Directions:

1. Start the pellet grill on SMOKE with the lid open until the fire is established. Set the temperature to 225 °F and preheat, lid closed, for 10 to 15 minutes.
2. Rinse and cut the ends off of the asparagus.
3. In a bowl, combine olive oil, chopped parsley, minced garlic, and the salt and pepper.
4. Season your asparagus with olive oil mixture.
5. Place the asparagus on a heavy-duty foil and fold the sides.
6. Smoke for 55 to 60 minutes or until soft (turn every 15 minutes).
7. Serve hot.

Nutrition: Calories: 352; Carbs: 6.7g; Fat: 36.2g; Fiber: 3.1g; Protein: 3.4g

40. Smoked Corn Cob with Spicy Rub

Preparation Time: 1 hour and 40 minutes

Servings: 4

Ingredients:

- 10 ears of fresh sweet corn on the cob
- 1/2 cup of macadamia nut oil
- Kosher salt and fresh ground black pepper to taste
- 1/2 tsp of garlic powder
- 1/2 tsp of hot paprika flakes
- 1/2 tsp of dried parsley
- 1/4 tsp of ground mustard

Directions:

1. Start the pellet grill on SMOKE with the lid open until the fire is established. Set the temperature to 350 °F and preheat, lid closed, for 10 to 15 minutes.
2. Combine macadamia nut oil with garlic powder, hot paprika flakes, dried parsley, and ground mustard.
3. Rub your corn with macadamia nut oil mixture and place on a grill rack.
4. Smoke corn for 80 to 90 minutes.
5. Serve hot.

Nutrition: Calories: 248; Carbs: 25g; Fat: 15g; Fiber: 3.2g; Protein: 4.4g

41. Smoked Sweet Pie Pumpkins

Preparation Time: 2 hours

Servings: 6

Ingredients:

- 4 small pie pumpkins
- avocado oil to taste

Directions:

1. Start the pellet grill on SMOKE with the lid open until the fire is established. Set the temperature to 250 °F and preheat, lid closed, for 10 to 15 minutes.
2. Cut pumpkins in half, top to bottom, and drizzle with avocado oil.
3. Place pumpkin halves on the smoker away from the fire.
4. Smoke pumpkins from 1 1/2 to 2 hours.
5. Remove pumpkins from smoked and allow to cool.
6. Serve to taste.

Nutrition: Calories: 167.1; Carbs: 10g; Fat: 14.2g; Fiber: 1g; Protein: 1.7g

42. Smoked Vegetable "Potpourri" (Pellet)

Preparation Time: 1 hour

Servings: 6

Ingredients:

- 2 large zucchini sliced
- 2 red bell peppers sliced
- 2 Russet potatoes sliced
- 1 red onion sliced
- 1/2 cup of olive oil
- Salt and ground black pepper to taste

Directions:

1. Start the pellet grill on SMOKE with the lid open until the fire is established. Set the temperature to 350 °F and preheat, lid closed, for 10 to 15 minutes.
2. In the meantime, rinse and slice all vegetables; pat dry on a kitchen paper.
3. Generously season with the salt and pepper, and drizzle with olive oil.
4. Place your sliced vegetables into grill basket or onto grill rack and smoke for 40 to 45 minutes.
5. Serve hot.

Nutrition: Calories: 330.1; Carbs: 29g; Fat: 21.6g; Fiber: 4g; Protein: 4.6g

BEEF RECIPES

43. Fully Grilled Steak

Preparation Time: 60 Minutes
Cooking Time: 15 Minutes
Servings: 2

Ingredients:

- 2 USDA Choice or Prime 1¼-1½ Inch New York Strip Steak (Approx. 12-14 ounces Each) Extra Virgin Olive Oil
- 4 teaspoons of Western Love or Salt and Pepper

Directions:

1. Remove the steak from the refrigerator, loosely cover with wrap about 45 minutes before returning to room temperature.
2. When the steak reaches room temperature, polish both sides with olive oil.
3. Season from each side of the steak with a teaspoon of rub or salt and pepper and let absorb at room temperature for at least 5 minutes before grilling.
4. Configure a wood pellet smoker and grill for direct cooking using a baking grate, set the temperature high, and preheat to at least 450 ° F using the pellets.
5. Place the steak on the grill and cook for 2-3 minutes until browned on one side.
6. On the same side, rotate the steak 90 degrees to mark the cross grill and cook for another 2-3 minutes. Turn the steak over and bake until the desired finish is achieved.
7. 3-5 minutes for medium rare (135 ° F internal temperature)

8. 6-7 minutes for medium (140 ° F internal temperature)
9. 8-10 minutes for medium wells (internal temperature 150 ° F)
10. Transfer the steak to a platter, loosen the tent with foil and leave for 5 minutes before serving.

Nutrition: Calories: 240 Carbs: 0g Fat: 15g Protein: 19g

44. Meat Chuck Short Rib

Preparation Time: 20 Minutes

Cooking Time: 5-6 Hours

Servings: 2

Ingredients:

- English cut 4 bone slab beef chuck short rib
- 3-4 cups of mustard yellow mustard or extra virgin olive oil
- 3-5 tablespoons of Western Love

Directions:

1. Cut the fat cap off the rib bone, leaving 1/4 inch fat, and remove the silvery skin.
2. Remove the membrane from the bone and move the spoon handle below the membrane to lift the piece of meat and season the meat properly. Grab the membrane using a paper towel and pull it away from the bone.
3. Apply mustard or olive oil to all sides of the short rib slab. By rubbing it, you can season all sides.
4. Using mesquite or hickory pellets, set the wood pellet smoker and grill to indirect heating and preheat to 225 ° F.
5. Insert a wood pellet smoker and grill or remote meat probe into the thickest part of the rib bone plank. If your grill does not have a meat probe or you do not have a remote meat probe, use an instant reading digital thermometer to read the internal temperature while cooking.
6. Place the short rib bone on the grill with the bone side down and smoke at 225 ° F for 5 hours.
7. If the ribs have not reached an internal temperature of at least 195 ° F after 5 hours, increase the pit temperature to 250 ° F until the internal temperature reaches 195 ° to 205 ° F.
8. Place the smoked short rib bone under the loose foil tent for 15 minutes before serving.

Nutrition: Calories: 357 Carbs: 0g Fat: 22g Protein: 37g

45. Texas Style Brisket Flat

Preparation Time: 45 Minutes (Additional Marinade, Optional)

Cooking Time: 5-6 Hours

Servings: 8

Ingredients:

- 6 ½ lbs. beef brisket flat
- ½ cup of roasted garlic flavored extra virgin olive oil
- ½ Cup Texas Style Brisket Love or Favorite Brisket Love

Directions:

1. Cut off the fat cap of the brisket and remove the silver skin.
2. Rub all sides of the trimmed meat with olive oil.
3. Put the rub to all sides of the brisket so that it is completely covered by the rub.
4. Wrap the brisket twice with plastic wrap and cool overnight to allow the meat to penetrate. Or, if needed, you can cook the brisket immediately.
5. Remove the brisket from the refrigerator and insert a wood pellet smoker and grill or remote meat probe in the thickest part of the meat. If your grill does not have a meat probe, features, or does not have a remote meat probe, use an instant reading digital thermometer to read the internal temperature while cooking.
6. Set wood pellet smoker and grill for indirect cooking using mesquite or oak pellets and preheat to 250 ° F.
7. Smoke the brisket at 250 ° F until the internal temperature reaches 160 ° F (about 4 hours).
8. Tke out the brisket from the grill, wrap it twice in sturdy aluminum foil, keep the meat probe in place and return to the smoking grill.
9. Raise the pit temperature to 325 ° F and cook the brisket for another 2 hours until the internal temperature reaches 205 ° F.
10. Remove the brisket with foil, wrap it with a towel and put it in the cooler. Let sit in the cooler for 2-4 hours before slicing into cereals and serving.

Nutrition: Calories: 180 Carbs: 2g Fat: 9g Protein: 23g

46. Reverse-Seared Tri-Tip

Preparation Time: 10 minutes

Cooking Time: 3 hours

Servings: 4

Ingredients:

- 1½ pounds tri-tip roast
- 1 batch Espresso Brisket Rub

Directions:

1. Supply your smoker with wood pellets and follow the manufacturer's specific start-up procedure. Let the grill reheat, with the lid closed to have a beautiful grill, to 180°F.
2. Season the tri-tip roast with the rub. With your two hands, work the rub into the meat.
3. Place the meat to roast directly on the grill grate and smoke until its internal temperature reaches 140°F.
4. Increase the grill's temperature to 450°F and continue to cook until the roast's internal temperature reaches 145°F. This same technique can be done over an open flame or in a cast-iron skillet with some butter.
5. Remove the tri-tip roast from the grill and let it rest 10 to 15 minutes, before slicing and serving.

Nutrition: Calories: 290 Carbs: 5g Fat: 18g Protein: 30g

47. George's Smoked Tri-Tip

Preparation Time: 25 minutes

Cooking Time: 5 hours

Servings: 4

Ingredients:

- 1½ pounds tri-tip roast
- Salt
- Freshly ground black pepper
- 2 teaspoons garlic powder
- 2 teaspoons lemon pepper
- ½ cup apple juice

Directions:

1. Supply your smoker with wood pellets and follow the manufacturer's specific start-up procedure. Allow your grill to preheat with the lid closed, to 180°F.
2. Season the tri-tip roast with salt, pepper, garlic powder, and lemon pepper. Using your two hands, work on the seasoning into the meat.
3. Place the meat to roast directly on the grill grate and smoke for 4 hours.
4. Pull the tri-tip from the grill and place it on enough aluminium foil to wrap it completely.
5. Increase the grill's temperature to 375°F.
6. Fold in three sides of the foil around the roast and add the apple juice. Fold in the last side, completely enclosing the tri-tip and liquid. Return the wrapped tri-tip to the grill and cook for 45 minutes more.
7. Remove the tri-tip roast from the grill and let it rest for 10 to 15 minutes, before unwrapping, slicing, and serving.

48. Pulled Beef

Preparation Time: 25 minutes

Cooking Time: 12 to 14 hours

Servings: 5 to 8

Ingredients:

- 1 (4-pound) top round roast
- 2 tablespoons yellow mustard
- 1 batch Espresso Brisket Rub
- ½ cup beef broth

Directions:

1. Supply your smoker with wood pellets and follow the manufacturer's specific start-up procedure. Allow your griller to preheat with the lid closed to have a quality food to 225°F.
2. Coat the top round roast all over with mustard and season it with the rub. Using your two hands, work the rub into the meat.
3. Place the meat to roast directly on the grill grate and smoke until its internal temperature reaches 160°F and a dark bark has formed.
4. Pull the roast from the grill and place it on enough aluminum foil to wrap it completely.
5. Increase the grill's temperature to 350°F.
6. Fold in three sides of the foil around the roast and add the beef broth. Fold in the last side, completely enclosing the roast and liquid. Return the wrapped roast to the grill and cook until its internal temperature reaches 195°F.
7. Pull the roast from the grill and place it in a cooler. Cover the cooler and let the roast rest for 1 or 2 hrs.

8. Remove your roast from the cooler and unwrap it. Pull apart the beef using just your fingers. Serve immediately.

Nutrition: Calories: 213 Carbs: 0g Fat: 16g Protein: 15g

49. Smoked Roast Beef

Preparation Time: 10 minutes

Cooking Time: 12 to 14 hours

Servings: 5 to 8

Ingredients:

- 1 (4-pound) top round roast
- 1 batch Espresso Brisket Rub
- 1 tablespoon butter

Directions:

1. Supply your smoker with wood pellets and follow the manufacturer's specific start-up procedure. Allow your griller to preheat with the lid closed, to 180°F.
2. Season the top round roast with the rub. Using your two hands, work the rub into the meat.
3. Place the meat to roast directly on the grill grate and smoke until its internal temperature reaches 140°F. Remove the roast from the grill.
4. Place a cast-iron skillet on the grill grate and increase the grill's temperature to 450°F. Place the roast in the skillet, add the butter, and cook until its internal temperature reaches 145°F, flipping once after about 3 minutes. (I recommend reverse-searing the meat over an open flame rather than in the cast-iron skillet, if your grill has that option.)
5. Remove the food you roast from the grill and let it rest for 10 to 15 minutes, before slicing and serving.

Nutrition: Calories: 290 Carbs: 3g Fat: 9g Protein: 50g

50. Smoked Beef Ribs

Preparation Time: 25 minutes

Cooking Time: 4 to 6 hours

Servings: 4 to 8

Ingredients:

- 2 (2- or 3-pound) racks beef ribs
- 2 tablespoons yellow mustard
- 1 batch Sweet and Spicy Cinnamon Rub

Directions:

1. Supply your smoker with wood pellets and follow the manufacturer's specific start-up procedure. Allow your griller to preheat with the lid closed, to 225°F.
2. Take off the membrane from the backside of the ribs. This can be done by cutting just through the membrane in an X pattern and working a paper towel between the membrane and the ribs to pull it off.
3. Coat the ribs all over with mustard and season them with the rub. Using your two hands, work with the rub into the meat.
4. Put your ribs directly on the grill grate and smoke until their internal

temperature reaches between 190°F and 200°F.

5. Remove the racks from the grill and cut them into individual ribs. Serve immediately.

Nutrition: Calories: 230 Carbs: 0g Fat: 17g Protein: 20g

51. Almond Crusted Beef Fillet

Preparation Time: 15 minutes

Cooking Time: 55 minutes

Servings: 4

Ingredients:

- ¼ cup chopped almonds
- 1 tablespoon Dijon mustard
- 1 Cup chicken broth
- Salt
- 1/3 cup chopped onion
- ¼ cup olive oil
- Pepper
- 2 tablespoons curry powder
- 3 pounds beef fillet tenderloin

Directions:

1. Rub the pepper and salt into the tenderloin.
2. Place the almonds, mustard, chicken broth, curry, onion, and olive oil into a bowl. Stir well to combine.
3. Take this mixture and rub the tenderloin generously with it.
4. Add wood pellets to your smoker and follow your cooker's startup procedure. Preheat your smoker, with your lid closed, until it reaches 450.
5. Lay on the grill, cover, and smoke for ten minutes on both sides.
6. Continue to cook until it reaches your desired doneness.
7. Take it all the grill and let it rest for at least ten minutes.

Nutrition: Calories: 118 Carbs: 3g Fat: 3g Protein: 20g

52. Bacon Cheese Steak Meatloaf

Preparation Time: 15 minutes

Cooking Time: 2 hours

Servings: 8

Ingredients:

- 2 chopped garlic cloves
- 1 chopped poblano chili
- 1 chopped medium onion
- 1 tablespoon canola oil
- 2 pounds extra lean ground beef
- ½ cup tiger sauce
- 2 cups breadcrumbs
- 1 beaten egg
- 2 cups shredded Swiss cheese
- 1 tablespoon A1 Steak sauce
- ½ pound cooked and crumbled bacon
- 2 tablespoons Montreal steak seasoning

Directions:

1. Using a pan on stove top, add in the oil and heat before adding the poblano, onion, and garlic. Allow these veggies to cook for three to five minutes, or until

the onion has become just barely translucent.

2. Add wood pellets to your smoker and follow your cooker's startup procedure. Preheat your smoker, with your lid closed, until it reaches 225.
3. Place the cooked veggies in a container along with the breadcrumbs, egg, Swiss cheese, bacon, steak sauce, steak seasoning, and ground beef. Make sure that you use your hands to mix everything together. They will do a better job than a spoon will. Once everything is very well incorporated, form it into a loaf shape.
4. Place the meatloaf in a cast iron skillet and set this on the grill. Lower the lid on your grill and allow the meatloaf to smoke for two hours. The meat is completely cooked once it has reached 165.
5. Once your meatloaf is cooked through, brush on the tiger sauce and take it off the grill and allow this to rest for ten minutes before you serve the meatloaf.

Nutrition: Calories: 385 Carbs: 2g Fat: 30g Protein: 23g

53. Herbed Beef Eye Fillet

Preparation Time: 15 minutes

Cooking Time: 8 hours

Servings: 6

Ingredients:

- Pepper
- Salt
- 2 tablespoons chopped rosemary
- 2 tablespoons chopped basil
- 2 tablespoons olive oil
- 3 cloves crushed garlic
- ¼ cup chopped oregano
- ¼ cup chopped parsley
- 2 pounds beef eye fillet

Directions:

1. Use salt and pepper to rub in the meat before placing in a container.
2. Place the garlic, oil, rosemary, oregano, basil, and parsley in a bowl. Stir well to combine.
3. Rub the fillet generously with this mixture on all sides. Let the meat to sit on the counter for 30 minutes.
4. Add wood pellets to your smoker and follow your cooker's startup procedure. Preheat your smoker, with your lid closed, until it reaches 450.
5. Lay the meat on the grill, cover, and smoke for ten minutes per side or your preferred tenderness.
6. Once it is done to your likeness, allow it to rest for ten minutes. Slice and enjoy.

Nutrition: Calories: 202 Carbs: 0g Fat: 8g Protein: 33g

54. Balsamic Vinegar Molasses Steak

Preparation Time: 15 minutes

Cooking Time: 8 hours

Servings: 4

Ingredients:

- Pepper
- Salt
- 1 tablespoon balsamic vinegar
- 2 tablespoons molasses
- 1 tablespoon red wine vinegar
- 1 cup beef broth
- 2 ½ pounds steak of choice

Directions:

1. Lay the steaks in a zip top bag.
2. Add the balsamic vinegar, red wine vinegar, molasses, and beef broth to a bowl. Combine thoroughly by stirring.
3. On the top of the steaks, drizzle this mixture.
4. Place into the refrigerator for eight hours.
5. Add wood pellets to your smoker and follow your cooker's startup procedure. Preheat your smoker, with your lid closed, until it reaches 350.
6. Take out the frozen steaks out of the refrigerator 30 minutes before you are ready to grill.
7. Place on the grill, cover, and smoke for ten minutes per side, or until meat is tender.
8. Place onto plates and let them rest ten minutes.

Nutrition: Calories: 164 Carbs: 6g Fat: 5g Protein: 22

55. Herbed Steaks

Preparation Time: 10 minutes

Cooking Time: 5 hours

Servings: 4

Ingredients:

- Pinch red pepper flakes
- ½ teaspoon coriander seeds
- 2 teaspoons green peppercorns
- 2 teaspoons black peppercorns
- 2 tablespoons chopped mint leaves
- ¼ cup olive oil
- 2 tablespoons peanut oil
- 3 pounds flank steak

Directions:

1. Sprinkle the flank steak with salt and rub generously. Lay the meat in a large zip-top bag.
2. Mix together the red pepper flakes, coriander, peppercorns, mint leaves, olive oil, and peanut oil.
3. Pour this mixture over the flank steak.
4. Place into the refrigerator for four hours.
5. Add wood pellets to your smoker and follow your cooker's startup procedure. Preheat your smoker, with your lid closed, until it reaches 450.
6. Take it out the flank steak out of the refrigerator 30 minutes before you are ready to grill it.
7. Put the flank steak onto the grill and grill ten minutes on each. You can grill

longer if you want the steak more well-done.

8. After removing from the grill and set for about ten minutes. Slice before serving.

Nutrition: Calories: 240 Carbs: 12g Fat: 11g Protein: 23g

56. Beer Honey Steaks

Preparation Time: 10 minutes
Cooking Time: 55 minutes
Servings: 4

Ingredients:

- Pepper
- Juice of one lemon
- 1 cup beer of choice
- 1 tablespoon honey
- Salt
- 2 tablespoons olive oil
- 1 teaspoon thyme
- 4 steaks of choice

Directions:

1. Season the steaks with pepper and salt.
2. Combine it together the olive oil, lemon juice, honey, thyme, and beer.
3. Rub the steaks with this mixture generously.
4. Add wood pellets to your smoker and follow your cooker's startup procedure. Preheat your smoker, with your lid closed, until it reaches 450.
5. Place the steaks onto the grill, cover, and smoke for ten minutes per side.
6. For about 10 minutes, let it cool after removing from the grill.

Nutrition: Calories: 245 Carbs: 8gFat: 5g Protein: 40g

57. La Rochelle Steak

Preparation Time: 10 minutes

Cooking Time: 5 hours

Servings: 4

Ingredients:

- 1 tablespoon red currant jelly
- ½ teaspoon salt
- 3 teaspoon curry powder
- 8 ounces pineapple chunks in juice
- 1 ½ pounds flank steak
- ¼ cup olive oil

Directions:

1. Put the flank steak into a large bag.
2. Mix the pepper, salt, red currant jelly, curry powder, pineapple chunks with juice, and olive oil together.
3. Pour this mixture over the flank steak.
4. Place into the refrigerator for four hours.
5. Add wood pellets to your smoker and follow your cooker's startup procedure. Preheat your smoker, with your lid closed, until it reaches 350.
6. Then you are ready to cook the steak, remove steak from refrigerator 30 minutes before ready to cook.
7. Lay the steaks on the grill, cover, and smoke for ten minutes on both sides, or done to your liking.
8. Remove your roasted food from the grill and allow to cool for about ten minutes.

Nutrition: Calories: 200 Carbs: 0g Fat: 7g Protein: 33g

58. Fennel and Thyme Veal Shoulder

Preparation Time: 15 minutes

Cooking Time: 55 minutes

Servings: 4

Ingredients:

- Pepper
- 2 tablespoons thyme
- 4 tablespoons olive oil
- 2 tablespoons chopped thyme
- 1 thinly sliced fennel bulb
- Salt
- 3 ½ pound veal shoulder roast

Directions:

1. Sprinkle the roast with pepper and salt and rub generously.
2. In a bowl, put the oil, wine, fennel, thyme, pepper, and salt. Stir well to combine.
3. Rub the roasted meat generously on all sides with this mixture.
4. Add wood pellets to your smoker and follow your cooker's startup procedure. Preheat your smoker, with your lid closed, until it reaches 450.
5. Cover after placing in the grill. For about 25 minutes cook the meat.
6. Take the roast off the grill and take an internal temperature. It should be no lower than 130.

Nutrition: Calories: 150 Carbs: 5g Fat: 5g Protein: 23g

59. Lemony Mustard Crusted Veal

Preparation Time: 10 minutes

Cooking Time: 2 hours 30 minutes

Servings: 2

Ingredients:

- Pepper
- Salt
- ¼ cup breadcrumbs
- 2 tablespoons water
- 1 teaspoon basil
- 1 pound veal round roast
- 1 tablespoon Dijon mustard
- 1 tablespoon lemon juice

Directions:

1. Lay the roast in a shallow roasting pan on a rack.
2. Mix together the pepper, thyme, basil, lemon juice, mustard, water, and breadcrumbs.
3. Spread this mixture over the roast being sure to get all sides.
4. Add wood pellets to your smoker and follow your cooker's startup procedure. Preheat your smoker, with your lid closed, until it reaches 450.
5. Place the roast onto the grill and cook for ten minutes per side until it is to your desired doneness.
6. Take off from the grill and allow to set for ten minutes.

Nutrition: Calories: 390 Carbs: 0g Fat: 15g Protein: 40g

60. Classic Burger

Preparation Time: 10 minutes

Cooking Time: 1 hour

Servings: 6

Ingredients:

- Pepper
- Salt
- 1 chopped onion
- ½ pound ground pork
- 1 tablespoon chopped parsley
- 4 tablespoon olive oil
- 1 ¼ pounds ground beef
- Toppings of choice

Directions:

1. Combine all together all of the ingredients, except the toppings.
2. Use your hands and mix the ingredients well until everything is thoroughly combined. Form into six patties.
3. Place into the refrigerator for 30 minutes.
4. Add wood pellets to your smoker and follow your cooker's startup procedure. Preheat your smoker, with your lid closed, until it reaches 425.
5. Grill the burgers, covered, for four minutes on each side.
6. Serve with toppings of your choice.

Nutrition: Calories: 237 Carbs: 27g Fat: 9g Protein: 11g

61. Green Burgers

Preparation Time: 10 minutes
Cooking Time: 35 minutes
Servings: 4

Ingredients:

- Pepper
- 2 pounds ground beef
- 1 tablespoon chopped cilantro
- 1 egg
- 1 pound frozen spinach, thawed and drained
- 3 cloves garlic
- 3 tablespoons olive oil
- 1 tablespoon chopped tarragon
- Salt
- 2 chopped green onions

Directions:

1. Wash and chop the green onion. Mix the onion and spinach together.
2. Add in the salt, pepper, cilantro, tarragon, oil, egg, garlic, and ground beef.
3. Use your two hands and mix all the ingredients until everything is thoroughly combined. Shape into six burgers.
4. Add wood pellets to your smoker and follow your cooker's startup procedure. Preheat your smoker, with your lid closed, until it reaches 380.
5. Cover the burgers after placing on the grill. Both sides should be cooked for 5 minutes.
6. Serve with toppings of choice.

Nutrition: Calories: 190 Carbs: 3g Fat: 10g Protein: 21g

62. Stuffed Peppers

Preparation Time: 10 minutes

Cooking Time: 2 hours

Servings: 4

Ingredients:

- 1 tablespoons garlic, minced
- 1 large diced onion
- 1 ½ cup grated cheddar, divided
- 1 teaspoon pepper
- 14 ounces can tomato paste
- 1 teaspoon seasoned salt
- ½ pound sausage
- ½ pound ground beef
- 1 poblano chili, seeded and chopped
- 4 large bell peppers, seeds, top, and core removed

Directions:

1. Lay the peppers inside of a disposable aluminum pan. If they won't stay upright on their own, you can wrap a foil ring around their base to keep them standing.
2. Add wood pellets to your smoker and follow your cooker's startup procedure. Preheat your smoker, with your lid closed, until it reaches 350.
3. Using the stovetop, heat a large pan. Brown the sausage and the beef for five to seven minutes. Drain the fat and crumble up the meat.
4. Mix in the garlic, pepper, and salt, cup of cheese, tomato paste, poblano, and onion. Stir until everything is well combined.
5. Place the meat mixture in the peppers. Put the pan onto the grill, cover, and let them smoke for an hour.

6. Top the peppers with the remaining cheese and allow them to smoke, covered, for another 15 minutes. Serve.

Nutrition: Calories: 243 Carbs: 28g Fat: 7g Protein: 19g

63. T-Bone with Blue Cheese Butter

Preparation Time: 10 minutes

Cooking Time: 1 hour

Servings: 4

Ingredients:

- 2 tablespoons garlic, minced
- 2 tablespoons salt
- 4 ounces T-bone steaks
- ½ cup crumbled blue cheese
- 1 tablespoon pepper
- 4 tablespoons room temperature butter

Directions:

1. Mix the blue cheese and the butter together and set to the side. Do not refrigerate the butter unless you are making it way in advance.
2. Add wood pellets to your smoker and follow your cooker's startup procedure. Preheat your smoker, with your lid closed, until it reaches 165.
3. Rub the steaks with the garlic, pepper, and salt.
4. Cover the steaks after putting them on. For about 30 minutes, allow it to smoke.
5. Turn the heat up to 450 and let the steaks smoke for 15 minutes if you like them medium-rare. Cook longer to reach your desired doneness.
6. Take the steaks out off of the grill and let them rest for 3-5 minutes. Serve them topped with some of the blue cheese butter.

Nutrition: Calories: 267 Carbs: 0g Fat: 16g Protein: 29g

64. Crispy Burnt Ends

Preparation Time: 5 minutes

Cooking Time: 8 hours

Servings: 4

Ingredients:

- ¾ cup favorite barbecue sauce, divided
- 3 tablespoons favorite dry rub
- 3 pounds chuck roast

Directions:

1. Add wood pellets to your smoker and follow your cooker's startup procedure. Preheat your smoker, with your lid closed, until it reaches 275.
2. Liberally rub your chuck roast with your favorite dry rub.
3. Lay the meat directly on the grill, cover, and allow it to smoke for around five minutes, or until it forms a dark bark on the surface and it reaches 165.
4. Use foil to wrap up the meat and continue to let it smoke, covered, for another hour or until it reaches 195.
5. For about fifteen to twenty minutes, let the roast rest after removing from the grill. Cut the roast into two-inch cubes.

6. Place the cubes on a disposable baking pan and toss them in a half.
7. Barbecue Sauce:
8. Put the pan on the grill, cover, and allow it to smoke for another 1 to 2 hours, or until they have become hot and bubbly. The remaining barbecue sauce could be added at the cooking's last thirty minutes. Serve.

Nutrition: Calories: 260 Carbs: 0g Fat: 19g Protein: 23g

65. Roast Beast

Preparation Time: 5 minutes

Cooking Time: 8 hours

Servings: 4

Ingredients:

- 2 tablespoons steak seasoning
- EVOO
- 4 pound rump roast
- 1 tablespoon garlic, minced

Directions:

1. Add wood pellets to your smoker and follow your cooker's startup procedure. Preheat your smoker, with your lid closed, until it reaches 425.
2. Rub the roast all over with a good amount of olive oil and then season with the steak seasoning and garlic.
3. Lay the meat on the grill and then sear every side for about two to five minutes. Set it off of the grill.
4. Turn the heat of the smoker to 225.
5. Lay the roast back on the grill, cover, and allow it to smoke for three to four hours. Depending on your desired doneness, it should reach a temperature of 120 to 155.
6. Use some foil to tent the roast after removing it from the heat. Allow it to roast and rest for ten minutes and slice against the grain. Enjoy.

Nutrition: Calories: 400 Carbs: 21g Fat: 8g Protein: 22g

66. BBQ Brisket

Preparation Time: 12 Hours

Cooking Time: 10 Hours

Servings: 8

Ingredients:

- 1 beef brisket, about 12 pounds
- Beef rub as needed

Directions:

1. Season beef brisket with beef rub until well coated, place it in a large plastic bag, seal it and let it marinate for a minimum of 12 hours in the refrigerator.
2. When ready to cook, switch on the Traeger grill, fill the grill hopper with hickory flavored wood pellets, power the grill on by using the control panel, select 'smoke' on the temperature dial, or set the temperature to 225 degrees F and let it preheat for a minimum of 15 minutes.
3. When the grill has preheated, open the lid, place marinated brisket on the grill grate fat-side down, shut the grill, and

smoke for 6 hours until the internal temperature reaches 160 degrees F.

4. Then wrap the brisket in foil, return it back to the grill grate and cook for 4 hours until the internal temperature reaches 204 degrees F.
5. When done, transfer brisket to a cutting board, let it rest for 30 minutes, then cut it into slices and serve.

Nutrition: Calories: 328 Fat: 21g Carbs: 0g Protein: 32g

67. Prime Rib Roast

Preparation Time: 24 Hours

Cooking Time: 4 Hours and 30 Minutes

Servings: 8

Ingredients:

- 1 prime rib roast, containing 5 to 7 bones
- Rib rub as needed

Directions:

1. Season rib roast with rib rub until well coated, place it in a large plastic bag, seal it and let it marinate for a minimum of 24 hours in the refrigerator.
2. When ready to cook, switch on the Traeger grill, fill the grill hopper with cherry flavored wood pellets, power the grill on by using the control panel, select 'smoke' on the temperature dial, or set the temperature to 225 degrees F and let it preheat for a minimum of 15 minutes.
3. When the grill has preheated, open the lid, place rib roast on the grill grate fat-side up, change the smoking temperature to 425 degrees F, shut the grill, and smoke for 30 minutes.
4. Then change the smoking temperature to 325 degrees F and continue cooking for 3 to 4 hours until roast has cooked to the desired level, rare at 120 degrees F, medium rare at 130 degrees F, medium at 140 degrees F, and well done at 150 degrees F.
5. When done, transfer roast rib to a cutting board, let it rest for 15 minutes, then cut it into slices and serve.

Nutrition: Calories: 248 Fat: 21.2g Carbs: 0g Protein: 28g

68. Grilled Filet Mignon

Servings: 3

Calories: 229

Cooking Time: 20 minutes

Ingredients:

- Salt
- Pepper
- Filet mignon - 3

Directions:

1. Preheat your grill to 450 degrees.
2. Season the steak with a good amount of salt and pepper to enhance its flavor.
3. Place on the grill and flip after 5 minutes.
4. Grill both sides for 5 minutes each.

5. Take it out when it looks cooked and serve with your favorite side dish.

Nutrition: Carbohydrates: 0 g; Protein: 23 g; Fat: 15 g; Sodium: 240 mg; Cholesterol: 82 mg

69. Kalbi Beef Ribs

Servings: 6

Calories: 355

Cooking Time: 23 minutes

Ingredients:

- Thinly sliced beef ribs - 2 ½ lbs
- Soy sauce - ½ cup
- Brown sugar - ½ cup
- Rice wine or mirin - ⅛ cup
- Minced garlic - 2 tbsp
- Sesame oil - 1 tbsp
- Grated onion - ⅛ cup

Directions:

1. In a medium-sized bowl, mix the mirin, soy sauce, sesame oil, brown sugar, garlic, and grated onion.
2. Add the ribs to the bowl to marinate and cover it properly with cling wrap. Put it in the refrigerator for up to 6 hours.
3. Once you remove the marinated ribs from the refrigerator, immediately put them on the grill. Close the grill quickly so no heat is lost. Also, make sure the grill is preheated well before you place the ribs on it.
4. Cook on one side for 4 minutes and then flip it. Cook the other side for 4 minutes.
5. Pull it out once it looks fully cooked. Serve it with rice or any other side dish

Nutrition: Carbohydrates: 22 g; Protein: 28 g; Fat: 6 g; Sodium: 1213 mg; Cholesterol: 81 mg

70. Beef Jerky

Servings: 10

Calories: 309

Cooking Time: 5 hours

Ingredients:

- Sirloin steaks - 3 lbs
- Soy sauce - 2 cups
- Brown sugar - ½ cup
- Pineapple juice - 1 cup
- Mayo - 2 tbsp
- Hoisin - 2 tbsp
- Onion powder - 2 tbsp
- Rice wine vinegar - 2 tbsp
- Minced garlic - 2 tbsp

Directions:

1. Mix all the ingredients to prepare the marinade.
2. Lock them all in a zip-lock bag with the beef. Make sure the bag is airtight.
3. Refrigerate the bag and let it soak in the marinade for 6 to 24 hours.

4. Remove it from the refrigerator and put it in a bowl 1 hour before cooking.
5. Start your grill on the smoke setting.
6. Put your marinated beef on the grill, leaving a ½-inch space between each of the slices.
7. Leave them on the grill for 4-5 hours. Turn them once after 2 or 2 ½ hours.
8. Remove from the grill when done and let the beef sit for 30 minutes to 1 hour to cool down.
9. After some time, put it in the storage or refrigerator.
10. Try your beef jerky before you store it!

Nutrition: Carbohydrates: 20 g; Protein: 34 g; Fat: 3 g; Sodium: 2832 mg; Cholesterol: 85 mg

71. Homemade Meatballs

Servings: 12

Calories: 453

Cooking Time: 1 hour 20 minutes

Ingredients:

- Ground beef - 2 lbs
- White bread - 2 slices
- Whole milk - ½ cup
- Salt - 1 tbsp
- Onion powder - ½ tbsp
- Italian seasoning - 2 tbsp
- Ground black pepper- ¼ tbsp
- Minced garlic - ½ tbsp

Directions:

1. Combine the whole milk, white bread, minced garlic, onion powder, Italian seasoning, and black pepper.
2. Add the ground beef and mix well.
3. Preheat your wood pellet grill on the 'smoke' option and leave the lid open for 4-5 minutes.
4. Line a baking sheet and start placing small balls on the sheet.
5. Smoke for 35 minutes and then flip the balls.
6. Let it stay from 35 more minutes.
7. Once it turns golden brown, serve hot!

Nutrition: Carbohydrates: 7 g; Protein: 42 g; Fat: 27 g; Sodium: 550 mg; Cholesterol: 137 mg

72. Rib Roast

Servings: 8

Calories: 721

Cooking Time: 2 hours 10 minutes

Ingredients:

- Boneless rib roast - 5 lbs
- Beef broth - 2 cups
- Celery - ½ chopped
- Carrots - ½ cup chopped
- Rosemary - 1 tbsp
- Onion - ½ cup
- Granulated garlic - 1 tbsp
- Onion powder - 1 ½ tbsp
- Kosher salt - 4 tbsp
- Black ground pepper - 1 tbsp

Directions:

- Preheat your wood pellet grill to 250 degrees.
- Take the beef out of the refrigerator at least 1 hour before cooking.
- Mix the pepper, salt, rosemary, onion powder, and garlic in a bowl.
- Coat the rib roast with this mix. Set the roast aside after coating it well.
- Mix the onions, celery, and carrots in a high-sided pan.
- Place the coated roast rib on top of the vegetables. Place it on the preheated grill.
- After 1 hour, pour the beef broth on the container.
- Cook until the temperature reaches 120 degrees.
- Pull out the roast and let it sit for 20 minutes.
- Skim off the fat and strain the juice from the bottom.

Nutrition: Carbohydrates: 3 g; Protein: 42 g; Fat: 60 g; Sodium: 2450 mg; Cholesterol: 207 mg

73. Grilled Hanger Steak

Servings: 6
Calories: 133
Cooking Time: 50 minutes

Ingredients:

- Hanger Steak - 1
- Salt
- Pepper
- For Bourbon Sauce
- Bourbon whiskey - 1/8 cup
- Honey - 1/8 cup
- Sriracha - 1 tbsp
- Garlic - 1/2 tbsp
- Salt - 1/4 tbsp

Directions:

- Preheat the grill to 225 degrees.
- Use pepper and salt to season the steak liberally.
- Place the steak on the grill and close the lid.
- Let it cook until the temperature goes down to the finish.
- Take an iron skillet and place it on the stove.
- Add some butter to the pan and place the steak on it.
- Cook on both sides for 2 minutes each.
- Remove the steak from the stove.
- Add the bourbon sauce ingredients to the pan.
- Cook and whisk for 3-4 minutes. Pour it over your steak.
- Serve with your favorite side dish or simply have it with the bourbon sauce.

Nutrition: Carbohydrates: 6 g; Protein: 10 g; Fat: 7 g; Sodium: 180 mg; Cholesterol: 36 mg

74. Smoked Peppered Beef Tenderloin

Servings: 4-6
Calories: 70
Cooking Time: 105 minutes

Ingredients:

- Cloves of minced garlic - 2
- Snake River trimmed beef tenderloin roast - 1
- Strong cold coffee or bourbon - 2 tbsp
- Dijon mustard - ½ cup
- Coarsely ground green and black peppercorns
- Jacobsen salt

Directions:

1. Lay the tenderloin gently on a large piece of clean plastic wrap.
2. Combine the garlic, bourbon, and mustard in a small bowl. Slather the mixture over the tenderloin evenly. Let it sit for an hour at room temperature.
3. Unwrap the plastic wrap. Season the tenderloin with the ground green and black peppercorns and salt generously on all sides.
4. Once it is ready to cook, preheat the pellet grill to 180 degrees for 15 minutes with the lid closed.
5. Arrange the tenderloin on the grill grate directly and smoke it for about 1 hour.
6. Increase the temperature of the grill to around 400 degrees. Roast the tenderloin properly for 20-30 minutes until the internal temperature of the meat reaches 130 degrees. The time depends on the overall thickness of the tenderloin. Be careful not to overcook the meat!
7. Let it rest for around 10 minutes prior to slicing it. Enjoy!

Nutrition: Carbohydrates: 3 g; Protein: 10 g; Fat: 1 g; Sodium: 590 mg; Cholesterol: 20 mg

75. Slow Roasted Shawarma

Servings: 6-8

Calories: 377.3

Cooking Time: 4 hours 55 minutes

Ingredients:

- Top sirloin - 5.5 lbs
- Lamb fat - 4.5 lbs
- Boneless, skinless chicken thighs- 5.5 lbs
- Pita bread
- Traeger rub - 4 tbsp
- Double skewer - 1
- Large yellow onions - 2
- Variety of topping options such as tomatoes, cucumbers, pickles, tahini, Israeli salad, fries, etc.
- Cast iron griddle

Directions:

1. Assemble the stack of shawarma the night before you wish to cook it.
2. Slice all the meat and fat into ½-inch slices. Place them into 3 bowls. If you partially freeze them, it will be much easier to slice them.
3. Season the bowl with the rub, massaging it thoroughly into the meat.
4. Place half a yellow onion on the bottom of the skewers to ensure a firm base. Add 2 layers at a time from each bowl. Try to make the entire stack symmetrical. Place the other 2 onions on top. Wrap them in plastic wrap and refrigerate overnight.
5. When the meat is ready to cook, preheat the pellet grill for about 15 minutes with the lid closed at a
6. temperature of 275 degrees Fahrenheit.
7. Lay the shawarma directly on the grill grate and cook it for at least 3-4 hours. Rotate the skewers at least once.
8. Remove them from the grill and increase its temperature to 445 degrees Fahrenheit. During the time when the grill is preheating, place a cast iron griddle directly on the grill grate and brush it with some olive oil.
9. Once the griddle is hot enough, place the shawarma directly on the cast iron. Sear it on each side for 5-10 minutes. Remove it from the grill and slice off the edges. Repeat the process with the remaining shawarma.
10. Serve in pita bread along with your favorite toppings, such as tomatoes, cucumbers, Israeli salad, fries, pickles, or tahini. Enjoy!

Nutrition: Carbohydrates: 4.6 g; Protein: 30.3 g; Fat: 26.3 g; Sodium: 318.7 mg; Cholesterol: 125.5 mg

76. Tomato Vinegar Beef Tenderloin

Servings: 8

Calories: 412

Cooking Time: about 55 minutes

Ingredients:

- Tomatoes—6, cored, seeds removed, and peeled
- Beef tenderloin—1, trimmed
- Balsamic vinegar—2 tbsps.
- Extra virgin olive oil—2/3 cup Rib rub—any of your choice
- Fresh thyme—1 tsp., minced

Directions:

1. Prepare your Wood Pellet Smoker-Grill by preheating it to a temperature of about 450°F. Close the top lid and leave for 12–18 minutes.
2. Coat the meat properly with olive oil, rib rub, pepper, and salt. Transfer the prepared meat to a shallow roasting pan.
3. Roast for about 18–20 minutes. Reduce the temperature to 350°F and keep roasting for another 18–20 minutes.
4. Remove and allow to cool for 6–10 minutes.
5. Cut thin slices and use thyme to garnish.
6. Make a mixture of balsamic vinegar, tomatoes, thyme leaves, and olive oil in your blender or food processor.
7. Pour it all over the sliced meat.
8. Enjoy.

Nutrition: Carbohydrate 0 g; Protein 32 g; Fat 30 g; Sodium 256 mg; Cholesterol 118 mg

77. Spicy Grilled Beef Steak

Servings: 6

Calories: 634

Cooking Time: about 1 hour and 22 minutes

Ingredients:

- Chili powder—2 tbsps.
- Beef rib eye—4 steaks
- Brown sugar—1 tsp.
- Worcestershire sauce—2 tbsps.
- Garlic cloves—2, minced
- Ground cumin—1 tsp.
- Olive oil—2 tbsps.
- Salt—1 tsp.

Directions:

1. Mix salt and mashed garlic in a small mixing bowl. Add Worcestershire sauce, chili powder, brown sugar, olive oil, and cumin.
2. Use this mixture to coat the steaks.
3. Put the coated steaks, as well as the rest of the rub, in a large zip seal bag. Let it marinate in the refrigerator for about 5–24 hours.
4. Prepare your Wood Pellet Smoker-Grill by preheating it to about 225°F. Close the top lid and leave for 12–18 minutes.
5. Smoke the steaks for about 50–60 minutes. Then, remove.

6. Increase the temperature to about 350°F and cook the steaks again to get an internal temperature of about 135°F.
7. Remove and allow the meat to cool.
8. Your dish is ready to be served.

Nutrition: Carbohydrate 34 g; Protein 67 g; Fat 13 g; Sodium 786 mg; Cholesterol 160 mg

78. Wood Pellet Smoked Meat Loaf

Preparation Time: 10 Minutes

Cooking time: 60 Minutes

Servings: 8

Ingredients:

- 1 and ½ lbs of ground beef
- ½ lb of sausage
- ½ Cup of bread crumbs
- ¾ Cup of plain yogurt
- ¼ Cup of milk
- 2 Large eggs
- 2 Teaspoons of chopped garlic
- ½ Cup of Parmesan cheese
- 1 Tablespoon of dried parsley
- 1 Teaspoon of dried oregano
- 1 and ½ teaspoons of kosher salt
- Your favorite BBQ rub 1
- Pinch of black pepper

Directions:

- Start your pellet smoker grill and turn it to a temperature of about 350°F to heat it up.
- In a medium size bowl, mix all your wet ingredients together; then place the seasonings in it and mix the sausage and the ground beef all together in a bowl Form a loaf of your mixture; then roll it into your favorite rub. Place the beef loaf on the pellet grill rack; then put the meat probe right into the center.
- Smoke the Meat loaf on High smoke for about 30 minutes at a temperature of 350°F until the internal temperature displays 160° F; it may take an hour
- Slice your meat loaf; then serve and enjoy its delicious taste.

Nutrition: Calories: 343, Fat: 25.6g, Carbohydrates: 10g, Protein: 17.2g, Dietary Fiber 0.5 g

79. Wood Pellet Corned Beef with Cabbage

Preparation Time: 10 Minutes

Cooking time: 30 Minutes

Servings: 4

Ingredients:

- A cut of corned Beef
- 2 Cups of water
- 5 to 6 red potatoes
- 1 Head of cabbage
- 3 Teaspoon of garlic salt
- 1 Teaspoon of ground black pepper
- 3 to 4 tablespoons of whole grain mustard
- 3 Tablespoons of melted butter

Directions:

- Start by rinsing the two sides of the corned beef under cold water for about 2 minutes in order to excess any excess of salt
- Coat both the sides of the corned beef with 2 tablespoons of mustard
- Add the water and the corned beef to an aluminum pan and smoke it at a temperature of about 220° F
- Remove the stem and remove the core of the cabbage; then quarter it
- Melt the butter; then stir in about 1 tablespoon of mustard and about 1 teaspoon of garlic salt
- Place the cabbage quarters into the aluminum pan in each of the corners and core it side up so that it looks like bowls
- Chop the potatoes into half and season it with about 2 teaspoons of garlic salt and about ½ teaspoon of pepper.
- Place the potatoes along the edges of the aluminum pan between the quarters of the cabbage
- Cover with the aluminum foil; then turn up your wood pellet grill to about 280°F and cook for about 2 additional hours until the internal temperature of meat reaches about 200 to 205° F
- Remove the aluminum foil and cook for about 15 minutes. Slice the meat; then serve and enjoy it!

Nutrition: Calories: 213.5, Fat: 15g, Carbohydrates: 8g, Protein: 7.9g, Dietary Fiber: 1.2g

80. Smoked Porterhouse Steak

Preparation Time: 1 hour

Servings: 2-4

The Meat:

- 2 porterhouse steaks (1 inch thick) (20 oz. or 1.25 lb.).
- The Mixture:
- Melted butter – 4 tbsp.
- Worcestershire Sauce – 2 tsp.
- Dijon Mustard – 1 tbsp.
- Coffee rub – 1 tsp.

The Fire:

- Wood pellet smoker, hickory wood pellets.

Directions:

- Start your wood pellet smoker on grill instructions when you are ready to cook.
- Set the temperature to smoke setting. Preheat for 5 minutes while keeping the lid closed.
- Mix the butter, Worcestershire sauce, and mustard until it is smooth. Brush the mixture on both sides of the steaks. Season the porterhouse steaks with coffee rub.
- Arrange the steaks on the grill grate and smoke them for about 30 minutes. Remove the steaks using tongs.
- Increase the heat to 450° F. Brush the steaks again with the butter sauce mixture that was prepared earlier.
- When the grill comes up to the temperature, place the steaks back on the grill grate and cook until it is done according to your choice. Whichever doneness you prefer i.e. rare, medium rare or well done.
- In case of medium rare, cook until the internal temperature is about 135° F. Before serving, let the steaks rest for 5 minutes.

81. Corned Beef Pastrami

Preparation Time: 6 hours

Servings: 5-6 Serving

The Meat:

- Pre-packed corned beef (silverback) (6 lbs.)

The Mixture:

- Coarse ground pepper – 3 tbsp.
- Garlic (very fine grains) – 2 tbsp.
- Onion Flakes – 2 tbsp.
- Chili Powder – 2 tbsp.
- Expresso rub – 2 tbsp.
- The Fire:
- Wood pellet smoker, pecan wood pellets.

Directions:

- First, the beef cut must be rinsed off thoroughly. Afterward, let it soak in cold water for a whole day. The water should be changed every six hours. In this way, more salt and cure are pulled from the meat.
- After it has been soaking for a whole day, take the beef out of the water. Dry it thoroughly with paper towel.
- Mix all of the rub ingredients in a bowl and then apply them to the beef. Set the beef aside while you preheat your smoker.
- Place the beef in the smoker with temperature up to 250 degrees. Cook until the internal temperature is 150 degrees. To reach this temperature, the beef should be cooking for about 3-4 hours.

- Now wrap the beef cut in a foil. Keep cooking until the internal temperature goes up to 185 -195 degrees. It should take about 2 more hours to reach this internal temperature.
- The corned beef pastrami is ready.

82. Smoked Rib Eye with Bourbon Butter

Preparation Time: 1 hour 30 minutes

Servings: 4-6 Serving

The Meat:

- 4 Ribeye steaks (1 inch thick).

The Mixture:

Fresh ground pepper – 1/2 tsp.

- Garlic (minced) – 1-2 cloves.
- Green Onion (finely minced) – 1 tbsp.
- Parsley (finely minced) – 1 tbsp.
- Salt– 1/2 tsp.
- Bourbon – 2 tbsp.
- Butter – 1/2 cup.
- Traeger prime rub.

The Fire:

- Wood pellet smoker, mesquite wood pellets.

Directions:

- In a mixing bowl, add butter, parsley, chives, bourbon, garlic, salt, and pepper. Stir all of the ingredients with a wooden spoon. Put the smoker on.
- Until the fire is established, keep the lid open. The fire should be established in about 4-5 minutes.
- Season the rib eye steaks with a prime rub. Arrange the steaks on the grill and smoke them for about an hour.
- Then temporarily remove the steaks from the smoker and set your smoker's temperature up to 450F. Place the steaks back on the grill. Give one side about 6-8 minutes and then turn it over.
- Keep the other side for the same amount of time. Otherwise, if you like your steak medium rare then keep cooking and turning over until the internal temperature is 135 F.
- Take the steak out of the smoker after they are cooked to your liking and immediately pat them with the bourbon butter sauce you made earlier.
- Let the meat rest for about 3 minutes before serving it.

LAMB RECIPES

83. Smoked Lamb Stew

Preparation Time: 2 hours

Servings: 4-6

Ingredients:

The Meat:

- Lamb (1/2 inch chunks) – (3 lbs.)

The Mixture:

- Beef Stock – 2 cups.
- 4 chopped Garlic cloves.
- Tomato paste – ¼ cup.
- Beer – 12 oz.
- 2 Bay leaves.
- Olive oil – 3 tbsp.
- 3 diced Large Carrots.
- Dried Thyme - 2 tsp.
- 1 diced turnip Peas - 2 cup
- 1 diced parsnip.
- Coarse kosher salt and pepper (as required).

The Fire:

Wood pellet smoker, oak wood pellets.

Directions:

1. Start your wood pellet grill on smoke. For 5 minutes keep the lid open until the fire is established.

2. Then take the temperature to 450 F. while keeping the lid closed preheat for 15 minutes more.
3. Use salt and pepper for seasoning the lamb. Heat oil in a wood pellet smoker and brown all sides separately for 6-8 minutes.
4. Now add all the lamb pieces in the Dutch oven. Add garlic and sauté for exactly 2 minutes. Then after that, put in tomato paste and cook for one more minute.
5. Then, collectively add beer, beef stock, thyme, bay leaves, salt, and pepper. Cook this now on a high heat.
6. The remaining vegetables are also browned while meat has been cooking. In the cooking pot, put in all the vegetables and let it cook for one hour.
7. Finally, the dish is ready to serve.

84. Ground Lamb Kebabs

Preparation Time: 1 hour

Servings: 2-4

The Meat:

- Ground Lamb – 1-1/2 lb.

The Mixture:

- Minced onions – 1/3 cup.
- Minced garlic – ½ cloves.
- Cilantro – 3 tbsp.
- Minced fresh mint – 1 tbsp.
- Ground cumin 1 tsp.
- Paprika – 1 tsp.
- Salt – 1 tsp.
- Ground coriander – ½ tsp.
- Cinnamon – ¼ tsp.
- Pita bread - to serve.

The Fire:

Wood pellet smoker, cherry wood pellets.

Directions:

1. First of all, take a large mixing bowl. Put in all of the ingredients except for the pita bread. Now start making meatballs out of the mixture. The meatballs should be about 2 inches in diameter.
2. Take a bamboo skewer for each of the meatballs. Now, wet your hands to easily mold the skewered meal. Mold them each into a cigar shape.
3. Leave it in the refrigerator for 30 minutes at least or preferably overnight. Set your wood pellet smoker on the smoke option and with the lid open the fire establish. It takes about 4-5 minutes.
4. Then after that set the temperature to 350F and preheat it for about 10-15 minutes. During preheating keep the lid closed. Place the kebabs on the grill.
5. After 30 minutes turn them over.
6. Also, if the internal temperature reads 160F then it is time to turn them over. Warm the bread before serving it with kebobs.

85. Lamb Lollipops with Mango Chutney

Preparation Time: 1 hour

Servings: 4-6

The Meat:

- 6 Lamb chops (frenched) – (3/4 inch thick).

The Mixture:

- Olive oil – 2 tbsp.
- Kosher salt – ½ tsp.
- For Mango chutney:
- 1 chopped mango.
- 3 Garlic cloves (chopped).
- ½ chopped habanero pepper.
- Salt – 1 tsp.
- Fresh lime juice – 1 tbsp.
- Chopped fresh cilantro – 3 sprigs.
- Pepper – ½ tsp.

The Fire:

Wood pellet smoker, apple wood pellets.

Directions:

1. Combine all the ingredients for the chutney and put them in a food processor. Pulse them till you get your desired consistency. Set this aside.
2. With the lid open, establish fire in your wood pellet smoker for about 4-5 minutes. Then preheat it at a high temperature.
3. Keep the lid closed for about 15 minutes of preheating. Put the lamb chops on a baking sheet and drizzle olive oil on it. Then season with salt and pepper. Allow them to sit for 10 minutes at room temperature. On the grill grate place your lamb chops. Now close the lid and grill for 5 minutes.
4. Then flip the chops over and allow them to smoke for another 3 minutes. If you want precise internal temperature, it should be 130 F to indicate that it is cooked through.
5. Remove the meat from the grill after that. Allow it to rest at least 10 minutes before serving them with the delicious chutney and sprinkled chopped mint.

86. Lamb Rack Wrapped In Apple Wood Walnut

Preparation Time: 25 Minutes
Cooking Time: 60 to 90 Minutes
Servings: 4

Ingredients:

- 3 tablespoons of Dijon mustard
- 2 pieces of garlic, chopped or 2 cups of crushed garlic
- ½ teaspoon of garlic
- ½ teaspoon kosher salt
- ½ teaspoon black pepper
- ½ teaspoon rosemary
- 1 (1½ pound) ram rack, French
- 1 cup crushed walnut

Directions:

1. Put mustard, garlic, garlic powder, salt, pepper and rosemary in a small bowl.
2. Spread the seasoning mix evenly on all sides of the lamb and sprinkle with crushed walnuts. Lightly press the

walnuts by hand to attach the nuts to the meat.

3. Wrap the walnut-coated lamb rack loosely in plastic wrap and refrigerate overnight to allow the seasoning to penetrate the meat.
4. Remove the walnut-covered lamb rack from the refrigerator and let it rest for 30 minutes to reach room temperature.
5. Set the wood pellet r grill for indirect cooking and preheat to 225 ° F using apple pellets.
6. Lay the grill directly on the rack with the lamb bone down.
7. Smoke at 225 ° F until the thickest part of the ram rack reaches the desired internal temperature. This is measured with a digital instantaneous thermometer near the time listed on the chart.
8. Place the mutton under a loose foil tent for 5 minutes before eating

Nutrition: Calories: 165 Carbs: 0g Fat: 8g Protein: 20g

87. Roasted Lamb Leg

Preparation Time: 20 Minutes
Cooking Time: 1.5 Hours to 2 Hours
Servings: 6

Ingredients:

- 1 boneless leg of a lamb
- ½ cup of roasted garlic flavored extra virgin olive oil
- ¼ cup dried parsley
- 3 garlics, chopped
- 2 tablespoons of a fresh lemon juice or 1 tablespoon of lemon zest (from 1 medium lemon)
- 2 tablespoons of dried oregano
- 1 tablespoon dried rosemary
- ½ teaspoon black pepper

Directions:

1. Remove the net from the lamb's leg. Cut off grease, silver skin, and large pieces of fat.
2. In a small bowl, mix olive oil, parsley, garlic, lemon juice or zest, oregano, rosemary, and pepper.
3. Spice the inside and outside surfaces of the lamb's boneless legs.
4. Secure the boneless lamb leg using a silicone food band or butcher twine. Use a band or twine to form and maintain the basic shape of the lamb
5. Wrap the lamb loosely in plastic wrap and refrigerate overnight to allow the seasoning to penetrate the meat.
6. Remove the rum from the refrigerator and leave at room temperature for 1 hour.
7. Set up a wood pellet smoker and grill for indirect cooking and preheat to 400 ° F using selected pellets.
8. Remove the wrap from the ram.
9. Insert a wood pellet smoker and grill meat probe or remote meat probe into the thickest part of the lamb. If your grill does not have a meat probe or you do not have a remote meat probe, use an instant reading digital thermometer to read the internal temperature while cooking. Roast the lamb at 400 ° F until the internal temperature of the thickest part reaches the desired finish.

10. Place the lamb under the loose foil tent for 10 minutes, then cut it against the grain and eat.

Nutrition: Calories: 200 Carbs: 1g Fat: 13g Protein: 20g

88. Greek Leg of Lamb

Preparation Time: 15 minutes
Cooking Time: 25 minutes
Servings: 6

Ingredients:

- 2 tablespoons finely chopped fresh rosemary
- 1 tablespoon ground thyme
- 5 garlic cloves, minced
- 2 tablespoons sea salt
- 1 tablespoon freshly ground black pepper
- Butcher's string
- 1 whole boneless (6- to 8-pound) leg of lamb
- ¼ cup extra-virgin olive oil
- 1 cup red wine vinegar
- ½ cup canola oil

Directions:

1. In a container, combine the rosemary, thyme, garlic, salt, and pepper; set aside.
2. Using butcher's string, tie the leg of lamb into the shape of a roast. Your butcher should also be happy to truss the leg for you.
3. Rub the lamb generously with the olive oil and season with the spice mixture. Put it to a plate, cover with plastic wrap, and refrigerate for 4 hours.
4. Remove the lamb from the refrigerator but do not rinse.
5. Supply your smoker with wood pellets and follow the manufacturer's specific start-up procedure. Preheat, with the lid closed, to 325°F.
6. In a small bowl, combine the red wine vinegar and canola oil for basting.
7. Place the lamb directly on the grill, close the lid, and smoke for 20 to 25 minutes per pound (depending on desired doneness), basting with the oil and vinegar mixture every 30 minutes. Lamb is generally served medium-rare to medium, so it will be done when a meat thermometer where inserted in the thickest part reads 140°F to 145°F.
8. Let the lamb meat rest for about 15 minutes before slicing to serve.

Nutrition: Calories: 130 Carbs: 2g Fat: 5g Protein: 19g

89. Smoked Christmas Crown Roast of Lamb

Preparation Time: 1 hour
Cooking Time: 2 hours
Servings: 4

Ingredients:

- 2 racks of lamb, trimmed, drenched, and tied into a crown
- 1¼ cups extra-virgin olive oil, divided
- 2 tablespoons chopped fresh basil
- 2 tablespoons chopped fresh rosemary
- 2 tablespoons ground sage
- 2 tablespoons ground thyme
- 8 garlic cloves, minced
- 2 teaspoons salt

- 2 teaspoons freshly ground black pepper

Directions:

1. Set the lamb out in the counter to take the chill off, about an hour.
2. In a container, combine 1 cup of olive oil, the basil, rosemary, sage, thyme, garlic, salt, and pepper.
3. Baste the entire crown with the herbed olive oil and wrap the exposed drenched bones in aluminum foil.
4. Supply your smoker with wood pellets and follow the manufacturer's specific start-up procedure. Preheat, with the lid closed, to 275°F.
5. Put the lamb directly on the grill, close the lid, and smoke for 1 hour 30 minutes to 2 hours, or wait until a meat thermometer inserted in the thickest part reads 140°F.
6. Remove the lamb from the heat, tent with foil, and let rest for about 15 minutes before serving. The temperature will rise about 5°F during the rest period, for a finished temperature of 145°F.

Nutrition: Calories: 206 Carbs: 4g Fat: 9g Protein: 32g

90. Succulent Lamb Chops

Preparation Time: 15 minutes
Cooking Time: 20 minutes
Servings: 4

Ingredients:

- For The Marinade
- ½ cup rice wine vinegar
- 1 teaspoon liquid smoke
- 2 tablespoons extra-virgin olive oil
- 2 tablespoons dried minced onion
- 1 tablespoon chopped fresh mint
- For The Lamb Chops
- 8 (4-ounce) lamb chops
- ½ cup hot pepper jelly
- 1 tablespoon Sriracha
- 1 teaspoon salt
- 1 teaspoon freshly ground black pepper

Directions:

1. In a small container, whisk together the rice wine vinegar, liquid smoke, olive oil, minced onion, and mint.
2. Place the lamb chops in an aluminum roasting pan. Pour the marinade over the meat, turning to coat thoroughly. Cover it with a plastic wrapper and marinate in the refrigerator for 2 hours.
3. Supply your smoker with wood pellets and follow the manufacturer's specific start-up procedure. Preheat, with the lid closed, to 165°F, or the "Smoke" setting.
4. Put your saucepan top of the stove then low heat, combine the hot pepper jelly and Sriracha and keep warm.
5. When you are going to cook the chops, remove them from the marinade and pat dry. Discard the marinade.

6. Season all the chops with a salt and pepper, then place them directly on the grill grate, close the lid, and smoke for 5 minutes to "breathe" some smoke into them.
7. Remove the chops from the grill. Increase the pellet cooker temperature to 450°F, or the "High" setting. Once your griller is up to temperature, place the chops on the grill and sear, cooking for 2 minutes per side to achieve medium-rare chops. A meat thermometer that usually inserted in the thickest part of the meat should read 145°F. Continue grilling, if necessary, to your desired doneness.
8. Serve the chops with the warm Sriracha pepper jelly on the side.

Nutrition: Calories: 277 Carbs: 0g Fat: 26g Protein: 18g

91. Roasted Rosemary Lamb

Preparation Time: 15 minutes
Cooking Time: 4 hours
Servings: 2

Ingredients:

- 1 lamb rack
- 2 rosemary sprigs, chopped
- Salt and pepper to taste
- 12 baby potatoes
- 1/2 cup butter
- 1 bunch asparagus
- 2 tablespoons olive oil

Directions:

1. Set your wood pellet grill to 225 degrees F. Sprinkle the lamb with the rosemary, salt and pepper.
2. In a baking pan, add the potatoes and coat with the butter. Add the lamb to the grill.
3. Place the pan with potatoes beside the lamb. Roast for 3 hours.
4. Coat the asparagus with the olive oil. In your last twenty minutes of cooking, stir the asparagus into the potatoes.
5. Serve the lamb with the asparagus and baby potatoes.

Nutrition: Calories: 197 Carbs: 3g Fat: 14g Protein: 15g

92. Grilled Lamb

Preparation Time: 10 minutes
Cooking Time: 1 hour
Servings: 6

Ingredients:

- 2 racks of lamb, fat trimmed
- 2 tablespoons Dijon mustard
- Steak seasoning
- 1 teaspoon fresh rosemary, chopped
- 1 tablespoon fresh parsley, chopped

Directions:

1. Coat the lamb with the mustard.
2. Sprinkle all sides with the seasoning, rosemary and parsley.
3. Set your wood pellet grill to 400 degrees F.
4. Sear the meat side of the lamb for 6 minutes.
5. Reduce temperature to 300 degrees F.
6. Grill it for about 20 minutes, turning once or twice.

7. Let rest for 10 minutes before slicing and serving.

Nutrition: Calories: 241 Carbs: 0g Fat: 17g Protein: 21g

93. Chipotle Lamb

Preparation Time: 15 minutes
Cooking Time: 2 hours 30 minutes
Servings: 6

Ingredients:

- 1 rack lamb ribs
- 3/4 cup olive oil
- Pepper to taste
- 1 tablespoon chipotle powder
- 3 cloves, garlic
- 1/4 cup Apple wood bacon rub
- 2 tablespoons rosemary, chopped
- 2 tablespoons thyme, chopped
- 2 tablespoons sage, chopped
- 2 tablespoons parsley

Directions:

1. Coat the lamb ribs with olive oil. Season with the pepper and chipotle powder.
2. Marinate for 15 minutes. Set your wood pellet grill to 275 degrees F.
3. Combine the rest of the ingredients. Spread the mixture on all sides of the lamb.
4. Cook thc lamb for 2 hours. Allow it to rest about ten minutes before carving and serving.

Nutrition: Calories: 210 Carbs: 0g Fat: 13g Protein: 22g

94. Hickory Rack of Lamb

Preparation Time: 10 minutes
Cooking Time: 2 hours
Servings: 3

Ingredients:

- 1 (3 pounds) rack of lamb (drenched)
- Marinade Ingredients:
- 1 lemon (juiced)
- 1 teaspoon ground black pepper
- 1 teaspoon thyme
- ¼ cup balsamic vinegar
- 1 teaspoon dried basil
- 2 tablespoons Dijon mustard
- 2 cloves garlic (crushed)

Rub Ingredients:

- ½ teaspoon cayenne pepper
- ½ teaspoon ground black pepper
- ¼ teaspoon Italian seasoning
- 1 teaspoon oregano
- 1 teaspoon dried mint
- 1 teaspoon paprika
- 1 teaspoon garlic powder
- 1 teaspoon onion powder
- 1 teaspoon dried parsley
- 1 teaspoon dried basil
- 1 teaspoon dried rosemary
- 4 tablespoons olive oil

Directions:

1. Put all the marinade ingredients in a empty container. Pour the marinade into a gallon zip-lock bag. Add the rack of lamb and massage the marinade into the rack. Seal the bag and place it in a refrigerator. Refrigerate for 8 hour or overnight.
2. When ready to roast, remove the rack of lamb from the marinade and let it sit for

about 2 hour or until it is at room temperature.
3. Meanwhile, combine all the rub ingredients except the olive oil in a mixing bowl.
4. Rub the rub mixture over the rack of lamb generously. Drizzle rack with the olive oil.
5. Start your grill on smoke with the lid opened until fire starts.
6. Close the lid and preheat grill to 225°F using hickory wood pellets.
7. Place the rack of your lamb on the grill grate, bone side down. Smoke it for about two hours or until the internal temperature of the meat reaches 140-145°F.
8. Take off the rack of lamb from the grill and let it rest for about 10 minutes to cool.

Nutrition: Calories: 800 Fat: 41.1g Carbs: 6.7g Protein 93.8g

95. Leg of Lamb

Preparation Time: 10 minutes
Cooking Time: 2 hours
Servings: 6

Ingredients:

- 1 (2 pounds) leg of lamb
- 1 teaspoon dried rosemary
- 2 teaspoon freshly ground black pepper
- 4 cloves garlic (minced)
- 2 teaspoon salt or more to taste
- ½ teaspoon paprika
- 1 teaspoon thyme
- 2 tablespoons olive oil
- 1 teaspoon brown sugar
- 2 tablespoons oregano

Directions:

1. Trim the meat of excess fat and remove silver-skin.
2. In a mixing bowl, combine the thyme, salt, sugar, oregano, paprika, black pepper, garlic and olive oil.
3. Generously, rub the mixture over the leg of lamb. Cover seasoned leg of lamb with foil and let it sit for 1 hour to marinate.
4. Start your grill on smoke and leave the lid open for 5 minutes, or until fire starts. Cover the lid and preheat grill to 250°F using hickory, maple or apple wood pellets.
5. Remove the foil and place the leg of lamb on a smoker rack. Place the rack on the grill and smoke the leg of lamb for about 4 hours, or until it reach the internal temperature of your meat 145°F. Take off the leg of lamb from the grill and let it rest for a few minutes to cool. Cut into sizes and serve.

Nutrition: Calories 334 Fat: 16g Carbs: 2.9g Protein 42.9g

96. Smoked Lamb chops

Preparation Time: 10 Minutes
Cooking Time: 50 Minutes
Servings: 4

Ingredients:

- 1 rack of lamb, fat trimmed
- 2 tablespoons rosemary, fresh
- 2 tablespoons sage, fresh
- 1 tablespoon garlic cloves, roughly chopped
- 1/2 tablespoon salt

- 1/2 tablespoon pepper, coarsely ground
- 1/4 cup olive oil
- 1 tablespoon honey

Directions:

1. Preheat your wood pellet smoker to 225°F using a fruitwood.
2. Put all your ingredients except the lamb in a food processor. Liberally apply the mixture on the lamb.
3. Place the lamb on the smoker for 45 minutes or until the internal temperature reaches 120°F.
4. Sear the lamb on the grill for 2 minutes per side. Let rest for 5 minutes before serving. Slice and enjoy.

Nutrition: Calories 704 Fat 56g Carbs 24g Protein 27g

97. Wood Pellet Smoked Lamb Shoulder

Preparation Time: 10 Minutes
Cooking Time: 1hour 30 Minutes
Servings: 7

Ingredients:

- For Smoked Lamb Shoulder
- 5 pound lamb shoulder, boneless and excess fat trimmed
- 2 tablespoons kosher salt
- 2 tablespoons black pepper
- 1 tablespoon rosemary, dried
- The Injection
- 1 cup apple cider vinegar
- The Spritz
- 1 cup apple cider vinegar
- 1 cup apple juice

Directions:

1. Preheat the wood pellet smoker with a water pan to 2250 F.
2. Rinse the lamb in cold water then pat it dry with a paper towel. Inject vinegar to the lamb.
3. Pat the lamb dry again and rub with oil, salt black pepper and rosemary. Tie with kitchen twine.
4. Smoke uncovered for 1 hour then spritz after every 15 minutes until the internal temperature reaches 1950 F.
5. Take off the lamb from the grill and place it on a platter. Let cool before shredding it and enjoying it with your favorite side.

Nutrition: Calories 243 Fat 19g Carbs 0g Protein 17g

98. Wood Pellet Smoked Pulled Lamb Sliders

Preparation Time: 10 Minutes
Cooking Time: 7 Hours
Servings: 7

Ingredients:
For the Lamb's shoulder

- 5 pound lamb shoulder, boneless
- 1/2 cup olive oil
- 1/4 cup dry rub
- 10 ounces spritz
- The Dry Rub
- 1/3 cup kosher salt
- 1/3 cup pepper, ground
- 1-1/3 cup garlic, granulated
- The Spritz
- 4 ounces Worcestershire sauce
- 6 ounces apple cider vinegar

Directions:

1. Preheat the wood pellet smoker with a water bath to 2500 F.
2. Trim any fat from the lamb then rub with oil and dry rub.
3. Place the lamb on the smoker for 90 minutes then spritz with a spray bottle every 30 minutes until the internal temperature reaches 1650 F.
4. Transfer the lamb shoulder to a foil pan with the remaining spritz liquid and cover tightly with foil.
5. Place back in the smoker and smoke until the internal temperature reaches 2000 F.
6. Remove from the smoker and let rest for 30 minutes before pulling the lamb and serving with slaw, bun, or aioli. Enjoy

Nutrition: Calories 339 Fat 22 Carbs 16g Protein 18g

99. Smoked Lamb Meatballs

Preparation Time: 10 Minutes
Cooking Time: 1 Hour
Servings: 5

Ingredients:

- 1 pound lamb shoulder, ground
- 3 garlic cloves, finely diced
- 3 tablespoons shallot, diced
- 1 tablespoon salt
- 1 egg
- 1/2 tablespoon pepper
- 1/2 tablespoon cumin
- 1/2 tablespoon smoked paprika
- 1/4 tablespoon red pepper flakes
- 1/4 tablespoon cinnamon, ground
- 1/4 cup panko breadcrumbs

Directions:

1. Set the wood pellet smoker to 2500 F using a fruitwood.
2. In a mixing bowl, combine all meatball ingredients until well mixed.
3. Form small-sized balls and place them on a baking sheet. Place the baking sheet in the smoker and smoke until the internal temperature reaches 1600 F.
4. Remove from the smoker and serve. Enjoy.

Nutrition: Calories 73 Fat 5.2g Carbs 1.5g Protein 4.9g

100. Crown Rack of Lamb

Preparation Time: 10 Minutes
Cooking Time: 30 Minutes
Servings: 6

Ingredients:

- 2 racks of lamb, drenched
- 1 tablespoon garlic, crushed
- 1 tablespoon rosemary, finely chopped
- 1/4 cup olive oil
- 2 feet twine

Directions:

1. Rinse the racks with cold water then pat them dry with a paper towel.
2. Lay the racks on a flat board then score between each bone, about ¼ inch down.
3. In a mixing bowl, mix garlic, rosemary, and oil then generously brush on the lamb.
4. Take each lamb rack and bend it into a semicircle forming a crown-like shape.

5. Use the twine to wrap the racks about 4 times starting from the base to the top. Make sure you tie the twine tightly to keep the racks together.
6. Preheat the wood pellet to 400-4500 F then place the lamb racks on a baking dish. Plac ethe baing dish on the pellet grill.
7. Cook for 10 minutes then reduce temperature to 3000 F. cook for 20 more minutes or until the internal temperature reaches 1300 F.
8. Remove the lamb rack from the wood pellet and let rest for 15 minutes.
9. Serve when hot with veggies and potatoes. Enjoy.

Nutrition: Calories 390 Fat 35g Carbs 0g Protein 17g

101.Wood Pellet Smoked Leg of Lamb

Preparation Time: 15 Minutes
Cooking Time: 3hourss
Servings: 6

Ingredients:

- 1 leg lamb, boneless
- 4 garlic cloves, minced
- 2 tablespoons salt
- 1 tablespoon black pepper, freshly ground
- 2 tablespoons oregano
- 1 tablespoon thyme
- 2 tablespoons olive oil

Directions:

1. Cut off any excess of fat from the lamb and tie the lamb using twine to form a nice roast.
2. In a mixing bowl, mix garlic, spices, and oil. Rub all over the lamb, wrap with a plastic bag then refrigerate for an hour to marinate.
3. Place the lamb on a smoker set at 2500 F. smoke the lamb for 4 hours or until the internal temperature reaches 1450 F.
4. Remove from the smoker and let rest to cool. Serve and enjoy.

Nutrition: Calories 356 Fat16g Carbs 3g Protein 49g

102. Wood Pellet Grilled Aussie Leg of Lamb Roast

Preparation Time: 30 Minutes
Cooking Time: 2 Hours
Serves: 8

Ingredients:

- 5 pounds Aussie leg of lamb, boneless
- Smoked Paprika Rub
- 1 tablespoon raw sugar
- 1 tablespoon kosher salt
- 1 tablespoon black pepper
- 1 tablespoon smoked paprika
- 1 tablespoon garlic powder
- 1 tablespoon rosemary, dried
- 1 tablespoon onion powder
- 1 tablespoon cumin
- 1/2 tablespoon cayenne pepper
- Roasted Carrots
- 1 bunch rainbow carrots
- Olive oil
- Salt
- Pepper

Directions:

1. Heat the wood pellet grill to 3750 F.
2. Trim any excess fat from the lamb.
3. Put all your rub ingredients and rub all over the lamb. Place the lamb on the grill and smoke for 2 hours.
4. Toss the carrots in oil, salt, and pepper then add to the grill after the lamb has cooked for 1 ½ hour.
5. Cook until the roast internal temperature reaches 1350 F. remove the lamb from the grill and cover with foil. Let rest for 30 minutes.
6. Remove the carrots from the grill once soft and serve with the lamb. Enjoy.

Nutrition: Calories 257 Fat 8g Carbs 6g Protein 37g

103. Simple Grilled Lamb Chops

Preparation Time: 10 Minutes
Cooking Time: 6 Minutes
Servings: 6

Ingredients:

- 1/4 cup distilled white vinegar
- 2 tablespoons salt
- 1/2 tablespoon black pepper
- 1 tablespoon garlic, minced
- 1 onion, thinly sliced
- 2 tablespoons olive oil
- 2 pounds lamb chops

Directions:

1. In a reseal able bag, mix vinegar, salt, black pepper, garlic, sliced onion, and oil until all salt has dissolved.
2. Add the lamb chops and toss until well coated. Place in the fridge to marinate for 2 hours.
3. Preheat the wood pellet grill to high heat.
4. Remove the lamb from the fridge and discard the marinade. Wrap any exposed bones with foil.
5. Grill your lamb meat for three minutes per side. You can also broil in a broiler for more crispness. Serve and enjoy

Nutrition: Calories 519 Fat 44.8g Carbs 2.3g Protein 25g

104. Wood Pellet Grilled Lamb with Brown Sugar Glaze

Preparation Time: 15 Minutes
Cooking Time: 10 Minutes
Servings: 4

Ingredients:

- 1/4 cup brown sugar
- 2 tablespoons ginger, ground
- 2 tablespoons tarragon, dried
- 1 teaspoon cinnamon, ground
- 1 tablespoons black pepper, ground
- 1 tablespoons garlic powder
- 1/2 tablespoons salt
- 4 lamb chops

Directions:

1. In a mixing bowl, mix sugar, ginger, dried tarragon, cinnamon, black pepper, garlic, and salt.
2. Rub the lamb chops with the seasoning and place it on a plate. Refrigerate for an hour to marinate.
3. Preheat the grill to high heat then brush the grill grate with oil.

4. Arrange the lamb chops on the grill grate in a single layer and cook for 5 minutes on each side. Serve and enjoy.

Nutrition: Calories 241 Fat 13.1g Carbs 15.8g Protein 14.6g

105. Grilled leg of lambs Steaks

Preparation Time: 10 Minutes
Cooking Time: 10 Minutes
Servings: 4

Ingredients:

- 4 lamb steaks, bone-in
- 1/4 cup olive oil
- 4 garlic cloves, minced
- 1 tablespoon rosemary, freshly chopped
- Salt and black pepper

Directions:

1. Put the lamb in a shallow container in a single layer. Top with oil, garlic cloves, rosemary, salt, and black pepper then flip the steaks to cover on both sides.
2. Let sit for 30 minutes to marinate.
3. Preheat the wood pellet grill to high and brush the grill grate with oil.
4. Place the lamb steaks on the grill grate and cook until browned and the internal is slightly pink. The internal temperature should be 1400 F.
5. Let rest for 5 minutes before serving. Enjoy.

Nutrition: Calories 327 Fat 21.9g Carbs 1.7g Protein 29.6g

106. Wood pellet Grilled Lamb Loin Chops

Preparation Time: 10 Minutes
Cooking Time: 10 Minutes
Servings: 6

Ingredients:

- 2 tablespoons herbs de Provence
- 1-1/2 tablespoons olive oil
- 2 garlic cloves, minced
- 2 tablespoons lemon juice
- 5 ounces lamb loin chops
- Salt and black pepper to taste

Directions:

1. In a small mixing bowl, mix herbs de Provence, oil, garlic, and juice. Rub the mixture on the lamb chops then refrigerate for an hour.
2. Preheat the wood pellet grill to medium-high then lightly oil the grill grate.
3. Seasoned the lamb chops using salt and black pepper.
4. Put your lamb chops on the griller and cook for 4 minutes on each side.
5. Take off the chops from the grill and place them in an aluminum covered plate. Let rest for 5 minutes before serving. Enjoy.

Nutrition: Calories 579 Fat 43.9g Carbs 0.7g Protein 42.5g

107. Spicy Chinese Cumin Lamb Skewers

Preparation Time: 20 Minutes
Cooking Time: 6 Minutes
Servings: 10

Ingredients:

- 1 pound lamb shoulder, cut into 1/2 inch pieces
- 10 skewers
- 2 tablespoons ground cumin
- 2 tablespoons red pepper flakes
- 1 tablespoon salt

Directions:

1. Thread the lamb pieces onto skewers.
2. Preheat the wood pellet grill to medium heat and lightly oil the grill grate.
3. Put the skewers on the griller grate and cook while turning occasionally. Sprinkle cumin, pepper flakes, and salt every time you turn the skewer.
4. Cook it for about six minutes or until nicely browned. Serve and enjoy.

Nutrition: Calories 77 Fat 5.2g Carbs 1.6g Protein 6.3g

108. Wood pellet grill Dale's Lamb

Preparation Time: 15 Minutes
Cooking Time: 50 Minutes
Servings: 8

Ingredients:

- 2/3 cup lemon juice
- 1/2 cup brown sugar
- 1/4 cup Dijon mustard
- 1/4 cup soy sauce
- 1/4 cup olive oil
- 2 garlic cloves, minced
- 1 piece ginger root, freshly sliced
- 1 tablespoon salt
- 1/2 tablespoon black pepper, ground
- 5 pounds leg of lamb, butterflied

Directions:

1. In a mixing bowl, mix lemon juice, sugar, Dijon mustard, sauce, oil, garlic cloves, ginger root, salt, and pepper.
2. Put the lamb in a container and pour the seasoning mixture over it. Cover the dish and put in a fridge to marinate for 8 hours.
3. Preheat a wood pellet grill to medium heat. Drain the marinade from the dish and bring it to boil in a small saucepan.
4. Reduce heat and let simmer while whisking occasionally.
5. Oil the grill grate and place the lamb on it. Cook for 50 minutes or until the internal temperature reaches 1450 F while turning occasionally.
6. Slice the lamb and cover with the marinade. Serve and enjoy.

Nutrition: Calories 451 Fat 27.2g Carbs 17.8g Protein 32.4g

109. Garlic and Rosemary Grilled Lamb Chops

Preparation Time: 10 Minutes
Cooking Time: 20 Minutes
Servings: 4

Ingredients:

- 2 pounds lamb loin, thick-cut
- 4 garlic cloves, minced
- 1 tablespoon kosher salt
- 1/2 tablespoon black pepper
- 1 lemon zest
- 1/4 cup olive oil

Directions:

1. In a small mixing bowl, mix garlic, lemon zest, oil, salt, and black pepper then pour the mixture over the lamb.
2. Flip the lamb chops to make sure they are evenly coated. Place the chops in the fridge to marinate for an hour.
3. Preheat the wood pellet grill to high heat then sear the lamb for 3 minutes on each side.
4. Lessen your heat and cook the chops for 6 minutes or until the internal temperature reaches 1500 F.
5. Remove the lamb from the grill and wrap it in a foil. Let it rest for 5 minutes before serving. Enjoy.

Nutrition: Calories 171.5 Fat 7.8g Carbs 0.4g Protein 23.2g

110.Grilled Lamb Chops

Preparation Time: 1 Hour
Cooking Time: 8 Minutes
Servings: 3

Ingredients:

- 2 garlic cloves, crushed
- 1 tablespoons rosemary leaves, fresh chopped
- 2 tablespoons olive oil
- 1 tablespoon lemon juice, fresh
- 1 tablespoon thyme leaves, fresh
- 1 tablespoon salt
- 9 lamb loin chops

Directions:

1. Add the garlic, rosemary, oil, juice, salt, and thyme in a food processor. Pulse until smooth.
2. Rub the marinade on the lamb chops both sides and let marinate for 1 hour in a fridge. Tke it from the fridge and let sit at room temperature for 20 minutes before cooking.
3. Preheat your wood pellet smoker to high heat. Smoke the lamb chops for 5 minutes on each side.
4. Sear the lamb chops for 3 minutes on each side. Remove from the grill and serve with a green salad.

Nutrition: Calories 1140 Fat 99g Carbs 1g Protein 55g

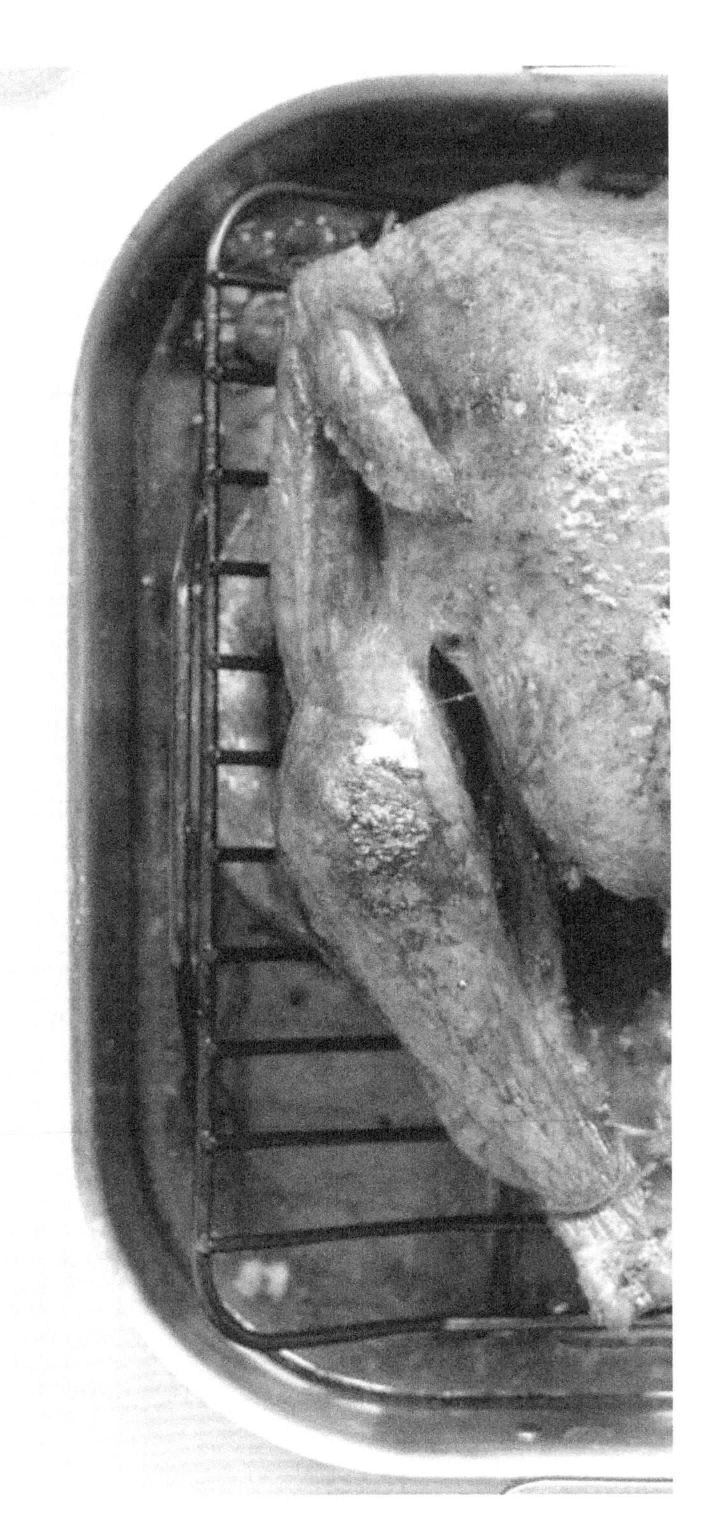

111. Tasty Smoked Chicken with Lemon Garlic Filling

Preparation Time: 15 minutes
Cooking Time: 3 hours
Servings: 10

Ingredients:

- Whole chicken (3-lb., 1.4-kg.)
- The Rub
- 3 tablespoons Lemon pepper
- ¾ teaspoon Salt
- 2 teaspoons Dried basils
- 1 ½ teaspoons Oregano
- 1 teaspoons Garlic powder
- 3 teaspoons Minced garlic
- 1 teaspoon Dried thyme
- The Filling
- 2 Fresh lemons
- 4 Garlic cloves
- ½ cup Chopped onion
- 4 sprigs Fresh thyme

Directions:

1. Combine lemon pepper with salt, dried basils, oregano, garlic powder, minced garlic, and dried thyme then mix well.
2. Apply the rub ingredients over the chicken skin then fill the chicken cavity with sliced fresh lemons, garlic cloves, chopped onion, and fresh thyme. Set aside.
3. Next, plug the wood pellet smoker then fill the hopper with the wood pellet. Turn the switch on.
4. Set the wood pellet smoker for indirect heat then adjust the temperature to 275°F (135°C).

5. When the wood pellet smoker is ready, place the seasoned chicken in it and smoke for 3 hours or until the internal temperature has reached 170°F (77°C).
6. Once it is done, remove the smoked chicken from the smoker and transfer it to a serving dish.
7. Cut the smoked chicken into slices and serve.

Nutrition: Calories: 218 Carbs: 6g Fat: 5g Protein: 37g

112. Cinnamon Apricot Smoked Chicken Thighs

Preparation Time: 15 minutes
Cooking Time: 1 hour 35 minutes
Servings: 10

Ingredients:

- Chicken thighs (3-lb., 1.4-kg.)
- The Rub
- 2 teaspoons Ground cinnamon
- 1 tablespoon Smoked paprika
- 3/4 tablespoon Cumin
- 3/4 teaspoon Ginger
- 3/4 teaspoon Salt
- 1 1/2 teaspoons Pepper
- 2 tablespoons Brown sugar
- a pinch Ground clove
- 1/4 teaspoon Cayenne
- The Glaze
- 1 cup Apricot marmalade
- 2 tablespoons Soy sauce
- 1 tablespoon Apricot syrup
- 1 tablespoon Cider vinegar
- 1/2 teaspoon Ground mustard

Directions:

1. Mix the entire rub ingredients—cinnamon, smoked paprika, cumin, ginger, salt, pepper, brown sugar, ground clove, and cayenne then stir until combined.
2. Rub the spices mixture over your chicken thighs then let it rest for a few minutes.
3. Next, plug the wood pellet smoker then fill the hopper with the wood pellet. Turn the switch on.
4. Set the wood pellet smoker for indirect heat then adjust the temperature to 275°F (135°C).
5. Arrange the seasoned chicken thighs in the wood pellet smoker and smoke for an hour and 30 minutes.
6. In the meantime, add apricot marmalade to a bowl then stir in soy sauce, apricot syrup, cider vinegar, and ground mustard to the bowl. Mix until incorporated and set aside.
7. Wait until the internal temperature of the smoked chicken thighs has reached 170°F (77°C) and coat the smoked chicken thighs with the apricot mixture.
8. Let the glazed smoked chicken thighs sit in the wood pellet smoker for 5 minutes then take them out.
9. Arrange the smoked chicken thighs on a serving dish then serve.

Nutrition: Calories: 270 Carbs: 24g Fat: 13g Protein: 15g

113. Chili Smoked Chicken Wings Bourbon

Preparation Time: 15 minutes
Cooking Time: 1 hour 15 minutes
Servings: 10

Ingredients:

- Chicken wings (3-lb., 1.4-kg.)
- The Rub
- 2 tablespoons Vegetable oil
- ½ cup Brown sugar
- 1 tablespoon Chili powder
- ½ teaspoon Ground cumin
- 1 teaspoon Black pepper
- ¾ teaspoon Salt
- The Sauce
- 2 teaspoons Olive oil
- ¼ cup Diced onion
- 1 tablespoon Minced garlic
- ½ cup Bourbon
- ¼ cup Tomato paste
- 1 ½ cups Ketchup
- ¼ cup Apple cider vinegar
- 3 tablespoons Worcestershire sauce
- 1 tablespoon Liquid smoke
- ¼ teaspoon Cayenne

Directions:

1. Place brown sugar, chili powder, ground cumin, black pepper, and salt in a bowl then drizzle vegetable oil over the spices. Mix well.
2. Apply the spice mixture on the chicken wings then let it rest for approximately 30 minutes.
3. Next, plug the wood pellet smoker then fill the hopper with the wood pellet. Turn the switch on.
4. Set the wood pellet smoker for indirect heat then adjust the temperature to 275°F (135°C).
5. Arrange the seasoned chicken wings in the wood pellet smoker and smoke for an hour.
6. In the meantime, preheat a saucepan over medium heat then pour olive oil into it.
7. Once your oil is hot, stir in diced onion together with minced garlic then sauté until wilted aromatic.
8. Add bourbon, tomato paste, ketchup, apple cider vinegar, Worcestershire sauce, liquid smoke, and cayenne to the saucepan then bring to a simmer. Remove from heat.
9. Check the internal temperature of the smoked chicken wings and once it reaches 170°F (77°C), remove from the wood pellet smoker.
10. Baste half of the sauce over the smoked chicken wings then wrap with aluminum foil. Let it rest for 15 minutes.
11. Unwrap the smoked chicken wings then transfer it to a serving dish. Serve and enjoy.

Nutrition: Calories: 240 Carbs: 0g Fat: 16g Protein: 22g

114. Sweet Smoked Chicken Breast with Aromatic Celery Seeds

Preparation Time: 10 minutes
Cooking Time: 1 hour 10 minutes
Servings: 6

Ingredients:

- Boneless chicken breast (4-lbs., 1.8-kg.)
- The Rub
- Olive oil – 3 tablespoons
- Brown sugar – ¼ cup
- Celery seeds – ¾ teaspoon
- Smoked paprika – 3 tablespoons
- Salt – 1 teaspoon
- Black pepper – ½ teaspoon
- Cayenne pepper – 1 ½ teaspoons
- Garlic powder – 1 tablespoon
- Onion powder – 1 tablespoon

Directions:

1. Combine brown sugar with celery seeds, smoked paprika, salt, black pepper, cayenne pepper, garlic powder, and onion powder. Mix well and set aside.
2. Rub the chicken breast with olive oil then sprinkle the dry spice mixture over the chicken breast. Let it rest for approximately 30 minutes.
3. Next, plug the wood pellet smoker then fill the hopper with the wood pellet. Turn the switch on.
4. Set the wood pellet smoker for indirect heat then adjust the temperature to 350°F (177°C)
5. Place the seasoned chicken breast in the wood pellet smoker and smoke for an hour.
6. Check the internal temperature of the smoked chicken breast and once it reaches 170°F (77°C), remove it from the wood pellet smoker.
7. Transfer the smoked chicken breast to a serving dish then cut into thick slices. Serve and enjoy.

Nutrition: Calories: 180 Carbs: 0g Fat: 10g Protein: 22g

115. Pineapple Cinnamon Smoked Chicken Thighs

Preparation Time: 10 minutes
Cooking Time: 1 hour 15 minutes
Servings: 8

Ingredients:

- Chicken thighs (3-lb., 1.4-kg.)
- The Marinade
- 1 ½ cups Ketchup
- 2 cups Pineapple chunks
- ½ cup White vinegar
- 2 teaspoons Garlic powder
- 3 tablespoons Olive oil
- 2 teaspoons Dried thyme
- 1 teaspoon Allspice
- 2 teaspoons Ground cinnamon
- 1 ½ teaspoons Cayenne powder
- ¾ teaspoons Salt

Directions:

1. Place ketchup, pineapple chunks, white vinegar, garlic powder, dried thyme, allspice, ground cinnamon, cayenne powder, and salt in a food processor then drizzle olive oil over the spices. Process until smooth.

2. Transfer the mixture to a saucepan then stir well.
3. Bring the sauce to a simmer then remove from heat. Let it cool.
4. Pour half of the mixture into a zipper-lock plastic bag then add the chicken thighs to the bag.
5. Seal and shake the plastic bag until the chicken thighs are completely coated with the spices then marinate for 4 hours to 8 hours. Store in the fridge to keep it fresh.
6. After several hours, remove the marinated chicken thighs from the fridge then thaw at room temperature.
7. Next, plug the wood pellet smoker then fill the hopper with the wood pellet. Turn the switch on.
8. Set the wood pellet smoker for indirect heat then adjust the temperature to 275°F (135°C).
9. Arrange the seasoned chicken thighs in the wood pellet smoker then smoke for an hour and 15 minutes.
10. Once the internal temperature of the smoked chicken thighs has reached 170°F (77°C), remove them from the wood pellet smoker.
11. Arrange the smoked chicken thighs on a serving dish then drizzle the remaining sauce on top. Serve and enjoy.

Nutrition: Calories: 297 Carbs: 15g Fat: 8g Protein: 0g

116. Wood Pellet Smoked Chicken Breasts

Preparation Time: 15 minutes
Cooking Time: 45 minutes
Servings: 4

Ingredients:

- 4 boneless and skinless chicken breasts.
- 1 tablespoon of olive oil.
- 2 tablespoons of brown sugar.
- 2 tablespoons of turbinate sugar.
- 1 teaspoon of celery seeds.
- 2 tablespoons of paprika.
- 2 tablespoons of kosher salt to taste.
- 1 teaspoon of black pepper to taste.
- 1 teaspoon of cayenne pepper.
- 2 tablespoons of garlic powder.
- 2 tablespoons of onion powder.

Directions:

1. Using a large mixing bowl, add in the celery seeds, paprika, cayenne pepper, sugars, garlic powder, onion powder, salt, and pepper to taste then mix properly to combine. Use paper towels to pat the chicken dry then rub all sides with the oil. Add some sprinkles of the mixed rub all over the chicken breast, wrap the chicken in a plastic bag then set aside in the fridge to rest for about fifteen to thirty minutes.
2. Preheat a Wood Pellet smoker and Grill (the smoker precisely) to smoke for about five minutes then turn the heat to 350 degrees and preheat for about fifteen minutes with the lid closed. Place the spiced/coated chicken on the grill then cook for about twelve to thirteen minutes.

3. Flip the chicken side to side, over and cook for another eight to ten minutes until it attains an internal temperature of 165 degrees F. once cooked, warp the chicken in aluminum foil and let rest for about three to five minutes. Slice and serve.

Nutrition: Calories 327 Fat 9g Carbohydrates 23g Protein 40g

117. Delicious Chicken Fritters

Preparation Time: 15 minutes
Cooking Time: 45 minutes
Servings: 8

Ingredients:

- 2 teaspoons of baking powder.
- 1 cup of shredded cheddar cheese.
- 1 1/2 lbs. of chicken breast.
- 2 eggs.
- 3/4 cup of almond flour.
- 1 1/2 teaspoon of lemon juice.
- 1 small and sliced.
- 3 tablespoons of mayonnaise.
- Olive oil.
- 2 tablespoons of chopped parsley.
- 2 teaspoons of chicken seasoning.
- 1 tablespoon of chopped scallions.
- 2 tablespoons of sour cream.
- 1 chopped yellow onion.
- 1/3 cup of almond milk.

Directions:

1. Preheat a Wood Pellet smoker and Grill to 425 degrees F, rub the chicken with olive oil then season with half of the chicken seasoning. Place the seasoned chicken on the preheated grill and grill for about twenty-five minutes until it attains an internal temperature of 165 degrees F.
2. Let the cooked chicken rest for a few minutes then pull into smaller pieces with a fork then add into a large mixing bowl. Add in other ingredients like the onion, tomato, eggs, parsley, milk, and cheese then mix everything to combine, set aside.
3. In another mixing bowl, add in the rest of the chicken seasoning, flour, and baking powder then mix properly to combine. Pour the mixture into the bowl containing the pulled chicken mixture then mix everything properly to combine. Cover the mixing bowl with a plastic wrap then refrigerate for about two hours.
4. Using another mixing bowl, add in the mayonnaise, sour cream, scallions, parsley, and lemon juice then mix properly to combine. This makes the serving dip. Feel free to store the dip the refrigerator until ready to be served.
5. Preheat the wood Pellet griddle to medium-low flame then grease the griddle with oil. Make fitters shape out of the chicken mixture, place the fritters on the preheated griddle, and cook for about three to four minutes. Flip the fritters over and cook for another three to four minutes then serve with the dip.

Nutrition: Calories 190 Fat 14.1g Carbohydrates 6.7g Protein 9.1g

118. Wood Pellet Grilled Chicken Satay

Preparation Time: 15 minutes
Cooking Time: 35 minutes
Servings: 4

Ingredients:

- Marinade
- 1 1/2 pounds of a boneless and skinless chicken breasts or thighs.
- 3/4 cup of coconut milk.
- 2 tablespoons of fish sauce.
- 2 tablespoons of soy sauce.
- 2 tablespoons of lime juice.
- 1/2 teaspoon of kosher salt to taste.
- 1/2 teaspoon of black pepper to taste.
- 1/2 teaspoon of garlic powder.
- 1/4 teaspoon of cayenne pepper.
- Dipping sauce
- 1/2 cup of coconut milk.
- 1/3 cup of peanut butter.
- 2 minced cloves of garlic.
- 1 tablespoon of soy sauce.
- 1 teaspoon of fish sauce.
- 1 tablespoon of lime juice.
- 1/2 tablespoon of swerve sweetener.
- 1 tablespoon of sriracha hot sauce.
- 1 cup of chopped cilantro.

Directions:

1. Properly slice the chicken as desired, preferably lengthwise then add to a Ziploc bag, set aside. Using a large mixing bowl, add in the milk, fish sauce, soy sauce, lime juice, garlic powder, cayenne pepper, salt, and pepper to taste then mix properly to combine. Pour the marinade into the resealable bag then shake properly to coat, refrigerate for about thirty minutes to three hours.
2. To make the dipping sauce, place all its ingredients in a mixing bowl then mix properly to combine, set aside. Preheat a Wood Pellet Smoker and Grill to 350 degrees F, thread the chicken onto skewers then place the skewers on the preheated grill.
3. Cook the chicken satay for about ten to fifteen minutes until it reads 1650 F. make sure you flip the chicken occasionally as you cook. Serve with the prepared dipping sauce and enjoy.

Nutrition: Calories 488 Fat 32g Carbohydrates 10g Protein 41g

119. Grilled BBQ Chicken Wings

Preparation Time: 10 minutes
Cooking Time: 2 hours 20 minutes
Servings: 5

Ingredients:

- 4 lbs. of chicken wings or drumettes.
- 1/2 cup of baking powder.
- 2 minced garlic cloves.
- 1 teaspoon of salt to taste.
- 1 teaspoon of black pepper to taste.
- 1 cup of barbecue sauce.
- Lime wedges for serving.

Directions:

1. Place the chicken wings in a plastic or Ziploc bag, add in the baking powder, garlic, pepper, and salt to taste then shake to coat. This will help you get crispy chicken wings when grilled. Let the chicken rest for about fifteen minutes then set aside.
2. Preheat the Wood Pellet Smoker and Grill to 275 to 300 degrees F, place chicken wings on the grill and cook for about six to ten minutes. Flip the

chicken side to side, over and cook for another five to six minutes until it is cooked through and attains an internal temperature of 185 degrees F.
3. Glaze the wings with the barbeque sauce, grill for another two minutes then serve. With lime wedges

Nutrition: Calories: 429 Carbohydrates 35g Fat: 30g Protein 31g

120. Smoked Whole Chicken with Dry Rub

Preparation Time: 10 minutes
Cooking Time: 1 hour 25 minutes
Servings: 4

Ingredients:

- 1 whole chicken.
- 1 can of beer or soda.
- 3 tablespoons of dry chicken rub.
- 2 tablespoon of olive oil.
- Lemon wedges for serving

Directions:

1. Turn the Wood Pellet Smoker and Grill to smoke for about five minutes then preheat to 350 degrees F. Use paper towels to pat the chicken dry then rub oil on the entire surface of the chicken. Coat the chicken with the dry rub then set aside.
2. Discard about half of the bear (you can choose to drink it up), add in a few tablespoons of the dry rub into the rest of the bear in the can then gently place the can into the chicken's cavity in such a way that the chicken stands upright.
3. Place the chicken on a baking sheet pan then place the pan on the grill gates. Cook the chicken for about one hour to one and a half hours until an inserted thermometer reads 165 degrees F. Once cooked, let the chicken cool for about fifteen minutes, slice, and serve with lemon wedges.

Nutrition: Calories 337 Fat 18g Carbohydrates 1.3g Protein 36g

121. Cajun Patch Cock Chicken

Preparation Time: 30 Minutes
Cooking Time: 2.5 Hours
Servings: 4

Ingredients:

- 4-5 pounds of fresh or thawed frozen chicken

- 4-6 glasses of extra virgin olive oil
- Cajun Spice Lab 4 tablespoons or Lucile Bloody Mary Mix Cajun Hot Dry Herb Mix Seasoning

Directions:

1. Place the chicken breast on a cutting board with the chest down.
2. Using kitchen or poultry scissors, cut along the side of the spine and remove.
3. Turn your chicken side to side, over and press down firmly on the chest to flatten it. Carefully loosen and remove the skin on the chest, thighs and drumsticks.
4. Rub olive oil freely under and on the skin. Season chicken in all directions and apply directly to the meat under the skin.
5. Wrap the chicken in plastic wrap and place in the refrigerator for 3 hours to absorb the flavor.
6. Use hickory, pecan pellets, or blend to configure a wood pellet smoker grill for indirect cooking and preheat to 225 ° F.
7. If the unit has a temperature meat probe input, such as a MAK Grills 2 Star, insert the probe into the thickest part of the breast.
8. Make chicken for 1.5 hours.
9. After one and a half hours at 225 ° F, raise the pit temperature to 375 ° F and roast until the inside temperature of the thickest part of the chest reaches 170 ° F and the thighs are at least 180 ° F.
10. Place the chicken under a loose foil tent for 15 minutes before carving.

Nutrition: Calories: 351 Carbs: 12g Fat: 15g Protein: 36g

122. Yan's Grilled Quarters

Preparation Time: 20 Minutes
Cooking Time: 1 To 1.5 Hours
Servings: 4

Ingredients:

- 4 fresh or thawed frozen chicken quarters
- 4-6 glasses of extra virgin olive oil
- 4 tablespoons of Yang's original dry lab

Directions:

1. Cut off excess skin and fat chicken. Carefully peel the chicken skin and rub olive oil above and below each chicken skin.
2. In Jean's original dry lab, apply seasonings to the top and bottom of the skin and the back of the chicken house.
3. Wrap the seasoned chicken in plastic wrap and store refrigerated for 2-4 hours to absorb flavor.
4. Configure a wood pellet smoker grill for indirect cooking and use the pellets to preheat to 325 ° F.
5. Place chicken on grill and cook at 325 ° F for 1 hour.
6. After one hour, raise the pit temperature to 400 ° F to finish the chicken and crisp the skin.
7. When the inside temperature of the thickest part of the thighs and feet reaches 180 ° F and the juice becomes clear, pull the crispy chicken out of the grill.
8. Let the crispy grilled chicken rest under a loose foil tent for 15 minutes before eating.

Nutrition: Calories: 250 Carbs: 0g Fat: 12g Protein: 21g

123. Roasted Tuscan Thighs

Preparation Time: 20 Minutes
Cooking Time: 40-60 Minutes
Servings: 4

Ingredients:

- 8 chicken thigh, with bone, with skin
- 3 extra virgin olive oils with roasted garlic flavor
- 3 cups of Tuscan or Tuscan seasoning per thigh

Directions:

1. Cut off excess skin on chicken thighs and leave at ¼ inch to shrink.
2. Carefully peel off the skin and remove large deposits of fat under the skin and behind the thighs.
3. Lightly rub olive oil behind and below the skin and thighs. A seasoning from Tuscan, seasoned on the skin of the thigh and the top and bottom of the back.
4. Wrap chicken thighs in plastic wrap, refrigerate for 1-2 hours, and allow time for flavor to be absorbed before roasting.
5. Set the wood pellet smoker grill for indirect cooking and use the pellets to preheat to 375 degrees Fahrenheit.
6. Depending on the grill of the wood pellet smoker, roast for 40-60 minutes until the internal temperature of the thick part of the chicken thigh reaches 180 ° F. Place the roasted Tuscan thighs under a loose foil tent for 15 minutes before serving.

Nutrition: Calories: 260 Carbs: 1g Fat: 20g Protein: 19g

124. Teriyaki Smoked Drumstick

Preparation Time: 15 Minutes
Cooking Time: 1.5 Hours to 2 Hours
Servings: 4

Ingredients:

- 3 cup teriyaki marinade and cooking sauce like Yoshida's original gourmet
- Poultry seasoning 3 teaspoons
- 1 teaspoon garlic powder
- 10 chicken drumsticks

Directions:

1. In a medium bowl, mix the marinade and cooking sauce with the chicken seasoning and garlic powder.
2. Peel off the skin of the drumstick to promote marinade penetration.
3. Put the drumstick in a marinade pan or 1 gallon plastic sealable bag and pour the marinade mixture into the drumstick. Refrigerate overnight.
4. Rotate the chicken leg in the morning.
5. Configure a wood pellet smoking grill for indirect cooking.
6. Place the skin on the drumstick and, while the grill is preheating, hang the drumstick on a poultry leg and wing rack to drain the cooking sheet on the counter. If you do not have a poultry leg and feather rack, you can dry the drumstick by tapping it with a paper towel.
7. Preheat wood pellet smoker grill to 180 ° F using hickory or maple pellets.
8. Make marinated chicken leg for 1 hour.

9. After 1 hour, raise the whole temperature to 350 ° F and cook the drumstick for another 30-45 minutes until the thickest part of the stick reaches an internal temperature of 180 ° F.
10. Place the chicken drumstick under the loose foil tent for 15 minutes before serving.

Nutrition: Calories: 280 Carbs: 0g Fat: 13g Protein: 35g

125. Smoked Bone In-Chicken Breast

Preparation Time: 20 Minutes
Cooking Time: 3 to 4 Hours
Servings: 6

Ingredients:

- 1 (8-10 pounds) boned chicken breast
- 6 tablespoons extra virgin olive oil
- 5 Yang original dry lab or poultry seasonings

Directions:

1. Remove excess fat and skin from chicken breast.
2. Slowly separate the skin of your chicken from the breast and leave the skin alone. Apply olive oil to the chest, under the skin and on the skin.
3. Rub or season carefully under the chest cavity, under the skin and on the skin.
4. Place the chicken breast in a V-rack for easy handling, or place it directly on a grill grate with the breast up.
5. Rest the chicken breasts on the kitchen counter at room temperature and preheat the wood pellet smoker grill.
6. Configure a wood pellet smoker grill for indirect cooking and preheat to 225 ° F using hickory or pecan pellets.
7. Smoke the boned chicken breast directly in a V rack or grill at 225 ° F for 2 hours.
8. After 2 hours of hickory smoke, raise the pit temperature to 325 ° F. Roast until the thickest part of the chicken breast reaches an internal temperature of 170 ° F and the juice is clear.
9. Place the hickory smoked chicken breast under a loose foil tent for 20 minutes, then scrape the grain.

Nutrition: Calories: 186 Carbs: 0g Fat: 4g Protein: 0g

126. Hickory Smoke Patchcock Chicken

Preparation Time: 20 Minutes
Cooking Time: 3-4 Hours
Servings: 6

Ingredients:

- 1 fresh or thawed frozen young chicken
- ¼ Extra virgin olive oil with cup roasted garlic flavor
- 6 poultry seasonings or original dry lab in January

Directions:

1. Using poultry scissors or a big butcher knife, carefully remove the chicken spine along both sides.
2. Press down on the sternum to flatten the patch-cocked chicken.
3. Remove excess fat and skin from the breast.
4. Slowly separate the skin of your chicken from the breast and leave the skin alone. Apply olive oil intrathoracic ally, under the skin and on the skin.
5. Sprinkle seasoning or dry rub with seasoning on chest cavity, under skin and on skin.
6. Configure a wood pellet smoking grill for indirect cooking and preheat to 225 ° F using hickory pellets.
7. Place the chicken skin down on a non-stick grill mat made of Teflon-coated fiberglass.
8. Suck the chicken at 225 ° F for 2 hours.
9. After 2 hours, raise the pit temperature to 350 ° F.
10. Roast chicken until the thickest part of the chest reaches an internal temperature of 170 ° F and the juice is clear.
11. Place the Hickory smoked roast chicken under a loose foil tent for 20 minutes before engraving.

Nutrition: Calories: 180 Carbs: 1g Fat: 16g Protein: 7g

127.Bacon Cordon Blue

Preparation Time: 30 Minutes
Cooking Time: 2 To 2.5 Hours
Servings: 6

Ingredients:

- 24 bacon slices
- 3 large boneless, skinless chicken breasts, butterfly
- 3 extra virgin olive oils with roasted garlic flavor
- 3 Yang original dry lab or poultry seasonings
- 12 slice black forest ham
- 12-slice provolone cheese

Directions:

1. Weave 4 slices of bacon tightly, leaving extra space on the edges. Bacon weave is used to interlock alternating bacon slices and wrap chicken cordon blue.
2. Slice or rub two chicken breast fillets with olive oil on both sides.
3. Scattered the seasoning mixture on both sides of the chicken breast.
4. Lay the seasoned chicken fillets on the bacon weave and slice one ham and one provolone cheese on each.
5. Repeat this process with another chicken fillet, ham and cheese. Fold chicken, ham and cheese in half.
6. Lay the bacon strips from the opposite corner to completely cover the chicken cordon blue.
7. Use a silicon food grade cooking band, butcher twine, and toothpick to secure the bacon strip in place.
8. Repeat this process for the remaining chicken breast and ingredients.
9. Using apple or cherry pellets, configure a wood pellet smoker grill for indirect cooking and preheat (180 ° F to 200 ° F) for smoking.
10. Inhale bacon cordon blue for 1 hour.
11. After smoking for 1 hour, raise the pit temperature to 350 ° F.

12. Bacon cordon blue occurs when the internal temperature reaches 165 ° F and the bacon becomes crispy.
13. Rest for 15 minutes under a loose foil tent before serving.

Nutrition: Calories: 250 Carbs: 11g Fat: 7g Protein: 34g

128. Lemon Cornish Chicken Stuffed With Crab

Preparation Time: 30 Minutes
Cooking Time: 1 Hour 30 Minutes
Servings: 4

Ingredients:

- 2 Cornish chickens (about 1¾ pound each)
- Half lemon, half
- 4 tablespoons western rub or poultry rub
- 2 cups stuffed with crab meat

Directions:

1. Rinse chicken thoroughly inside and outside, tap lightly and let it dry.
2. Carefully loosen the skin on the chest and legs. Rub the lemon under and over the skin and into the cavity. Rub the western lab under and over the skin on the chest and legs. Carefully return the skin to its original position.
3. Wrap the Cornish hen in plastic wrap and refrigerate for 2-3 hours until flavor is absorbed.
4. Prepare crab meat stuffing according to the instructions. Make sure it is completely cooled before packing the chicken. Loosely fill the cavities of each hen with crab filling.
5. Tie the Cornish chicken legs with a butcher's leash to put the filling.
6. Set wood pellet smoker grill for indirect cooking and preheat to 375 ° F with pellets.
7. Place the stuffed animal on the rack in the baking dish. If you do not have a rack that is small enough to fit, you can also place the chicken directly on the baking dish.
8. Roast the chicken at 375 ° F until the inside temperature of the thickest part of the chicken breast reaches 170 ° F, the thigh reaches 180 ° F, and the juice is clear.
9. Test the crab meat stuffing to see if the temperature has reached 165 ° F.
10. Place the roasted chicken under a loose foil tent for 15 minutes before serving.

Nutrition: Calories: 275 Carbs: 0g Fat: 3g Protein: 32g

129. Pellet Smoked Chicken Burgers

Preparation Time: 15 minutes
Cooking Time: 1 hour 10 minutes
Servings: 6

Ingredients:

- 2 pounds ground chicken breast
- 2/3 cup of finely chopped onions
- 1 tablespoon of cilantro, finely chopped
- 2 tablespoons fresh parsley, finely chopped
- 2 tablespoons of olive oil
- 1/2 teaspoon of ground cumin

- 2 tablespoons of lemon juice freshly squeezed
- 3/4 teaspoon of salt and red pepper to taste

Directions:

1. In a bowl add all ingredients; mix until combined well.
2. Form the mixture into 6 patties.
3. Start your pellet grill on SMOKE (oak or apple pellets) with the lid open until the fire is established. Set the temperature to 350°F and preheat, lid closed, for 10 to 15 minutes.
4. Smoke the chicken burgers for 45 - 50 minutes or until cooked through, turning every 15 minutes.
5. Your burgers are ready when internal temperature reaches 165 °F. Serve hot.

Nutrition: Calories: 221 Carbs: 2.12g Fat: 8.5g Protein: 32.5g

130. Perfect Smoked Chicken Patties

Preparation Time: 15 minutes
Cooking Time: 55 minutes
Servings: 6

Ingredients:

- 2 pounds ground chicken breast
- 2/3 cup minced onion
- 1 Tablespoon cilantro (chopped)
- 2 Tablespoons fresh parsley, finely chopped
- 2 Tablespoons olive oil
- 1/8 teaspoon crushed red pepper powdered for the taste
- 1/2 teaspoon ground cumin
- 2 Tablespoons fresh lemon juice
- 3/4teaspoonkosher salt
- 2 teaspoons paprika
- Hamburger buns for serving

Directions:

1. In a bowl combine all ingredients from the list.
2. Using your hands, mix well. Form mixture into 6 patties. Refrigerate until ready to grill (about 30 minutes).
3. Start your pellet grill on SMOKE with the lid open until the fire is established). Set the temperature to 350°F and preheat, lid closed, for 10 to 15 minutes.
4. Arrange chicken patties on the grill rack and cook for 35 to 40 minutes turning once.
5. Serve hot with hamburger buns and your favorite condiments.

Nutrition: Calories: 258 Carbs: 2.5g Fat: 9.4g Protein: 39g

131. Smoked Chicken Breast with Honey Garlic

Preparation Time: 10 minutes
Cooking Time: 4 hours 20 minutes
Servings: 4

Ingredients:

- 4 chicken breasts boneless and skinless
- 8 cloves of garlic finely chopped
- 1/4 can honey
- 2 tablespoons olive oil
- Herbs and spices to a taste
- Salt and freshly ground pepper to taste

Directions:

1. In a skillet, heat oil and sauté garlic until tender.
2. Remove from heat and stir in honey. Add salt and pepper, and herbs and spices as desired.
3. Place chicken breasts in a baking dish and cover evenly with the garlic and honey mixture.
4. Preheat pellet smoker to 225°F.
5. Smoke chicken directly on the rack for 4 hours or until internal temperature reaches 165°F (the total cooking time will depend on the cut of the meat).
6. Remove chicken from the smoker and let rest for 10 minutes. Serve warm.

Nutrition: Calories: 341.66 Carbs: 3g Fat: 13g Protein: 50.5g

132. Smoked Chicken Breasts in Lemon Marinade

Preparation Time: 10 minutes
Cooking Time: 1 hour
Servings: 6

Ingredients:

- 4 pounds boneless chicken breast
- 6 lemons sliced, without seeds
- 2 tablespoons olive oil
- 1 tablespoon of garlic minced
- 2 tablespoons of onion finely chopped
- 1 teaspoon of allspice
- Salt and powdered black pepper for taste
- 1/2 cup water

Directions:

1. In a large bowl, place chicken pieces and lemon rings.
2. In a separate bowl, combine all remaining ingredients.
3. Pour the spice mixture over chicken and mix thoroughly. Refrigerate for 4 hours.
4. Start your pellet grill on SMOKE with the lid open until the fire is established). Put the temperature to 250°F and allow to preheat, lid closed, for 10 to 15 minutes.
5. Place marinated chicken into the smoker, and cook for 35 to 45 minutes.
6. Your chicken is ready when internal temperature reaches 165 °F.
7. Allow to rest for 15 minutes, slice and serve.

Nutrition: Calories: 618 Carbs: 12.7g Fat: 73g Protein: 11.5g

133. Smoked Chicken Breasts with Dried Herbs

Preparation Time: 10 minutes
Cooking Time: 55 minutes
Servings: 4

Ingredients:

- 4 chicken breasts boneless
- 1/4 cup garlic-infused olive oil
- 2 clove garlic minced
- 1/4 teaspoon of dried sage
- 1/4 teaspoon of dried lavender
- 1/4 teaspoon of dried thyme
- 1/4 teaspoon of dried mint
- 1/2 tablespoons dried crushed red pepper
- Kosher salt to taste

Directions:

1. Place the chicken breasts in a shallow plastic container.
2. In a bowl, combine all remaining ingredients, and pour the mixture over the chicken breast and refrigerate for one hour.
3. Remove the chicken breast from the sauce (reserve sauce) and pat dry on kitchen paper.
4. Start your pellet grill on SMOKE (hickory pellet) with the lid open until the fire is established). Put the temperature to 250°F and preheat, lid closed, for 10 to 15 minutes.
5. Place chicken breasts on the smoker. Close pellet grill lid and cook for about 30 to 40 minutes or until chicken breasts reach 165°F. Serve hot with reserved marinade.

Nutrition: Calories: 391 Carbs: 0.7g Fat: 3.21g Protein: 20.25g

134. Smoked Chicken Burgers with Feta Cheese

Preparation Time: 10 minutes
Cooking Time: 1 hour
Servings: 6

Ingredients:

- 2 pounds of minced chicken meat
- Zest of 1 lemon
- 1 tablespoon olive oil
- 2 teaspoons oregano, fresh chopped
- 1/2 teaspoons of fresh thyme and marjoram finely chopped
- 1 teaspoon fresh parsley finely chopped
- Salt and ground pepper to taste
- 1 cup Feta cheese crumbled
- Olive oil for brushing

Directions:

1. Combine minced meat, lemon zest, olive oil, oregano, thyme and salt and pepper to taste.
2. Wet your heads and knead the meat mixture.
3. Cut Feta into small cubes and start making the meatballs.
4. Take about half a tablespoon of minced meat, roll in the shape of a circle, press in the middle with our thumb, place a piece of cheese there, "close" and gently roll into balls.
5. Start your pellet grill on SMOKE (hickory or apple pellets) with the lid open until the fire is established. Set the temperature to 380°F and preheat, lid closed, for 10 to 15 minutes.
6. Place burgers into the smoker and cook for 35 to 40 minutes.
7. Your chicken burgers are ready when internal temperature reaches 165 °F. Serve hot.

Nutrition: Calories: 455.7 Carbs: 10g Fat: 24.54g Protein: 27.g

135. Smoked Chicken on Pellet Grill

Preparation Time: 10 minutes
Cooking Time: 4 hours
Servings: 6

Ingredients:

- 1 large chicken
- 3/4 cup fresh butter
- 1 teaspoon fresh basil finely chopped
- 1 teaspoon fresh thyme finely chopped
- 1 teaspoon parsley finely chopped
- Salt and ground pepper

- 1/2 cup white wine

Directions:

1. Wash the chicken, patted dry with a paper towel and tie cross its legs.
2. In a small saucepan melt the butter over medium heat.
3. Add basil, parsley and thyme, wine and salt and pepper; stir well and remove from heat.
4. Brush the chicken generously with the herbed butter mixtures.
5. Preheat your smoker or pellet grill to 180°F, lid closed, for 10 to 15 minutes.
6. Smoke the chicken 4 hours.
7. When the chicken reach an internal temperature of 160°F in the thigh meat, take them out of the smoker.
8. Chop the chicken and serve hot.

Nutrition: Calories: 452 Carbs: 1.2g Fat: 37g Protein: 22.7g

136. Grilled Chicken Salad

Servings: 6

Preparation time: 20 minutes

Cooking time: 1 hour and 15 minutes

Ingredients:

- 1 whole chicken
- 3 tablespoons poultry rub
- For the Salad:
- 4 green onions, chopped
- 1 cup red grapes, halved
- 4 celery stalks, chopped
- 1 cup green grapes, halved
- ¾ teaspoon salt
- ½ teaspoon ground black pepper
- 2 tablespoons brown sugar
- ½ cup sour cream
- ½ cup mayonnaise
- 1 lemon, juiced

Directions:

1. Open hopper of the smoker, add dry pallets, make sure ash-can is in place, then open the ash damper, power on the smoker and close the ash damper.
2. Set the temperature of the smoker to 250 degrees F, let preheat for 30 minutes or until the green light on the dial blinks that indicate smoker has reached to set temperature.
3. Meanwhile, break the whole chicken into thighs, legs, breasts, and wings and then season with poultry rub until well coated.
4. Place chicken pieces on the smoker grill, shut with lid and smoke for 35 to 45 minutes or until thoroughly cooked and the internal temperature of chicken reaches to 165 degrees F.
5. When done, transfer chicken pieces to a cutting board, let it rest for 5 minutes, then separate bones from the meats, discard bones and skin, and shred chicken with two forks.
6. Prepare the salad and for this, place onion, grapes and celery in a large salad bowl, drizzle with lemon juice and sprinkle with salt, black pepper, and sugar.
7. Whisk together sour cream and mayonnaise, add into the salad bowl along with chicken pieces and stir the salad gently until mixed.
8. Serve immediately.

Nutrition: Calories: 351.7; Total Fat: 10.2 g; Saturated Fat: 1.4 g; Protein: 43.8 g; Carbs: 20.5 g; Fiber: 4.6 g; Sugar: 7 g

137. Chicken Fajitas

Servings: 10
Preparation time: 10 minutes
Cooking time: 40 minutes

Ingredients:

- 2 pounds chicken breast, sliced
- 1 large orange bell pepper, deseeded, sliced
- 1 large white onion, peeled, sliced
- 1 large red bell pepper, deseeded, sliced
- 1/2 teaspoon onion powder
- 1/2 teaspoon garlic powder
- 1 teaspoon salt
- 2 tablespoons Chile Margarita Seasoning
- 2 tablespoons olive oil

Directions:

1. Open hopper of the smoker, add dry pallets, make sure ash-can is in place, then open the ash damper, power on the smoker and close the ash damper.
2. Set the temperature of the smoker to 450 degrees F, switch smoker to open flame cooking mode, press the open flame 3, remove the grill grates and the batch, replace batch with direct flame insert, then return grates on the grill in the lower position, place a large sheet pan lined with aluminum foil and let preheat for 30 minutes or until the green light on the dial blinks that indicate smoker has reached to set temperature.
3. Meanwhile, place chicken in a large bowl, add all the peppers, drizzle with oil, sprinkle with onion powder, garlic powder, salt, and seasoning and toss until mixed.
4. Add seasoned chicken and peppers along with onion in the sheet pan on the smoker grill, spread the ingredients in a single layer, then shut with lid and smoke for 10 minutes or until chicken is no longer pink.
5. When done, transfer chicken to a dish and serve with toasted tortillas.

Nutrition: Calories: 122; Total Fat: 1.5 g; Saturated Fat: 0 g; Protein: 22 g; Carbs: 4 g; Fiber: 1.5 g; Sugar: 0 g

138. Parmesan Chicken Wings

Servings: 6

Preparation time: 1 hour and 10 minutes

Cooking time: 45 minutes

Ingredients:

- 2 pounds chicken wings, trimmed
- 2 tablespoons butter, unsalted, melted
- ¼ cup grated parmesan cheese

For the Marinade:

- 1 ½ tablespoon minced garlic
- 2 tablespoons chicken seasoning
- 2 tablespoons parsley, chopped

- 2 tablespoons Dijon mustard
- 1 lemon, juiced
- ¼ cup olive oil

Directions:

- Place all the ingredients for the marinade in a small bowl and then stir until combined.
- Place chicken wings in a large plastic bag, pour in prepared marinade, then seal the bag, turn it upside down to coat the chicken wings with the marinade and let marinate in the refrigerator for 1 hour.
- When ready to cook, open hopper of the smoker, add dry pallets, make sure ash-can is in place, then open the ash damper, power on the smoker and close the ash damper.
- Set the temperature of the smoker to 350 degrees F, let preheat for 30 minutes or until the green light on the dial blinks that indicate smoker has reached to set temperature.
- Remove chicken wings from the marinade, place them on the smoker grill, shut with lid and smoke for 15 minutes or until the internal temperature of chicken reach to 165 degrees F, flipping the chicken wings halfway through.
- When done, transfer the chicken wings to a large bowl, add butter and cheese and toss until well coated.
- Serve straight away.

Nutrition: Calories: 510; Total Fat: 40 g; Saturated Fat: 14 g; Protein: 35 g; Carbs: 3 g; Fiber: 0 g; Sugar: 0 g

139. BBQ Chicken

Cooking Time: 3 Hours

Servings: 4

The Heat: Hardwood Apple

Ingredients:

- 2 pounds of the whole chicken
- 6 chilies, Thai chilis
- 1 teaspoon of paprika, sweet
- 1 scotch bonnet
- 2 tablespoons of brown sugar
- Salt, to taste
- 1 onion, chopped
- 5 garlic cloves, minced
- 4 cups of olive oil

Directions:

1. Take a food processor and pulse Thai chili, paprika, bonnet, brown sugar, onion, garlic and salt along with olive oil.
2. Now marinate the chicken by smothering it with the puree.
3. Let it sit for a few hours in the refrigerator for marinating.
4. When ready to cook, set the temperate for 300 degrees F and close the lid and preheat the grill for 20 minutes.
5. Next, place the chicken on the grill, breast side up, and smoke for 3 hours.
6. Once, the internal temperate reaches 165 degrees F.
7. Once the time completes, flip to cook from the other side.
8. Remove it from the grill and allow it to cool for 10 minutes.
9. Serve and enjoy.

140. Balsamic Vinegar Chicken Breasts

Cooking Time: 3 Hours
Servings: 4

The Heat: Hardwood Apple or Cheery

Ingredients:

- 6 tablespoons of olive oil
- 1 cup balsamic vinegar
- 3 cloves of garlic cloves, minced
- 1 teaspoon of basil, fresh
- 1 teaspoon of chili powder
- Salt and black pepper, to taste
- 2 pounds of chicken breast, boneless and skinless

Directions:

1. In a zip lock bag, add oil, balsamic vinegar, basil leaves, chili powder, garlic cloves, salt, and black pepper.
2. Now, place the chicken in the zip lock bag and mix well.
3. Marinate the chicken in the sauce, for 3 hours in the refrigerator.
4. Now, preheat the grill for 20 minutes at 225 degrees F .
5. Place the chicken onto the grill, and smoke for 3 hours.
6. Once the internal temperature reached 150 degrees, remove it from the grill, and then let it get cool for 10 minutes before serving.
7. Serve and enjoy.

141. Buffalo Wings

Cooking Time: 53 Minutes
Servings: 8

The Heat: Hardwood Hickory or Cherry

Ingredient for Chicken Wings:

- 4 pounds of Chicken Wings
- 2 teaspoons of Corn Starch
- 2 tablespoons of buffalo wings rub
- Salt, To Taste

Ingredients for Buffalo Sauce:

- 1/3 cup Spicy Mustard
- 1 Cup Franks Red Hot Sauce
- 8 tablespoons of Unsalted Butter

Side:

- 1 cup Blue cheese dressing

Directions:

1. Preheat the grill to 375 degrees F, for 15 minutes.
2. Meanwhile, the grill is preheating add wings to a large bowl and sprinkle corn starch, spice rub, and salt.
3. Mix the ingredients well.
4. When the grill is heated, place the wings on the grill and cook for 38 minutes, or until the internal temperature reaches 165 degrees F.
5. Meanwhile, in a bowl, mix all the buffalo sauce ingredients.

6. Put the sauce over the wings and then cook the wings for additional 15 minutes with the lid closed.
7. Serve the wings with blue cheese dressing.
8. Enjoy.

142. Herbed Smoked Hen

Cooking Time: 50 Minutes
Servings: 5

The Heat: Hardwood Hickory

Ingredients:

- 12 cups of filtered water
- 3 cups of beer nonalcoholic
- Sea Salt, to taste
- ⅓ Cup brown sugar
- 2 tablespoons of rosemary
- ½ teaspoon of sage
- 2.5 pounds of a whole chicken, trimmed and giblets removed
- 6 tablespoons of butter
- 2 tablespoons of Olive oil, for basting
- 1/3 cup Italian seasoning
- 1 tablespoon garlic powder
- 1 tablespoon of lemon zest

Directions:

1. Pour water in a large cooking pot and then add sugar and salt to the water.
2. Boil the water until the sugar and salt dissolve.
3. Now to the boiling water, add rosemary and sage.
4. Boil it until aroma comes.
5. Now pour the beer into the water and then submerge the chicken into the boiling water.
6. Turn off the heat and refrigerate the chicken for a few hours.
7. After few hours removed it from the brine, and then pat dry with the paper towel.
8. Let the chicken sit for a few minutes at room temperature.
9. Now rub the chicken with the butter and massage it completely for fine coating.
10. Season the chicken with garlic powder, lemon zest, and Italian seasoning.
11. Preheat the Electrical smoker at 270 degrees Fahrenheit until the smoke started to build.
12. Baste the chicken with olive oil and put it on the grill grate.
13. Cook the chicken with the lid closed, for 30 to 40 minutes, or until the internal temperature reaches 165 degrees F.
14. Serve and enjoy.

143. Smoked Chicken Thighs

Cooking Time: 2 Hours
Servings: 4

The Heat: Applewood

Ingredients:

- 2.5 pounds of chicken thighs
- 4 tablespoons soy sauce
- 4 teaspoons sesame oil
- 2 garlic cloves
- 1-inch ginger, grated
- 1 small white onion, chopped
- ½ tablespoon thyme

- 2 teaspoons allspice, powder
- ½ teaspoon cinnamon
- ½ teaspoon crushed red chili peppers, powder

Directions:

1. Take a food processor and add soya sauce, sesame oil, garlic cloves, ginger, onions, thyme, allspice powder, cinnamon, and red chili peppers.
2. Blend the mixture into a smooth paste.
3. Coat the chicken thighs with the blended paste, and marinate in a zip-lock plastic bag for 2 hours in the refrigerator.
4. Preheat the smoker at 225 degrees F, until the smoke started to form.
5. Place the chicken directly onto the grill grate, and cook the chicken for 2 hours.
6. Use a thermometer to read the internal temperature of the chicken.
7. Once the temperature reaches 145 degrees Fahrenheit, the chicken is ready to be served
8. Remove the chicken from the gill great, and let it sit at the room temperature for 20 minutes, before serving.
9. Serve and enjoy.

144. Maple Glazed Whole Chicken

Cooking Time: 3 Hours
Servings: 4

The Heat: Hardwood Apple

Ingredients for The Rub:

- Black pepper and salt, to taste
- 3 garlic cloves, minced
- 3 teaspoons of onion powder
- 1.5 teaspoons of ginger, minced
- ½ teaspoon of five-spice powder
- Basic Ingredients:
- 2.5 pounds whole chicken
- 4 tablespoons of melted butter
- 1 cup of grapefruit juice
- 2.5 cups chicken stock
- Ingredients for The Glaze:
- 6 teaspoons of coconut milk
- 3 tablespoons of sesame oil
- 3 tablespoons of maple syrup
- 1 tablespoon of lemon juice
- 4 tablespoons of melted butter

Directions:

- In a small cooking pot, pour the coconut milk and add sesame oil, maple syrup, melted butter, and lemon juice.
- Cook the mixture for a few minutes until all the ingredients are combined well, the glaze is ready.
- Reserve some of the mixture for further use.
- Take a separate cooking pot and add chicken stock, butter, and grapefruit juice.
- Simmer the mixture for a few minutes, and then add the chicken to this liquid.
- Submerge the chicken completely in the brain and let it sit for a few hours for marinating.
- In a separate bowl, combine all the rub ingredients.

- After a few hours pass, take out the chicken from the liquid and pat dry with a paper towel.
- Now cover the chicken with the rub mixture.
- Preheat the smoker grill for 20 minutes at 225 degrees Fahrenheit.
- Cherry or apple wood chip can be used to create the smoke.
- Place chicken onto the smoker grill grate and cook for 3 hours by closing the lid.
- After every 30 minutes, baste the chicken with the maple glaze.
- Once the internal temperature of the chicken reaches 165 degrees Fahrenheit the chicken is ready to be served.
- Remove the chicken from the grill grate and baste it with the glaze and additional butter on top.
- Let the chicken sit at the room temperature for 10 minutes before cutting and serving.

145. Sriracha Chicken Wings

Cooking Time: 2 Hours
Servings: 4

The Heat: Cherry Wood

Ingredients:

- 2 pounds of chicken wings
- 2 teaspoons garlic powder
- Sea salt, to taste
- Freshly ground black pepper, to taste
- Ingredients for The Sauce:
- 1/3 cup raw honey
- 1/3 cup Sriracha sauce
- 2 tablespoons coconut amino
- 3 limes, juice only

Directions:

- Take a large mixing bowl and combine the sauce ingredients including Sriracha sauce, raw honey, coconut amino, and lime juice.
- Rub the chicken with salt, black pepper, and garlic powder.
- Preheat the smoker grill for 30 minutes at 225 degrees F.
- Put the chicken directly onto the grill grate and smoke with the close lid for 2 hours.
- When the internal temperature reaches 150 degrees F the chicken is ready.
- Remove the chicken from the grill grate and dumped into the sauce bowl.
- Toss the chicken wings well for the fine coating.
- Serve and Enjoy.

146. Rosemary Chicken

Servings: 6

Preparation time: 4 hours and 10 minutes

Cooking time: 1 hour and 5 minutes

Ingredients:

- 4 pounds chicken thighs, boneless
- 2 teaspoons garlic powder
- 2 teaspoons salt
- 1/2 cup brown sugar
- 1 teaspoon ground black pepper

- 4 teaspoons fresh rosemary, chopped
- 1/4 cup soy sauce
- 1/2 cup apple cider vinegar
- 1/2 cup Worcestershire sauce
- 1/2 cup olive oil
- 1/2 of a lemon, juiced
- 1/4 cup Dijon mustard

Directions:

- Place all the ingredients in a small bowl, except for chicken and then stir well until combined.
- Place chicken thighs in a large plastic bag, pour in the prepared mixture, seal the bag, turn it upside down to coat the chicken pieces and let marinate in the refrigerator for a minimum for 4 hours.
- When ready to cook, open hopper of the smoker, add dry pallets, make sure ash-can is in place, then open the ash damper, power on the smoker and close the ash damper.
- Set the temperature of the smoker to 350 degrees F, let preheat for 30 minutes or until the green light on the dial blinks that indicate smoker has reached to set temperature.
- Remove chicken thighs from the marinade, place them on the smoker grill, shut with lid and smoke for 35 minutes or until thoroughly cooked and the internal temperature of chicken reaches to 165 degrees F.
- When done, transfer chicken to a dish and serve straight away.

Nutrition: Calories: 109; Total Fat: 5.5 g; Saturated Fat: 1.1 g; Protein: 13.7 g; Carbs: 0.6 g; Fiber: 0.2 g; Sugar: 0.1 g

147. Asian Wings

Servings: 4

Preparation time: 15 minutes

Cooking time: 2 hours

Ingredients:

- 2 pounds chicken wings
- 1 teaspoon minced garlic
- ½ teaspoon ginger powder
- 1 tablespoon honey
- 2 tablespoons soy sauce
- 1 teaspoon sesame oil
- 3 tablespoons hoisin sauce
- 1 tablespoon sesame seeds, toasted

Directions:

- Open hopper of the smoker, add dry pallets, make sure ash-can is in place, then open the ash damper, power on the smoker and close the ash damper.
- Set the temperature of the smoker to 225 degrees F, let preheat for 30 minutes or until the green light on the dial blinks that indicate smoker has reached to set temperature.
- Then place chicken wings on the smoker grill, shut with lid and smoke for 1 hour and 30 minutes.
- Meanwhile, place remaining ingredients except for sesame seeds in a small bowl and stir until combined.
- When chicken wings are smoked, transfer the smoked chicken wings in a large bowl and preheat the smoker at 375 degrees F.

- Add prepared mixture to the chicken wings, toss until well coated, then place chicken wings on a cookie sheet lined with aluminum foil and sprinkle with sesame seeds.
- Place the cookie sheet containing chicken wings on the grill grate, shut with lid and smoke for 30 minutes or until the internal temperature of chicken wings reach to 165 degrees F.
- When done, transfer chicken wings to a dish and serve straight away.

Nutrition: Calories: 165.7; Total Fat: 10.6 g; Saturated Fat: 1.2 g; Protein: 14.6 g; Carbs: 2.5 g; Fiber: 0.4 g; Sugar: 0.3 g

148. Orange Chicken Wings

Servings: 4

Preparation time: 1 hour and 10 minutes

Cooking time: 1 hour and 15 minutes

For the Sauce:

- 1 orange, zested
- 1 tablespoon corn starch
- 1 teaspoon ground ginger
- ½ teaspoon salt
- 1/4 teaspoon ground white pepper
- 1/3 cups brown sugar
- 1 tablespoon chili garlic paste
- 2 tablespoons soy sauce
- 1 cup orange juice, fresh
- 1/4 cup chicken stock
- For the Wings:
- 2 pounds chicken wings
- 2 tablespoons salt

Directions:

- Spread the chicken wings on a wire rack placed on a sheet pan and lined with paper towels, then pat dry the chicken wings, sprinkle them with salt and place them in the refrigerator for 1 hour.
- When ready to smoke, open hopper of the smoker, add dry pallets, make sure ash-can is in place, then open the ash damper, power on the smoker and close the ash damper.
- Set the temperature of the smoker to 350 degrees F, let preheat for 30 minutes or until the green light on the dial blinks that indicate smoker has reached to set temperature.
- Place chicken wings on the smoker grill, shut with lid and smoke for 45 minutes or until their skin is golden brown and the internal temperature of chicken wings reach to 170 degrees F.
- Meanwhile, prepare the orange sauce and for this, pour the chicken stock in a bowl, add corn starch, stir well and set aside until required.
- Add remaining ingredients for the sauce in a saucepan, whisk well until combined, then place it over medium heat and bring the sauce to simmer.

- Then whisk in corn starch-chicken stock mixture until mixed and continue simmering the sauce for 5 to 10 minutes or until the sauce has thickened; remove the pan from heat and set aside until required.
- When chicken wings are done, transfer them in a large bowl, pour in prepared orange sauce and toss until the chicken wings are well covered.
- Serve straight away.

Nutrition: Calories: 220; Total Fat: 8 g; Saturated Fat: 1 g; Protein: 12 g; Carbs: 25 g; Fiber: 1 g; Sugar: 12 g

149. Buffalo Wings (2nd Version)

Servings: 6

Preparation time: 10 minutes

Cooking time: 1 hour and 30 minutes

Ingredients:

- 5 pounds chicken wings
- 1 cup butter, unsalted
- 4 cups olive oil
- 2 cups Frank's hot sauce

Directions:

- Open hopper of the smoker, add dry pallets, make sure ash-can is in place, then open the ash damper, power on the smoker and close the ash damper.
- Set the temperature of the smoker to 450 degrees F, switch smoker to open flame cooking mode, press the open flame 3, remove the grill grates and the batch, replace batch with direct flame insert, then return grates on the grill in the lower position and let preheat for 30 minutes or until the green light on the dial blinks that indicate smoker has reached to set temperature.
- Then place the chicken wings on the smoker grill, shut with lid and smoke for 30 to 45 minutes or until the internal temperature of chicken reach to 115 degrees F.
- Meanwhile, place a small saucepan over medium heat, pour in hot sauce and bring to simmer.
- Then add butter, stir slowly until butter is melted completely and mixed in the sauce; remove the pan from the heat and set aside until required.
- Place a Dutch oven over medium heat, pour in the oil and let heat until it reaches to 375 degrees F temperature.
- When chicken wings have reached to 115 degrees, remove them from the smoker, then add them into the hot oil in the Dutch oven and fry for 3 to 5 minutes per side or until nicely golden brown and crispy.
- Transfer fried chicken wings in a large bowl, pour in prepared hot sauce and toss until well coated.

- Transfer chicken wings to a dish and serve straight away.

Nutrition: Calories: 73; Total Fat: 4 g; Saturated Fat: 1 g; Protein: 6 g; Carbs: 0 g; Fiber: 0 g; Sugar: 0 g

TURKEY RECIPES

150. Herb Roasted Turkey

Preparation Time: 15 Minutes

Cooking Time: 3 Hours And 30 Minutes

Servings: 12

Ingredients:

- 14 pounds turkey, cleaned
- 2 tablespoons chopped mixed herbs
- Pork and poultry rub as needed
- ¼ teaspoon ground black pepper
- 3 tablespoons butter, unsalted, melted
- 8 tablespoons butter, unsalted, softened
- 2 cups chicken broth

Directions:

1. Clean the turkey by removing the giblets, wash it inside out, pat dry with paper towels, then place it on a roasting pan and tuck the turkey wings by tiring with butcher's string.
2. Switch on the Traeger grill, fill the grill hopper with hickory flavored wood pellets, power the grill on by using the control panel, select 'smoke' on the temperature dial, or set the temperature to 325 degrees F and let it preheat for a minimum of 15 minutes.
3. Meanwhile, prepared herb butter and for this, take a small bowl, place the

softened butter in it, add black pepper and mixed herbs and beat until fluffy.

4. Place some of the prepared herb butter underneath the skin of turkey by using a handle of a wooden spoon, and massage the skin to distribute butter evenly.
5. Then rub the exterior of the turkey with melted butter, season with pork and poultry rub, and pour the broth in the roasting pan.
6. When the grill has preheated, open the lid, place roasting pan containing turkey on the grill grate, shut the grill and smoke for 3 hours and 30 minutes until the internal temperature reaches 165 degrees F and the top has turned golden brown.
7. When done, transfer turkey to a cutting board, let it rest for 30 minutes, then carve it into slices and serve.

Nutrition: Calories: 154.6 Fat: 3.1 g Carbs: 8.4 g Protein: 28.8 g

151. Turkey Legs

Preparation Time: 10 Minutes

Cooking Time: 5 Hours

Servings: 4

Ingredients:

- 4 turkey legs
- For the Brine:
- ½ cup curing salt
- 1 tablespoon whole black peppercorns
- 1 cup BBQ rub
- ½ cup brown sugar
- 2 bay leaves
- 2 teaspoons liquid smoke
- 16 cups of warm water
- 4 cups ice
- 8 cups of cold water

Directions:

1. Prepare the brine and for this, take a large stockpot, place it over high heat, pour warm water in it, add peppercorn, bay leaves, and liquid smoke, stir in salt, sugar, and BBQ rub and bring it to a boil.
2. Remove pot from heat, bring it to room temperature, then pour in cold water, add ice cubes and let the brine chill in the refrigerator.
3. Then add turkey legs in it, submerge them completely, and let soak for 24 hours in the refrigerator.
4. After 24 hours, remove turkey legs from the brine, rinse well and pat dry with paper towels.
5. When ready to cook, switch on the Traeger grill, fill the grill hopper with hickory flavored wood pellets, power the grill on by using the control panel, select 'smoke' on the temperature dial, or set the temperature to 250 degrees F and let it preheat for a minimum of 15 minutes.
6. When the grill has preheated, open the lid, place turkey legs on the grill grate, shut the grill, and smoke for 5 hours until nicely browned and the internal temperature reaches 165 degrees F. Serve immediately.

Nutrition: Calories: 416 Fat: 13.3 g Carbs: 0 g Protein: 69.8 g

152. Turkey Breast

Preparation Time: 12 Hours
Cooking Time: 8 Hours
Servings: 6

Ingredients:

For the Brine:

- 2 pounds turkey breast, deboned
- 2 tablespoons ground black pepper
- ¼ cup salt
- 1 cup brown sugar
- 4 cups cold water

For the BBQ Rub:

- 2 tablespoons dried onions
- 2 tablespoons garlic powder
- ¼ cup paprika
- 2 tablespoons ground black pepper
- 1 tablespoon salt
- 2 tablespoons brown sugar
- 2 tablespoons red chili powder
- 1 tablespoon cayenne pepper
- 2 tablespoons sugar
- 2 tablespoons ground cumin

Directions:

1. Prepare the brine and for this, take a large bowl, add salt, black pepper, and sugar in it, pour in water, and stir until sugar has dissolved.
2. Place turkey breast in it, submerge it completely and let it soak for a minimum of 12 hours in the refrigerator.
3. Meanwhile, prepare the BBQ rub and for this, take a small bowl, place all of its ingredients in it and then stir until combined, set aside until required.
4. Then remove turkey breast from the brine and season well with the prepared BBQ rub.
5. When ready to cook, switch on the Traeger grill, fill the grill hopper with apple-flavored wood pellets, power the grill on by using the control panel, select 'smoke' on the temperature dial, or set the temperature to 180 degrees F and let it preheat for a minimum of 15 minutes.
6. When the grill has preheated, open the lid, place turkey breast on the grill grate, shut the grill, change the smoking temperature to 225 degrees F, and smoke for 8 hours until the internal temperature reaches 160 degrees F.
7. When done, transfer turkey to a cutting board, let it rest for 10 minutes, then cut it into slices and serve.

Nutrition: Calories: 250 Fat: 5 g Carbs: 31 g Protein: 18 g

153. Apple wood-Smoked Whole Turkey

Preparation Time: 10 minutes
Cooking Time: 5 hours
Servings: 6

Ingredients:

- 1 (10- to 12-pound) turkey, giblets removed
- Extra-virgin olive oil, for rubbing
- ¼ cup poultry seasoning
- 8 tablespoons (1 stick) unsalted butter, melted
- ½ cup apple juice
- 2 teaspoons dried sage

- 2 teaspoons dried thyme

Directions:

1. Supply your smoker with wood pellets and follow the manufacturer's specific start-up procedure. Preheat, with the lid closed, to 250°F.
2. Rub the turkey with oil and season with the poultry seasoning inside and out, getting under the skin.
3. In a bowl, combine the melted butter, apple juice, sage, and thyme to use for basting.
4. Put the turkey in a roasting pan, place on the grill, close the lid, and grill for 5 to 6 hours, basting every hour, until the skin is brown and crispy, or until a meat thermometer inserted in the thickest part of the thigh reads 165°F.
5. Let the turkey meat rest for about 15 to 20 minutes before carving.

Nutrition: Calories: 180 Carbs: 3g Fat: 2g Protein: 39g

154. Savory-Sweet Turkey Legs

Preparation Time: 10 minutes
Cooking Time: 5 hours
Servings: 4

Ingredients:

- 1 gallon hot water
- 1 cup curing salt (such as Morton Tender Quick)
- ¼ cup packed light brown sugar
- 1 teaspoon freshly ground black pepper
- 1 teaspoon ground cloves
- 1 bay leaf
- 2 teaspoons liquid smoke
- 4 turkey legs
- Mandarin Glaze, for serving

Directions:

1. In a huge container with a lid, stir together the water, curing salt, brown sugar, pepper, cloves, bay leaf, and liquid smoke until the salt and sugar are dissolved; let come to room temperature.
2. Submerge the turkey legs in the seasoned brine, cover, and refrigerate overnight.
3. When ready to smoke, remove the turkey legs from the brine and rinse them; discard the brine.
4. Supply your smoker with wood pellets and follow the manufacturer's specific start-up procedure. Preheat, with the lid closed, to 225°F.
5. Arrange the turkey legs on the grill, close the lid, and smoke for 4 to 5 hours, or until dark brown and a meat thermometer inserted in the thickest part of the meat reads 165°F.
6. Serve with Mandarin Glaze on the side or drizzled over the turkey legs.

Nutrition: Calories: 190 Carbs: 1g Fat: 9g Protein: 24g

155. Marinated Smoked Turkey Breast

Preparation Time: 15 minutes
Cooking Time: 4 hours
Servings: 6

Ingredients:

- 1 (5 pounds) boneless chicken breast
- 4 cups water
- 2 tablespoons kosher salt
- 1 teaspoon Italian seasoning
- 2 tablespoons honey
- 1 tablespoon cider vinegar
- Rub:
- ½ teaspoon onion powder
- 1 teaspoon paprika
- 1 teaspoon salt
- 1 teaspoon ground black pepper
- 1 tablespoons brown sugar
- ½ teaspoon garlic powder
- 1 teaspoon oregano

Directions:

1. In a huge container, combine the water, honey, cider vinegar, Italian seasoning and salt.
2. Add the chicken breast and toss to combine. Cover the bowl and place it in the refrigerator and chill for 4 hours.
3. Rinse the chicken breast with water and pat dry with paper towels.
4. In another mixing bowl, combine the brown sugar, salt, paprika, onion powder, pepper, oregano and garlic.
5. Generously season the chicken breasts with the rub mix.
6. Preheat the grill to 225°F with lid closed for 15 minutes. Use cherry wood pellets.
7. Arrange the turkey breast into a grill rack. Place the grill rack on the grill.
8. Smoke for about 3 to 4 hours or until the internal temperature of the turkey breast reaches 165°F.
9. Remove the chicken breast from heat and let them rest for a few minutes. Serve.

Nutrition: Calories 903 Fat: 34g Carbs: 9.9g Protein 131.5g

156. Maple Bourbon Turkey

Preparation Time: 15 minutes
Cooking Time: 3 hours
Servings: 8
Ingredients:

- 1 (12 pounds) turkey
- 8 cup chicken broth
- 1 stick butter (softened)
- 1 teaspoon thyme
- 2 garlic clove (minced)
- 1 teaspoon dried basil
- 1 teaspoon pepper
- 1 teaspoon salt
- 1 tablespoon minced rosemary
- 1 teaspoon paprika
- 1 lemon (wedged)
- 1 onion
- 1 orange (wedged)
- 1 apple (wedged)
- Maple Bourbon Glaze:
- ¾ cup bourbon
- 1/2 cup maple syrup
- 1 stick butter (melted)
- 1 tablespoon lime

Directions:

1. Wash the turkey meat inside and out under cold running water.
2. Insert the onion, lemon, orange and apple into the turkey cavity.
3. In a mixing bowl, combine the butter, paprika, thyme, garlic, basil, pepper, salt, basil and rosemary.
4. Brush the turkey generously with the herb butter mixture.
5. Set a rack into a roasting pan and place the turkey on the rack. Put a 5 cups of chicken broth into the bottom of the roasting pan.
6. Preheat the grill to 350°F with lid closed for 15 minutes, using maple wood pellets.
7. Place the roasting pan in the grill and cook for 1 hour.
8. Meanwhile, combine all the maple bourbon glaze ingredients in a mixing bowl. Mix until well combined.
9. Baste the turkey with glaze mixture. Continue cooking, basting turkey every 30 minutes and adding more broth as needed for 2 hours, or until the internal temperature of the turkey reaches 165°F.
10. Take off the turkey from the grill and let it rest for a few minutes. Cut into slices and serve.

Nutrition: Calories 1536 Fat 58.6g Carbs: 24g Protein 20.1g

157. Thanksgiving Turkey

Preparation Time: 15 minutes
Cooking Time: 4 hours
Servings: 6

Ingredients:

- 2 cups butter (softened)
- 1 tablespoon cracked black pepper
- 2 teaspoons kosher salt
- 2 tablespoons freshly chopped rosemary
- 2 tablespoons freshly chopped parsley
- 2 tablespoons freshly chopped sage
- 2 teaspoons dried thyme
- 6 garlic cloves (minced)
- 1 (18 pound) turkey

Directions:

1. In a mixing bowl, combine the butter, sage, rosemary, 1 teaspoon black pepper, 1 teaspoon salt, thyme, parsley and garlic.
2. Use your fingers to loosen the skin from the turkey.
3. Generously, Rub butter mixture under the turkey skin and all over the turkey as well. 4. Season turkey generously with herb mix. 5. Preheat the grill to 300°F with lid closed for 15 minutes.
4. Place the turkey on the grill and roast for about 4 hours, or until the turkey thigh temperature reaches 160°F.
5. Take out the turkey meat from the grill and let it rest for a few minutes. Cut into sizes and serve.

Nutrition: Calories 278 Fat 30.8g Carbs: 1.6g Protein 0.6g

158. Spatchcock Smoked Turkey

Preparation Time: 15 minutes
Cooking Time: 4 hours 3 minutes
Servings: 6

Ingredients:

- 1 (18 pounds) turkey
- 2 tablespoons finely chopped fresh parsley
- 1 tablespoon finely chopped fresh rosemary
- 2 tablespoons finely chopped fresh thyme
- ½ cup melted butter
- 1 teaspoon garlic powder
- 1 teaspoon onion powder
- 1 teaspoon ground black pepper
- 2 teaspoons salt or to taste
- 2 tablespoons finely chopped scallions

Directions:

1. Remove the turkcy giblets and rinse turkey, in and out, under cold running water.
2. Place the turkey on a working surface, breast side down. Use a poultry shear to cut the turkey along both sides of the backbone to remove the turkey back bone.
3. Flip the turkey over, back side down. Now, press the turkey down to flatten it.
4. In a mixing bowl, combine the parsley, rosemary, scallions, thyme, butter, pepper, salt, garlic and onion powder.
5. Rub butter mixture over all sides of the turkey.
6. Preheat your grill to HIGH (450°F) with lid closed for 15 minutes.
7. Place the turkey directly on the grill grate and cook for 30 minutes. Reduce the heat to 300°F and cook for an additional 4 hours, or until the internal temperature of the thickest part of the thigh reaches 165°F.
8. Take out the turkey meat from the grill and let it rest for a few minutes. Cut into sizes and serve.

Nutrition: Calories: 780 Fat: 19g Carbs: 29.7g Protein 116.4g

159. Hoisin Turkey Wings

Preparation Time: 15 minutes
Cooking Time: 1 hour
Servings: 8

Ingredients:

- 2 pounds turkey wings
- ½ cup hoisin sauce
- 1 tablespoon honey
- 2 teaspoons soy sauce
- 2 garlic cloves (minced)
- 1 teaspoons freshly grated ginger
- 2 teaspoons sesame oil
- 1 teaspoons pepper or to taste
- 1 teaspoons salt or to taste
- ¼ cup pineapple juice
- 1 tablespoon chopped green onions
- 1 tablespoon sesame seeds
- 1 lemon (cut into wedges)

Directions:

1. In a huge container, combine the honey, garlic, ginger, soy, hoisin sauce, sesame oil, pepper and salt. Put all the

mixture into a zip lock bag and add the wings. Refrigerate for 2 hours.
2. Remove turkey from the marinade and reserve the marinade. Let the turkey rest for a few minutes, until it is at room temperature.
3. Preheat your grill to 300°F with the lid closed for 15 minutes.
4. Arrange the wings into a grilling basket and place the basket on the grill.
5. Grill for 1 hour or until the internal temperature of the wings reaches 165°F.
6. Meanwhile, pour the reserved marinade into a saucepan over medium-high heat. Stir in the pineapple juice.
7. Wait to boi then reduce heat and simmer for until the sauce thickens.
8. Brush the wings with sauce and cook for 6 minutes more. Remove the wings from heat.
9. Serve and garnish it with green onions, sesame seeds and lemon wedges.

Nutrition: Calories: 115 Fat: 4.8g Carbs: 11.9g Protein 6.8g

160. Turkey Jerky

Preparation Time: 15 minutes
Cooking Time: 4 hours
Servings: 6

Ingredients:

- Marinade:
- 1 cup pineapple juice
- ½ cup brown sugar
- 2 tablespoons sriracha
- 2 teaspoons onion powder
- 2 tablespoons minced garlic
- 2 tablespoons rice wine vinegar
- 2 tablespoons hoisin
- 1 tablespoon red pepper flakes
- 1 tablespoon coarsely ground black pepper flakes
- 2 cups coconut amino
- 2 jalapenos (thinly sliced)
- Meat:
- 3 pounds turkey boneless skinless breasts (sliced to ¼ inch thick)

Directions:

1. Pour the marinade mixture ingredients in a container and mix until the ingredients are well combined.
2. Put the turkey slices in a gallon sized zip lock bag and pour the marinade into the bag. Massage the marinade into the turkey. Seal the bag and refrigerate for 8 hours.
3. Remove the turkey slices from the marinade.
4. Activate the pellet grill for smoking and leave lip opened for 5 minutes until fire starts.
5. Close the lid and preheat your pellet grill to 180°F, using hickory pellet.
6. Remove the turkey slices from the marinade and pat them dry with a paper towel.
7. Arrange the turkey slices on the grill in a single layer. Smoke the turkey for about 3 to 4 hours, turning often after the first 2 hours of smoking. The jerky should be dark and dry when it is done.
8. Remove the jerky from the grill and let it sit for about 1 hour to cool. Serve immediately or store in refrigerator.

Nutrition: Calories: 109 Carbs: 12g Fat: 1g Protein: 14g

161. Smoked Turkey Legs

Preparation Time: 20 Minutes
Cooking Time: 5 Hours
Servings: 4-5

Ingredients:

- 1 Gallon of warm water
- ½ Gallon of cold water
- 4 Cups of ice
- 1 Cup of BBQ rub
- ½ Cup of curing salt
- ½ Cup of brown sugar
- 1 Tablespoon of crushed allspice berries
- 1 Tablespoon of whole black peppercorns
- 2 Bay leaves
- 2 Teaspoons of liquid smoke
- 4 to 5 turkey legs

Directions:

1. In a huge pot or saucepan, pour a gallon of warm water, the rub, the curing salt, the brown sugar, the allspice, the peppercorns, the bay leaves and the liquid smoke.
2. Bring the mixture to a boil over a high heat to dissolve the salt granules and let cool to a room temperature
3. Add some cold water or ice; then chill in the refrigerator and add the turkey legs
4. After about 24 hours, drain the turkey legs and discard the brine; then rinse the brine off the turkey legs with the cold water
5. Dry with paper towels; then brush off any solid spices
6. Set the temperature to about 250° F and preheat the pellet smoker with the lid closed for about 15 minutes.
7. Lay the turkey legs on the grill grate and smoke for about 4 to 5 hours
8. Serve immediately the turkey; then serve and enjoy!

Nutrition: Calories: 120 Fat: 3g Carbs: 0g Protein: 22g

162. Smoked Whole Turkey

Preparation Time: 20 Minutes
Cooking Time: 8 Hours
Servings: 6

Ingredients:

- 1 Whole Turkey of about 12 to 16 lb
- 1 Cup of your Favorite Rub
- 1 Cup of Sugar
- 1 Tablespoon of minced garlic
- ½ Cup of Worcestershire sauce
- 2 Tablespoons of Canola Oil

Directions:

1. Thaw the Turkey and remove the giblets
2. Pour in 3 gallons of water in a non-metal bucket of about 5 gallons
3. Add the BBQ rub and mix very well
4. Add the garlic, the sugar and the Worcestershire sauce; then submerge the turkey into the bucket.
5. Refrigerate the turkey in the bucket for an overnight.
6. Place the Grill on a High Smoke and smoke the Turkey for about 3 hours
7. Switch the grilling temp to about 350 degrees F; then push a metal meat

thermometer into the thickest part of the turkey breast

8. Cook for about 4 hours; then take off the wood pellet grill and let rest for about 15 minutes
9. Slice the turkey, then serve and enjoy your dish!

Nutrition: Calories: 165 Fat: 14g Carbs: 0.5g Protein: 15.2g

163. Smoked Turkey Breast

Preparation Time: 10 Minutes
Cooking Time: 1 Hour 30 minutes
Servings: 6

Ingredients:

- For The Brine
- 1 Cup of kosher salt
- 1 Cup of maple syrup
- ¼ Cup of brown sugar
- ¼ Cup of whole black peppercorns
- 4 Cups of cold bourbon
- 1 and ½ gallons of cold water
- 1 Turkey breast of about 7 pounds
- For The Turkey
- 3 Tablespoons of brown sugar
- 1 and ½ tablespoons of smoked paprika
- 1 ½ teaspoons of chipotle chili powder
- 1 ½ teaspoons of garlic powder
- 1 ½ teaspoons of salt
- 1 and ½ teaspoons of black pepper
- 1 Teaspoon of onion powder
- ½ teaspoon of ground cumin
- 6 Tablespoons of melted unsalted butter

Directions:

1. Before beginning; make sure that the bourbon; the water and the chicken stock are all cold
2. Now to make the brine, combine altogether the salt, the syrup, the sugar, the peppercorns, the bourbon, and the water in a large bucket.
3. Remove any pieces that are left on the turkey, like the neck or the giblets
4. Refrigerate the turkey meat in the brine for about 8 to 12 hours in a reseal able bag
5. Remove the turkey breast from the brine and pat dry with clean paper towels; then place it over a baking sheet and refrigerate for about 1 hour
6. Preheat your pellet smoker to about 300°F; making sure to add the wood chips to the burner
7. In a bowl, mix the paprika with the sugar, the chili powder, the garlic powder, the salt, the pepper, the onion powder and the cumin, mixing very well to combine.
8. Carefully lift the skin of the turkey; then rub the melted butter over the meat
9. Rub the spice over the meat very well and over the skin
10. Smoke the turkey breast for about 1 ½ hours at a temperature of about 375°

Nutrition: Calories: 94 Fat: 2g Carbs: 1g Protein: 18g

164. Whole Turkey

Preparation Time: 10 Minutes
Cooking Time: 7 Hours And 30 Minutes
Servings: 10

Ingredients:

- 1 frozen whole turkey, giblets removed, thawed
- 2 tablespoons orange zest
- 2 tablespoons chopped fresh parsley
- 1 teaspoon salt
- 2 tablespoons chopped fresh rosemary
- 1 teaspoon ground black pepper
- 2 tablespoons chopped fresh sage
- 1 cup butter, unsalted, softened, divided
- 2 tablespoons chopped fresh thyme
- ½ cup water
- 14.5-ounce chicken broth

Directions:

1. Open hopper of the smoker, add dry pallets, make sure ash-can is in place, then open the ash damper, power on the smoker and close the ash damper.
2. Set the temperature of the smoker to 180 degrees F, let preheat for 30 minutes or until the green light on the dial blinks that indicate smoker has reached to set temperature.
3. Meanwhile, prepare the turkey and for this, tuck its wings under it by using kitchen twine.
4. Place ½ cup butter in a bowl, add thyme, parsley, and sage, orange zest, and rosemary, stir well until combined and then brush this mixture generously on the inside and outside of the turkey and season the external of turkey with salt and black pepper.
5. Place turkey on a roasting pan, breast side up, pour in broth and water, add the remaining butter in the pan, then place the pan on the smoker grill and shut with lid.
6. Smoke the turkey for 3 hours, then increase the temperature to 350 degrees F and continue smoking the turkey for 4 hours or until thoroughly cooked and the internal temperature of the turkey reaches to 165 degrees F, basting turkey with the dripping every 30 minutes, but not in the last hour.
7. When you are done, take off the roasting pan from the smoker and let the turkey rest for 20 minutes.
8. Carve turkey into pieces and serve.

Nutrition: Calories: 146 Fat: 8 g Protein: 18 g Carbs: 1 g

165. Herbed Turkey Breast

Preparation Time: 8 Hours And 10 Minutes
Cooking Time: 3 Hours
Servings: 12

Ingredients:

- 7 pounds turkey breast, bone-in, skin-on, fat trimmed
- 3/4 cup salt
- 1/3 cup brown sugar
- 4 quarts water, cold
- For Herbed Butter:
- 1 tablespoon chopped parsley
- ½ teaspoon ground black pepper
- 8 tablespoons butter, unsalted, softened
- 1 tablespoon chopped sage

- ½ tablespoon minced garlic
- 1 tablespoon chopped rosemary
- 1 teaspoon lemon zest
- 1 tablespoon chopped oregano
- 1 tablespoon lemon juice

Directions:

1. Prepare the brine and for this, pour water in a large container, add salt and sugar and stir well until salt and sugar has completely dissolved.
2. Add turkey breast in the brine, cover with the lid and let soak in the refrigerator for a minimum of 8 hours.
3. Then remove turkey breast from the brine, rinse well and pat dry with paper towels.
4. Open hopper of the smoker, add dry pallets, make sure ash-can is in place, then open the ash damper, power on the smoker and close the ash damper.
5. Set the temperature of the smoker to 350 degrees F, let preheat for 30 minutes or until the green light on the dial blinks that indicate smoker has reached to set temperature.
6. Meanwhile, take a roasting pan, pour in 1 cup water, then place a wire rack in it and place turkey breast on it.
7. Prepare the herb butter and for this, place butter in a heatproof bowl, add remaining ingredients for the butter and stir until just mix.
8. Loosen the skin of the turkey from its breast by using your fingers, then insert 2 tablespoons of prepared herb butter on each side of the skin of the breastbone and spread it evenly, pushing out all the air pockets.
9. Place the remaining herb butter in the bowl into the microwave wave and heat for 1 minute or more at high heat setting or until melted.
10. Then brush melted herb butter on the outside of the turkey breast and place roasting pan containing turkey on the smoker grill.
11. Shut the smoker with lid and smoke for 2 hours and 30 minutes or until the turkey breast is nicely golden brown and the internal temperature of turkey reach to 165 degrees F, flipping the turkey and basting with melted herb butter after 1 hour and 30 minutes smoking.
12. When done, transfer the turkey breast to a cutting board, let it rest for 15 minutes, then carve it into pieces and serve.

Nutrition: Calories: 97 Fat: 4 g Protein: 13 g Carbs: 1 g

166. Jalapeno Injection Turkey

Preparation Time: 15 Minutes
Cooking Time: 4 Hours And 10 Minutes
Servings: 4

Ingredients:

- 15 pounds whole turkey, giblet removed
- ½ of medium red onion, peeled and minced
- 8 jalapeño peppers
- 2 tablespoons minced garlic
- 4 tablespoons garlic powder
- 6 tablespoons Italian seasoning
- 1 cup butter, softened, unsalted
- ¼ cup olive oil
- 1 cup chicken broth

Directions:

1. Open hopper of the smoker, add dry pallets, make sure ash-can is in place, then open the ash damper, power on the smoker and close the ash damper.
2. Make the temperature of the smoker up to 200 degrees F, let preheat for 30 minutes or until the green light on the dial blinks that indicate smoker has reached to set temperature.
3. Meanwhile, place a large saucepan over medium-high heat, add oil and butter and when the butter melts, add onion, garlic, and peppers and cook for 3 to 5 minutes or until nicely golden brown.
4. Pour in broth, stir well, let the mixture boil for 5 minutes, then remove pan from the heat and strain the mixture to get just liquid.
5. Inject turkey generously with prepared liquid, then spray the outside of turkey with butter spray and season well with garlic and Italian seasoning.
6. Place turkey on the smoker grill, shut with lid, and smoke for 30 minutes, then increase the temperature to 325 degrees F and continue smoking the turkey for 3 hours or until the internal temperature of turkey reach to 165 degrees F.
7. When done, transfer turkey to a cutting board, let rest for 5 minutes, then carve into slices and serve.

Nutrition: Calories: 131 Fat: 7 g Protein: 13 g Carbs: 3 g

167. Smoked Turkey Mayo with Green Apple

Preparation Time: 20 minutes
Cooking Time: 4 hours 10 minutes
Servings: 10

Ingredients:

- Whole turkey (4-lbs., 1.8-kg.)
- The Rub
- Mayonnaise – ½ cup
- Salt – ¾ teaspoon
- Brown sugar – ¼ cup
- Ground mustard – 2 tablespoons
- Black pepper – 1 teaspoon
- Onion powder – 1 ½ tablespoons
- Ground cumin – 1 ½ tablespoons
- Chili powder – 2 tablespoons
- Cayenne pepper – ½ tablespoon
- Old Bay Seasoning – ½ teaspoon
- The Filling
- Sliced green apples – 3 cups

Directions:

1. Place salt, brown sugar, brown mustard, black pepper, onion powder, ground cumin, chili powder, cayenne pepper, and old bay seasoning in a bowl then mix well. Set aside.
2. Next, fill the turkey cavity with sliced green apples then baste mayonnaise over the turkey skin.
3. Sprinkle the dry spice mixture over the turkey then wrap with aluminum foil.
4. Marinate the turkey for at least 4 hours or overnight and store in the fridge to keep it fresh.
5. On the next day, remove the turkey from the fridge and thaw at room temperature.

6. Meanwhile, plug the wood pellet smoker then fill the hopper with the wood pellet. Turn the switch on.
7. Set the wood pellet smoker for indirect heat then adjust the temperature to 275°F (135°C).
8. Unwrap the turkey and place in the wood pellet smoker.
9. Smoke the turkey for 4 hours or until the internal temperature has reached 170°F (77°C).
10. Remove the smoked turkey from the wood pellet smoker and serve.

Nutrition: Calories: 340 Carbs: 40g Fat: 10g Protein: 21g

168. Buttery Smoked Turkey Beer

Preparation Time: 15 minutes
Cooking Time: 4 hours
Servings: 6

Ingredients:

- Whole turkey (4-lbs., 1.8-kg.)
- The Brine
- Beer – 2 cans
- Salt – 1 tablespoon
- White sugar – 2 tablespoons
- Soy sauce – ¼ cup
- Cold water – 1 quart
- The Rub
- Unsalted butter – 3 tablespoons
- Smoked paprika – 1 teaspoon
- Garlic powder – 1 ½ teaspoons
- Pepper – 1 teaspoon
- Cayenne pepper – ¼ teaspoon

Directions:

1. Pour beer into a container then add salt, white sugar, and soy sauce then stir well.
2. Put the turkey into the brine mixture cold water over the turkey. Make sure that the turkey is completely soaked.
3. Soak the turkey in the brine for at least 6 hours or overnight and store in the fridge to keep it fresh.
4. On the next day, remove the turkey from the fridge and take it out of the brine mixture.
5. Wash and rinse the turkey then pat it dry.
6. Next, plug the wood pellet smoker then fill the hopper with the wood pellet. Turn the switch on.
7. Set the wood pellet smoker for indirect heat then adjust the temperature to 275°F (135°C).
8. Open the beer can then push it in the turkey cavity.
9. Place the seasoned turkey in the wood pellet smoker and make a tripod using the beer can and the two turkey-legs.
10. Smoke the turkey for 4 hours or until the internal temperature has reached 170°F (77°C).
11. Once it is done, remove the smoked turkey from the wood pellet smoker and transfer it to a serving dish.

Nutrition: Calories: 229 Carbs: 34g Fat: 8g Protein: 3g

169. Barbecue Chili Smoked Turkey Breast

Preparation Time: 15 minutes
Cooking Time: 4 hours 20 minutes
Servings: 8

Ingredients:

- Turkey breast (3-lb., 1.4-kg.)
- The Rub
- Salt – ¾ teaspoon
- Pepper – ½ teaspoon
- The Glaze
- Olive oil – 1 tablespoon
- Ketchup – ¾ cup
- White vinegar – 3 tablespoons
- Brown sugar – 3 tablespoons
- Smoked paprika – 1 tablespoons
- Chili powder – ¾ teaspoon
- Cayenne powder – ¼ teaspoon

Directions:

1. Score the turkey breast at several places then sprinkle salt and pepper over it.
2. Let the seasoned turkey breast rest for approximately 10 minutes.
3. In the meantime, plug the wood pellet smoker then fill the hopper with the wood pellet. Turn the switch on.
4. Set the wood pellet smoker for indirect heat then adjust the temperature to 275°F (135°C).
5. Place the seasoned turkey breast in the wood pellet smoker and smoke for 2 hours.
6. In the meantime, combine olive oil, ketchup, white vinegar, brown sugar, smoked paprika; chili powder, garlic powder, and cayenne pepper in a saucepan then stir until incorporated. Wait to simmer then remove from heat.
7. After 2 hours of smoking, baste the sauce over the turkey breast and continue smoking for another 2 hours.
8. Once the internal temperature of the smoked turkey breast has reached 170°F (77°C) remove from the wood pellet smoker and wrap with aluminum foil.
9. Let the smoked turkey breast rest for approximately 15 minutes to 30 minutes then unwrap it.
10. Cut the smoked turkey breast into thick slices then serve.

Nutrition: Calories: 290 Carbs: 2g Fat: 3g Protein: 63g

170. Hot Sauce Smoked Turkey Tabasco

Preparation Time: 20 minutes
Cooking Time: 4 hours 15 minutes
Servings: 8

Ingredients:

- Whole turkey (4-lbs., 1.8-kg.)
- The Rub
- Brown sugar – ¼ cup
- Smoked paprika – 2 teaspoons
- Salt – 1 teaspoon
- Onion powder – 1 ½ teaspoons
- Oregano – 2 teaspoons
- Garlic powder – 2 teaspoons
- Dried thyme – ½ teaspoon
- White pepper – ½ teaspoon
- Cayenne pepper – ½ teaspoon
- The Glaze
- Ketchup – ½ cup

- Hot sauce – ½ cup
- Cider vinegar – 1 tablespoon
- Tabasco – 2 teaspoons
- Cajun spices – ½ teaspoon
- Unsalted butter – 3 tablespoons

Directions:

1. Rub the turkey with 2 tablespoons of brown sugar, smoked paprika, salt, onion powder, garlic powder, dried thyme, white pepper, and cayenne pepper. Let the turkey rest for an hour.
2. Plug the wood pellet smoker then fill the hopper with the wood pellet. Turn the switch on.
3. Set the wood pellet smoker for indirect heat then adjust the temperature to 275°F (135°C).
4. Place the seasoned turkey in the wood pellet smoker and smoke for 4 hours.
5. In the meantime, place ketchup, hot sauce, cider vinegar, Tabasco, and Cajun spices in a saucepan then bring to a simmer.
6. Remove the sauce from heat and quickly add unsalted butter to the saucepan. Stir until melted.
7. After 4 hours of smoking, baste the Tabasco sauce over the turkey then continue smoking for 15 minutes.
8. Once the internal temperature of the smoked turkey has reached 170°F (77°C), remove from the wood pellet smoker and place it on a serving dish.

Nutrition: Calories: 160 Carbs: 2g Fat: 14g Protein: 7g

171. Ginger Sage Smoked Turkey Legs

Preparation Time: 10 minutes
Cooking Time: 3 hours 15 minutes
Servings: 8

Ingredients:

- Turkey legs (3.5-lb., 1.6-kg.)
- The Rub
- Vegetable oil – ¼ cup
- Onion powder – 2 tablespoons
- Smoked paprika – 1 tablespoon
- Garlic powder – ½ tablespoon
- Salt – ¾ teaspoon
- White pepper – ½ teaspoon
- Ground ginger – 1 teaspoon
- Ground sage – ½ teaspoon

Directions:

1. Combine onion powder with smoked paprika, garlic powder, salt, white pepper, ground ginger, and ground sage in a bowl then mix well.
2. Drizzle vegetable oil over the spices then stir until becoming a paste.
3. Rub the turkey legs with the spice mixture then let it rest for approximately 15 minutes.
4. Next, plug the wood pellet smoker then fill the hopper with the wood pellet. Turn the switch on.
5. Set the wood pellet smoker for indirect heat then adjust the temperature to 275°F (135°C).
6. Place the seasoned turkey in the wood pellet smoker and smoke for 3 hours.
7. Wait until the internal temperature of the smoked turkey legs has reached 170°F (77°C) and remove it from the wood pellet smoker.

8. Arrange the smoked turkey legs on a serving dish.

Nutrition: Calories: 190 Carbs: 1g

Fat: 9g Protein: 24g

172. Cured Turkey Drumstick

Preparation time: 20 minutes
Cooking time: 2.5 hours to 3 hours
Servings: 3

Ingredients:

- 3 fresh or thawed frozen turkey drumsticks
- 3 tablespoons extra virgin olive oil
- Brine component
- 4 cups of filtered water
- ¼Cup kosher salt
- ¼ cup brown sugar
- 1 teaspoon garlic powder
- Poultry seasoning 1 teaspoon
- 1/2 teaspoon red pepper flakes
- 1 teaspoon pink hardened salt

Directions:

1. Put the salt water ingredients in a 1 gallon sealable bag. Add the turkey drumstick to the salt water and refrigerate for 12 hours.
2. After 12 hours, remove the drumstick from the saline, rinse with cold water, and pat dry with a paper towel.
3. Air dry the drumstick in the refrigerator without a cover for 2 hours.
4. Remove the drumsticks from the refrigerator and rub a tablespoon of extra virgin olive oil under and over each drumstick.
5. Set the wood pellet or grill for indirect cooking and preheat to 250 degrees Fahrenheit using hickory or maple pellets.
6. Place the drumstick on the grill and smoke at 250 ° F for 2 hours.
7. After 2 hours, increase grill temperature to 325 ° F.
8. Cook the turkey drumstick at 325 ° F until the internal temperature of the thickest part of each drumstick is 180 ° F with an instant reading digital thermometer.
9. Place a smoked turkey drumstick under a loose foil tent for 15 minutes before eating.

Nutrition: Calories: 278 Carbs: 0g Fat: 13g Protein: 37g

173. Tailgate Smoked Young Turkey

Preparation Time: 20 Minutes
Cooking Time: 4 To 4 Hours 30 Minutes
Servings: 6

Ingredients:

- 1 fresh or thawed frozen young turkey
- 6 glasses of extra virgin olive oil with roasted garlic flavor
- 6 original Yang dry lab or poultry seasonings

Directions:

1. Remove excess fat and skin from turkey breasts and cavities.
2. Slowly separate the skin of the turkey to its breast and a quarter of the leg, leaving the skin intact.

3. Apply olive oil to the chest, under the skin and on the skin.
4. Gently rub or season to the chest cavity, under the skin and on the skin.
5. Set up tailgate wood pellet smoker grill for indirect cooking and smoking. Preheat to 225 ° F using apple or cherry pellets.
6. Put the turkey meat on the grill with the chest up.
7. Suck the turkey for 4-4 hours at 225 ° F until the thickest part of the turkey's chest reaches an internal temperature of 170 ° F and the juice is clear.
8. Before engraving, place the turkey under a loose foil tent for 20 minutes

Nutrition: Calories: 240 Carbs: 27g Fat: 9g Protein: 15g

174. Roast Turkey Orange

Preparation Time: 30 Minutes
Cooking Time: 2 hours 30 minutes
Servings:

Ingredients:

- 1 Frozen Long Island turkey
- 3 tablespoons west
- 1 large orange, cut into wedges
- Three celery stems chopped into large chunks
- Half a small red onion, a quarter
- Orange sauce:
- 2 orange cups
- 2 tablespoons soy sauce
- 2 tablespoons orange marmalade
- 2 tablespoons honey
- 3 teaspoons grated raw

Directions:

1. Remove the jibble from the turkey's cavity and neck and retain or discard for another use. Wash the duck and pat some dry paper towel.
2. Remove excess fat from tail, neck and cavity. Use a sharp scalpel knife tip to pierce the turkey's skin entirely, so that it does not penetrate the duck's meat, to help dissolve the fat layer beneath the skin.
3. Add the seasoning inside the cavity with one cup of rub or seasoning.
4. Season the outside of the turkey with the remaining friction or seasoning.
5. Fill the cavity with orange wedges, celery and onion. Duck legs are tied with butcher twine to make filling easier. Place the turkey's breast up on a small rack of shallow roast bread.
6. To make the sauce, mix the ingredients in the saucepan over low heat and cook until the sauce is thick and syrupy. Set aside and let cool.
7. Set the wood pellet smoker grill for indirect cooking and use the pellets to preheat to 350 ° F.
8. Roast the turkey at 350 ° F for 2 hours.
9. After 2 hours, brush the turkey freely with orange sauce.
10. Roast the orange glass turkey for another 30 minutes, making sure that the inside temperature of the thickest part of the leg reaches 165 ° F.
11. Place turkey under loose foil tent for 20 minutes before serving.
12. Discard the orange wedge, celery and onion. Serve with a quarter of turkey with poultry scissors.

Nutrition: Calories: 216 Carbs: 2g Fat: 11g Protein: 34g

PORK RECIPES

175. Baby Back Ribs

Preparation Time: 6 Hours And 10 Minutes
Cooking Time: 5 Hours And 50 Minutes
Servings: 6

Ingredients:

- 2 racks of baby back ribs
- 1 tablespoon onion powder
- 1 tablespoon garlic powder
- 4-ounce salt
- 1 tablespoon ground black pepper
- 1 teaspoon chipotle powder
- 2-ounce mild chili powder
- 4-ounce brown sugar
- 1 tablespoon ground cumin
- 1 tablespoon ground coriander

Directions:

1. Prepare spice mix and for this, stir together all the ingredients except for ribs, then sprinkle generously on both sides of ribs until evenly coated and met marinate for a minimum of 6 hours or overnight.
2. When ready to cook, open hopper of the smoker, add dry pallets, make sure ash-can is in place, then open the ash damper, power on the smoker and close the ash damper.

3. Set the temperature of the smoker to 350 degrees F, let preheat for 30 minutes, then set it to 200 degrees F and continue preheating for 20 minutes or until the green light on the dial blinks that indicate smoker has reached to set temperature.
4. Place marinated pork ribs on the smoker grill, shut with lid and smoke for 5 hours or until thoroughly cooked, brushing ribs with marinade every 15 minutes.
5. When done, transfer ribs to a cutting board, let rest for 5 minutes, then cut into pieces and serve straight away.

Nutrition: Calories: 661.8 Fat: 42.4 g Protein: 28.4 g Carbs: 48.6 g

176. Rosemary Pork Tenderloin

Preparation Time: 10 Minutes
Cooking Time: 1 hour 20 Minutes
Servings: 2

Ingredients:

- 1.5-pound pork tenderloin, fat trimmed
- 2 tablespoons minced garlic
- ¼ teaspoon ground black pepper
- 1 tablespoon Dijon mustard
- 1 tablespoon olive oil
- 6 sprigs of rosemary, fresh

Directions:

1. Open hopper of the smoker, add dry pallets, make sure ash-can is in place, then open the ash damper, power on the smoker and close the ash damper.
2. Set the temperature of the smoker to 350 degrees F, let preheat for 30 minutes, then set it to 375 degrees F and continue preheating for 20 minutes or until the green light on the dial blinks that indicate smoker has reached to set temperature.
3. Meanwhile, stir together garlic, black pepper, mustard, and oil until smooth paste comes together and then coat pork with this paste evenly.
4. Cut a kitochen string into six 10-inch long pieces, then place them parallel to each, about 2-inch apart, lay 3 sprigs horizontally across the kitchen string, place seasoned pork on it, cover the top with remaining sprigs and tie the strings to secure the sprigs around the tenderloin.
5. Place pork on the smoker grill, shut with lid, smoke for 15 minutes, then flip the pork tenderloin and continue smoking for another 15 minutes or until the internal temperature of pork reach to 145 degrees F.
6. When done, transfer pork tenderloin to a cutting board, let rest for 5 minutes, then remove all the strong and cut pork into even slices.

Nutrition: Calories: 480 Fat: 23 g Protein: 47 g Carbs: 13 g

177. Pulled Pork

Preparation Time: 25 Minutes
Cooking Time: 10 Hours And 50 Minutes
Servings: 2

Ingredients:

- 8-pound pork shoulder, fat trimmed
- 1 tablespoon garlic powder
- 1 tablespoon salt
- 1 teaspoon ground black pepper
- 1 teaspoon chipotle powder
- 1 teaspoon red chili powder
- 1 teaspoon dried thyme
- 2 tablespoons Dijon mustard

Directions:

1. Open hopper of the smoker, add dry pallets, make sure ash-can is in place, then open the ash damper, power on the smoker and close the ash damper.
2. Set the temperature of the smoker to 350 degrees F, let preheat for 30 minutes, then set it to 225 degrees F and continue preheating for 20 minutes or until the green light on the dial blinks that indicate smoker has reached to set temperature.
3. Meanwhile, rinse and pat dry pork and then coat with mustard.
4. Stir together garlic powder, salt, black pepper, chipotle powder, red chili powder, and thyme and rub this mixture on pork.
5. Place pork on the smoker grill, fat-side up, shut with lid and smoke until the internal temperature of pork reach to 160 degrees F.
6. Remove pork from the smoker and then wrap it with aluminum foil.
7. Then set the temperature of the smoker to 240 degrees F, return wrapped pork into the smoker and continue smoking the pork until the internal temperature of pork reach to 195 degrees F.
8. When you are done, transfer the pork to a cutting board, let rest for 15 minutes, then unwrap the pork and shred the meat with two forks.
9. Toss shredded pork into its juices and serve with ciabatta buns and barbecue sauce.

Nutrition: Calories: 745.8 Fat: 54.4g Protein: 53.3 g Carbs: 8.9 g

178. Honey Glazed Ham

Preparation Time: 25 Minutes
Cooking Time: 2 Hours And 50 Minutes
Servings: 10

Ingredients:

- 8 pounds bone-in ham
- 20 whole cloves
- 1/4 cup corn syrup
- 1 cup smoked honey
- 1 stick of butter, unsalted, softened

Directions:

1. Open hopper of the smoker, add dry pallets, make sure ash-can is in place, then open the ash damper, power on the smoker and close the ash damper.
2. Set the temperature of the smoker to 350 degrees F, let preheat for 30 minutes, then set it to 325 degrees F and continue preheating for 20 minutes or until the green light on the dial

blinks that indicate smoker has reached to set temperature.

3. Meanwhile, score ham using a sharp knife, then smear the meat with butter, stuff with cloves and place ham in an aluminum foil-lined baking pan.
4. Whisk together honey and corn syrup, pour three-fourth of this mixture over ham, and then place baking pan on the smoker.
5. Shut with lid and smoke ham for 1 hour and 30 minutes or 2 hours until thoroughly cooked, and the internal temperature of ham reaches to 140 degrees F, basting ham with remaining honey mixture every 15 minutes.
6. When done, remove the pan from the grill, let rest for 15 minutes and then slice to serve.

Nutrition: Calories: 120 Fat: 5 g Protein: 16 g Carbs: 1 g

179. Sweet & Salty Pork Belly

Preparation Time: 20 Minutes
Cooking Time: 55 Minutes
Servings:

Ingredients:

- 1-pound pork belly slices, thick-cut
- 2 teaspoons garlic powder
- 1 ½ teaspoon salt
- 1/2 cup brown sugar
- 2 teaspoons paprika
- ½ teaspoon chipotle powder

Directions:

1. Open hopper of the smoker, add dry pallets, make sure ash-can is in place, then open the ash damper, power on the smoker and close the ash damper.
2. Set the temperature of the smoker to 350 degrees F, let preheat for 30 minutes or until the green light on the dial blinks that indicate smoker has reached to set temperature.
3. Meanwhile, place ¼ cup sugar in a shallow dish, add garlic powder, salt, paprika, chipotle pepper, stir until mixed, then rub this mixture on all sides of slices until evenly coated.
4. Place pork slices on a rimmed cookie sheet lined with aluminum foil, place it on the smoker grill, shut with lid and smoke for 10 minutes.
5. Then sprinkle the remaining sugar over pork slices and continue smoking for 15 minutes or until pork is nicely browned and sugar has caramelized.
6. When done, transfer pork belly slices to a serving dish, let cool for 10 minutes and serve.

Nutrition: Calories: 346 Fat: 43 g Protein: 29 g Carbs: 0 g

180. Sweet Bacon Wrapped Smokes

Preparation Time: 40 Minutes
Cooking Time: 45 Minutes
Servings: 6

Ingredients:

- 14-ounce cocktail sausages
- 1-pound bacon strips, halved
- 1/2 cup brown sugar

Directions:

1. Place bacon strips on clean working space, roll them using a rolling pin until each strip is of even thickness, then wrap each strip of bacon around the sausage and secure with a toothpick.
2. Place wrapped sausages in a casserole dish in a single layer, sprinkle with sugar until covered entirely and let them rest in the refrigerator for 30 minutes.
3. Meanwhile, open hopper of the smoker, add dry pallets, make sure ash-can is in place, then open the ash damper, power on the smoker and close the ash damper.
4. Set the temperature of the smoker to 350 degrees F, let preheat for 30 minutes or until the green light on the dial blinks that indicate smoker has reached to set temperature.
5. Lay wrapped sausage on a cookie sheet lined with parchment sheet, place the cookie sheet on the smoker grill, shut with lid and smoke for 30 minutes or until thoroughly cooked.

Nutrition: Calories: 270 Fat: 27 g Protein: 9 g Carbs: 18 g

181. Prosciutto Wrapped Asparagus

Preparation Time: 10 Minutes

Cooking Time: 1 Hour 5 Minutes

Servings: 6

Ingredients:

- 2 bunches of asparagus
- 4-ounce prosciutto
- ½ tablespoon salt
- ½ tablespoon ground pepper
- 2 tablespoons apple cider vinegar, divided
- 2 tablespoons olive oil
- 3 tablespoons toasted pine nuts
- 1 lemon, zested

Directions:

1. Open hopper of the smoker, add dry pallets, make sure ash-can is in place, then open the ash damper, power on the smoker and close the ash damper.
2. Set the temperature of the smoker to 350 degrees F, let preheat for 30 minutes, then set it to 400 degrees F and continue preheating for 20 minutes or until the green light on the dial blinks that indicate smoker has reached to set temperature.
3. Meanwhile, rinse asparagus, pat dry with paper towels, then cut off bottom third off of stalks and wrap 4 to 5 asparagus stalks with a piece of prosciutto.
4. Drizzle oil over asparagus bunch, drizzle with 1 tablespoon vinegar, season with salt, black pepper, and lemon zest and then place them on a baking sheet.

5. Place baking sheet on the smoker grill, then shut with lid and smoke for 5 minutes.
6. Then shake the baking pan to turn asparagus bunches, drizzle with 1 tablespoon vinegar, return the baking pan on the smoker grill and continue smoking for 5 to 8 minutes or until thoroughly cooked.
7. When done, transfer asparagus bunches to a serving dish, scatter with pine nuts and serve.

Nutrition: Calories: 56.2 Fat: 3.7 g Protein: 4 g Carbs: 2.7 g

182. Lemon Pepper Pork Tenderloin

Preparation Time: 2 Hours And 20 Minutes

Cooking Time: 1 Hour 10 Minutes

Servings: 6

Ingredients:

- 2 pounds pork tenderloin, fat trimmed
- ½ teaspoon minced garlic
- 1/2 teaspoon salt
- 1/4 teaspoon ground black pepper
- 2 Lemons, zested
- 1 teaspoon minced parsley
- 1 teaspoon lemon juice
- 2 tablespoons olive oil

Directions:

1. Prepare the marinade and for this, place all the ingredients except for pork in a small bowl and stir until mixed.
2. Place pork tenderloin in a large plastic bag, pour in prepared marinade, seal the plastic bag, then turn it upside down to coat the pork and marinate for a minimum of 2 hours.
3. When ready to smoke, open hopper of the smoker, add dry pallets, and make sure ash-can is in place, then open the ash damper, power on the smoker and close the ash damper.
4. Set the temperature of the smoker to 350 degrees F, let preheat for 30 minutes, then set it to 375 degrees F and continue preheating for 20 minutes or until the green light on the dial blinks that indicate smoker has reached to set temperature.
5. Remove the pork tenderloin from the marinade, place it on the smoker grill, shut with lid and smoke for 20 minutes or until thoroughly cooked and the internal temperature of pork reach to 120 degrees F; flipping pork halfway through.
6. When you are done, transfer the pork meat to a cutting board, let rest for 10 minutes, and then slice to serve.

Nutrition: Calories: 144.5 Fat: 8.8 g Protein: 13.2 g Carbs: 3.1 g

183. Bacon Wrapped Jalapeno Poppers

Preparation Time: 10 Minutes
Cooking Time: 95 Minutes
Servings: 4

Ingredients:

- 16-ounce bacon, not thick-sliced
- 10 jalapeno peppers
- 20-ounce crushed pineapple with juice
- 8-ounce cream cheese, softened
- Barbecue sauce as needed

Directions:

1. Open hopper of the smoker, add dry pallets, make sure ash-can is in place, then open the ash damper, power on the smoker and close the ash damper.
2. Set the temperature of the smoker to 350 degrees F, let preheat for 30 minutes, then set it to 275 degrees F and continue preheating for 20 minutes or until the green light on the dial blinks that indicate smoker has reached to set temperature.
3. Meanwhile, cut each pepper lengthwise and then remove and discard its seeds.
4. Put a cream cheese in a shallow container, beat with an immersion blender until fluffy, add pineapples, mix well until combined and then stuff the mixture into jalapeno halved, leveling the top with a spatula.
5. Then wrap each stuffed jalapeno pepper with a slice of bacon and place it on a large baking sheet greased with oil; prepare the remaining wrapped peppers in the same manner.
6. Place baking sheet on the smoker grill, shut with lid and smoke for 45 minutes or until bacon is crispy.
7. Baste each pepper with a barbecue sauce and continue smoking for 5 minutes.
8. When done, transfer peppers to a dish and serve straight away.

Nutrition: Calories: 280 Fat: 26 g Protein: 9 g Carbs: 4 g

184. Bacon

Preparation Time: 10 Minutes
Cooking Time: 55 Minutes
Servings: 6

Ingredients:

- 1-pound bacon slices, thick-cut

Directions:

1. Open hopper of the smoker, add dry pallets, make sure ash-can is in place, then open the ash damper, power on the smoker and close the ash damper.
2. Set the temperature of the smoker to 375 degrees F, let preheat for 30 minutes or until the green light on the dial blinks that indicate smoker has reached to set temperature.
3. Meanwhile, take a large baking sheet, line it with parchment paper and place bacon slices on it in a single layer.
4. Place baking sheet on the smoker grill, shut with lid, and smoke for 20 minutes, then flip the bacon and continue smoking for 5 minutes or until bacon is no longer floppy.

5. When done, transfer bacon to a dish lined with paper towels to soak excess fat and then serve.

Nutrition: Calories: 80 Fat: 7 g Protein: 5 g Carbs: 0 g

185. Apple Wood Paprika Chili Smoked Pulled Pork

Preparation Time: 20 minutes
Cooking Time: 6 hours 10 minutes
Servings: 8

Ingredients:

- Pork Butt (4-lbs., 1.8-kg.)
- The Rub
- Smoked paprika – 2 ½ tablespoons
- Salt – 1 ½ teaspoons
- White sugar – 2 ½ tablespoons
- Ground cumin – 1 tablespoon
- Chili powder – 1 tablespoon
- Pepper – ¾ tablespoon
- Cayenne pepper – 1 ½ tablespoons
- The Sauce
- Yellow mustard – ¾ cup
- Chili powder – ½ tablespoon
- Brown sugar – 3 tablespoons
- Water – 3 tablespoons
- Soy sauce – ¾ teaspoons
- Unsalted butter – 1 tablespoon
- Liquid smoke – ¾ tablespoon

Directions:

1. Combine the rub ingredients—smoked paprika; salt, white sugar, ground cumin, chili powder, pepper, and cayenne pepper in a bowl then mix well.
2. Apply the rub mixture over the pork butt then set aside.
3. Next, plug the wood pellet smoker then fill the hopper with the wood pellet. Turn the switch on.
4. Set the wood pellet smoker for indirect heat then adjust the temperature to 250°F (121°C).
5. Wait until the wood pellet smoker reaches the desired temperature then place the seasoned pork butt in it.
6. Smoke the pork butt 3 hours or until the internal temperature of the smoked pork butt has reached 165°F (74°C).
7. Take out the smoked pork butt out of the wood pellet smoker then wrap with aluminum foil.
8. After that, return the wrapped smoked pork butt to the wood pellet smoker and continue smoking for another 3 hours.
9. In the meantime, place the entire sauce ingredients—yellow mustard, chili powder, brown sugar, water, soy sauce, unsalted butter, and liquid smoke in a saucepan then bring to a simmer. Remove from heat and set aside.
10. Once the internal temperature of the smoked pork butt has reached 205°F (96°C), take it out of the wood pellet smoker and let it rest for approximately 15 minutes.
11. Unwrap the smoked pork butt then using a fork shred the smoked pork butt.
12. Place the shredded smoked pork butt in a serving dish then drizzle the sauce on top. Mix well.

Nutrition: Calories: 230 Carbs: 1g Fat: 22g Protein: 8g

186. Bourbon Honey Glazed Smoked Pork Ribs

Preparation Time: 15 minutes
Cooking Time: 5 hours
Servings: 10

Ingredients:

- Pork Ribs (4-lbs., 1.8-kg.)
- The Marinade
- Apple juice – 1 ½ cups
- Yellow mustard – ½ cup
- The Rub
- Brown sugar – ¼ cup
- Smoked paprika – 1 tablespoon
- Onion powder – ¾ tablespoon
- Garlic powder – ¾ tablespoon
- Chili powder – 1 teaspoon
- Cayenne pepper – ¾ teaspoon
- Salt – 1 ½ teaspoons
- The Glaze
- Unsalted butter – 2 tablespoons
- Honey – ¼ cup
- Bourbon – 3 tablespoons

Directions:

1. Place apple juice and yellow mustard in a bowl then stir until combined.
2. Apply the mixture over the pork ribs then marinate for at least an hour.
3. In the meantime, combine brown sugar with smoked paprika, onion powder, garlic powder, chili powder, black pepper, cayenne pepper, and salt then mix well.
4. After an hour of marinade, sprinkle the dry spice mixture over the marinated pork ribs then let it rest for a few minutes.
5. Next, plug the wood pellet smoker then fill the hopper with the wood pellet. Turn the switch on.
6. Set the wood pellet smoker for indirect heat then adjust the temperature to 250°F (121°C).
7. When the wood pellet smoker is ready, place the seasoned pork ribs in the wood pellet smoker and smoke for 3 hours.
8. Meanwhile, place unsalted butter in a saucepan then melt over very low heat.
9. Once it is melted, remove from heat then add honey and bourbon to the saucepan. Stir until incorporated and set aside.
10. After 3 hours of smoking, baste the honey bourbon mixture over the pork ribs and wrap with aluminum foil.
11. Return the wrapped pork ribs to the wood pellet smoker and continue smoking for the next 2 hours.
12. Once the internal temperature of the smoked pork ribs reaches 145°F (63°C), remove the smoked pork ribs from the wood pellet smoker.
13. Unwrap the smoked pork ribs and serve.

Nutrition: Calories: 313 Carbs: 5g Fat: 20g Protein: 26g

187. Lime Barbecue Smoked Pork Shoulder Chili

Preparation Time: 20 minutes
Cooking Time: 6 hours 10 minutes
Servings: 8

Ingredients:

- Pork Shoulder (3.5-lb., 1.6-kg.)
- The Rub
- Brown sugar – 3 tablespoons
- Garlic powder – 1 tablespoon
- Smoked paprika -1 tablespoon
- Ground cumin – 1 tablespoon
- Salt – 1 teaspoon
- Chili powder – 1 ½ teaspoons
- Black pepper – 1 teaspoon
- The Glaze
- Red chili flakes – 1 tablespoon
- Vegetable oil – 2 tablespoons
- Minced garlic – 1 tablespoon
- Ground coriander – 1 ½ teaspoons
- Tomato ketchup – 1 ½ cups
- White sugar – ¼ cup
- Apple juice – ½ cup
- The Topping
- Fresh limes - 2

Directions:

1. Place brown sugar, garlic powder, smoked paprika, ground cumin, salt, chili powder, and black pepper in a bowl then stir until combined.
2. Rub the spices mixture over and side by side of the pork shoulder then let it rest for approximately an hour.
3. In the meantime, pour vegetable oil into a saucepan then preheat over medium heat.
4. Once the oil is hot, stir in minced garlic and sauté until wilted and aromatic. Remove the saucepan from heat.
5. Stir in red chili flakes, ground coriander, and white sugar into the saucepan then pour apple juice and tomato ketchup over the sauce. Mix well and set aside.
6. Next, plug the wood pellet smoker then fill the hopper with the wood pellet. Turn the switch on.
7. Set the wood pellet smoker for indirect heat then adjust the temperature to 250°F (121°C).
8. Place the seasoned pork shoulder in the wood pellet smoker and smoke for 3 hours. The internal temperature should be 165°F (74°C).
9. Take the pork shoulder out of the wood pellet smoker then place it on a sheet of aluminum foil.
10. Baste the glaze over the pork shoulder then arrange sliced limes over the pork shoulder.
11. Wrap the pork shoulder with aluminum foil then return it to the wood pellet smoker.
12. Smoke the wrapped smoked pork shoulder for another 3 hours or until the internal temperature has reached 205°F (96°C).
13. Once it is done, remove the smoked pork shoulder from the wood pellet smoker and let it rest for approximately 15 minutes.
14. Unwrap the smoked pork shoulder and place it on a serving dish.

Nutrition: Calories: 220 Carbs: 1g Fat: 18g Protein: 16g

188. Chili Sweet Smoked Pork Tenderloin

Preparation Time: 10 minutes
Cooking Time: 3 hours 30 minutes
Servings: 8

Ingredients:

- Pork Tenderloin (3-lb., 1.4-kg.)
- The Rub
- Apple juice – 1 cup
- Honey – ½ cup
- Brown sugar – ¾ cup
- Dried thyme – 2 tablespoons
- Black pepper – ½ tablespoon
- Chili powder – 1 ½ teaspoons
- Italian seasoning – ½ teaspoon
- Onion powder – 1 teaspoon

Directions:

1. Pour apple juice into a container then stir in honey, brown sugar, dried thyme, black pepper, chili powder, Italian seasoning, and onion powder. Mix well.
2. Rub the pork tenderloin with the spice mixture then let it rest for an hour.
3. Next, plug the wood pellet smoker then fill the hopper with the wood pellet. Turn the switch on.
4. Set the wood pellet smoker for indirect heat then adjust the temperature to 250°F (121°C).
5. When the wood pellet smoker has reached the desired temperature, place the seasoned pork tenderloin in the wood pellet smoker and smoke for 3 hours.
6. After 3 hours of smoking, increase the temperature of the wood pellet smoker to 350°F (177°C) and continue smoking the pork tenderloin for another 30 minutes.
7. Once the internal temperature of the smoked pork tenderloin has reached 165°F (74°C), remove it from the wood pellet smoker and transfer to a serving dish.
8. Cut the smoked pork tenderloin into thick slices then serve.

Nutrition: Calories: 318 Carbs: 7g Fat: 10g Protein: 8g

189. Gingery Maple Glazed Smoked Pork Ribs

Preparation Time: 15 minutes
Cooking Time: 5 hours 10 minutes
Servings: 8

Ingredients:

- Pork Ribs (4-lbs., 1.8-kg.)
- The Spices
- Apple juice – ¼ cup
- Brown sugar – 3 tablespoons
- Salt – 1 teaspoon
- Black pepper – 1 tablespoon
- Onion powder – ½ tablespoon
- Oregano – 1 tablespoon
- Cayenne pepper – ½ teaspoon
- Chili powder – 1 tablespoon
- The Glaze
- Maple syrup – ¼ cup
- Ginger – 1 teaspoon
- Apple cider vinegar – 1 teaspoon
- Mustard – ½ teaspoon

Directions:

1. Combine brown sugar with salt, black pepper, garlic powder, onion powder, oregano, cayenne pepper then mix well.
2. Baste the pork ribs with the apple juice then sprinkle the dry spice mixture over the pork ribs. Set aside.
3. Next, plug the wood pellet smoker then fill the hopper with the wood pellet. Turn the switch on.
4. Set the wood pellet smoker for indirect heat then adjust the temperature to 250°F (121°C).
5. Place the seasoned pork ribs in the wood pellet smoker and smoke for 5 hours.
6. In the meantime, combine maple syrup with ginger, apple cider vinegar, and mustard. Stir until incorporated and set aside.
7. After 5 hours of smoking, check the smoked pork ribs and once the internal temperature of the smoked pork ribs has reached 145°F (63°C), remove it from the wood pellet smoker.
8. Baste maple syrup mixture over the smoked pork ribs and quickly wrap with aluminum foil.
9. Let the wrapped smoked pork ribs rest for approximately 10 minutes then unwrap and serve it.

Nutrition: Calories: 301 Carbs: 9g Fat: 20g Protein: 23g

190. Tasty Grilled Pork Chops

Preparation Time: 15 minutes
Cooking Time: 1 hour 30 minutes
Servings: 4

Ingredients:

- 4 pork chops.
- 1/4 cup of olive oil.
- 1 1/2 tablespoons of brown sugar.
- 2 teaspoons of Dijon mustard.
- 1 1/2 tablespoons of soy sauce.
- 1 teaspoon of lemon zest.
- 2 teaspoons of chopped parsley.
- 2 teaspoons of chopped thyme.
- 1/2 teaspoon of salt to taste.
- 1/2 teaspoon of pepper to taste.
- 1 teaspoon of minced garlic.

Directions:

1. Using a small mixing bowl, add in all the ingredients on the list aside from the pork chops then mix properly to combine. This makes the marinade. Place the chops into a Ziploc bag, pour in the prepared marinade then shake properly to coat. Let the pork chops marinate in the refrigerator for about one to eight hours.
2. Next, preheat a Wood Pellet Smoker and Grill to 300 degrees F, place the marinated pork chops on the grill and cook for about six to eight minutes. Flip it side by side of the meat over and cook for an additional six to eight minutes until it attains an internal temperature of 165 degrees F.
3. Once cooked, let the pork chops rest for about five minutes, slice and serve.

Nutrition: Calories 313 Carbohydrate 5g Protein 30g Fat 14g

191. Delicious Barbeque and Grape Jelly Pork Chops

Preparation Time: 10 minutes
Cooking Time: 30 minutes
Servings: 4

Ingredients:

- 4 boneless pork chops.
- 1/2 cup of barbeque sauce.
- 1/4 cup of grape jelly.
- 2 minced cloves of garlic.
- 1/2 teaspoon of ground black pepper to taste.

Directions:

1. Using a small mixing bowl, add in the barbeque sauce, grape jelly, garlic, and pepper to taste then mix properly to combine. Using a reseal able plastic bag, add in the pork chops alongside with half of the prepared marinade then shake properly to coat. Wait until the pork marinate in the refrigerator for about four to eight hours.
2. Preheat a Wood Pellet Smoker and Grill to 350 degrees F, place the marinated pork chops on the grill, and grill for about six to eight minutes. Flip the pork over, blast with the reserved marinade then grill for an additional six to eight hours until it is cooked through and attains an internal temperature of 145 degrees F.
3. Once cooked, let the pork rest for about five minutes, slice and serve with your favorite sauce.

Nutrition: Calories 302 Carbohydrates 22g Protein 29g Fat 9g

192. Bacon Wrapped Jalapeno Poppers

Preparation Time: 10 minutes
Cooking Time: 1 hour 45 minutes
Servings: 8

Ingredients:

- 8 jalapeno peppers.
- 8 ounces of softened cream cheese.
- 1 minced green onion.
- 3/4 teaspoon of garlic powder.
- 1/2 cup of shredded cheddar cheese.
- 16 bacon slices.

Directions:

1. Using a very sharp knife to cut off the steam of each jalapeno then slice lengthwise. Scoop out the seeds, discard then set the jalapenos aside. Using a small mixing bowl, add in the cream cheese, green onion, garlic powder, and cheese then mix properly to combine.
2. Stuff each jalapeno with the cheese mixture then wrap up with a piece of bacon. Make sure to secure the bacon wraps with toothpicks. Place the stuffed jalapenos into the refrigerator and let rest for about one hour. This step will prevent the cheese from melting when grilling.
3. Preheat Wood Pellet Smoker and Grill to 275 degrees F, add in the jalapenos and grill for about forty-five minutes. Just check the doneness of the jalapenos halfway through the cooking process. Serve.

Nutrition: Calories 345 Carbohydrates 3g Protein 11g Fat 32g

193. Barbeque Baby Back Ribs

Preparation Time: 15 minutes
Cooking Time: 1 hour 30 minutes
Servings: 6

Ingredients:

- 2 racks baby back ribs.
- 3/4 cup of chicken broth.
- 3/4 cup of soy sauce.
- 1 cup of sugar.
- 6 tablespoons of cider vinegar.
- 6 tablespoons of olive oil.
- 3 minced garlic cloves.
- 2 teaspoons of salt to taste.
- 1 tablespoon of paprika.
- 1/2 teaspoon of chili powder.
- 1/2 teaspoon of pepper to taste.
- 1/4 teaspoon of garlic powder.
- A dash of cayenne pepper.
- Barbecue sauce.

Directions:

1. Using a large mixing bowl, add in half of the sugar, soy sauce, vinegar, oil, and garlic then mix properly to combine. This makes the marinade. Place the pork ribs in a Ziploc bag, pour in about 2/3 of the prepared marinade then sake properly to coat. Allow the ribs marinate in the refrigerator for overnight.
2. Using another mixing bowl, add in the rest of the sugar, salt, and seasonings on the list then mix properly to combine. Rub the ribs with the mixture, coating all sides then set aside. Preheat a Wood Pellet Smoker Grill to 250 degrees F, place the ribs on the preheated grill and grill for about two hours.
3. Blast the ribs with the reserved marinade and cook for an additional one hour. Once cooked, let rest for about five to ten minutes, slice, and serve.

Nutrition: Calories 647 Fat 41g Carbohydrate 30g Protein 37g

194. Delicious Grilled Pulled Pork

Preparation Time: 15 minutes
Cooking Time: 8 hours 15 minutes
Servings: 6

Ingredients:

- 1 (5 to 6 pounds) boneless pork butt.
- For the Rub:
- 2 tablespoons of paprika.
- 2 teaspoons of salt to taste.
- 2 teaspoons of dried oregano.
- 2 teaspoons of garlic powder.
- 2 teaspoons of dried thyme.
- 1/2 teaspoon of ground red pepper to taste.
- 1/2 teaspoon of ground black pepper to taste.

Directions:

1. Using a small mixing bowl, add in the paprika, oregano, garlic powder, thyme, red pepper, black pepper, and salt to taste then mix properly to combine. Place the pork meat into a large mixing bowl, add in the prepared rub then toss to coat. Cover the mixing bowl with a plastic wrap then let the

pork marinate for about one to three hours in the refrigerator.

2. Next, preheat a Wood Pellet Smoker and Grill to 255 degrees F, place pork on the smoker, and cook for about six hours. Wrap the pork in two pieces of aluminum foil, increase the grill temperature to 250 degrees F and cook the pork for an additional two hours until it is cooked through and tender.
3. Make sure the pork reads an internal temperature of 204 degrees F. Once cooked, let the pork cool for a few minutes, un-warp the foil then shred with a fork. Serve.

Nutrition: Calories 859 Fat 52g Carbohydrate 2g Protein 91g

195. Grilled Honey Pork Loin

Preparation Time: 10 minutes
Cooking Time: 1 hour 15 minutes
Servings: 8

Ingredients:

- 1 (3 lbs.) boneless pork loin.
- 2/3 cup of soy sauce.
- 1 teaspoon of ground ginger.
- 3 crushed garlic cloves.
- 1/4 cup of packed brown sugar.
- 1/3 cup of honey.
- 1 1/2 tablespoons of sesame oil.
- Vegetable oil.

Directions:

1. Using a very sharp kitchen knife to trim the fat off the pork loin then add into a Ziploc bag, set aside. Using a small mixing bowl, add in the soy sauce, ginger, and garlic then mix properly to combine. Pour the mixture into the bag containing the pork loin then shake properly to coat. Place the pork in the refrigerator to marinate for about three hours. Make sure you turn the meat occasionally.
2. Using another mixing bowl, add in the sugar, honey, and sesame oil then mix to combine. Preheat a Wood Pellet Smoker and Grill to 250 to 275 degrees F, place a pan on the griddle, pour in the sugar mixture then cook for few minutes until the sugar dissolves, set aside. Put the pork loin on the grill and cook for about one hour.
3. Next, brush the loin with the sugar mixture and cook for an additional forty-five minutes until the internal temperature reads 145 degrees F. Once cooked, let the pork rest for about five minutes, slice, and serve.

Nutrition: Calories 260.5 Fat 11.2g Carbohydrate 13.5g Protein 26g

196. The Best Pork Shoulder Steak

Preparation Time: 10 minutes
Cooking Time: 50 minutes
Servings: 4

Ingredients:

- 4 pork shoulder steaks.
- 1/3 cup of olive oil.
- 1 tablespoon of apple cider vinegar.
- 1 teaspoon of salt to taste.
- ½ teaspoon of black pepper to taste.

- 1 sliced onion.
- 2 tablespoons of chopped parsley.
- 1/2 teaspoon of chopped thyme.
- 1 teaspoon of ground cumin.
- 1 teaspoon of paprika.
- 1 teaspoon of oregano.

Directions:

1. Start by pounding the pork to an even thickness using a meat mallet then trim off any present fat. Place the pork in a Ziploc bag, add in other ingredients like the oil, vinegar, onion, parsley, thyme, cumin, paprika, oregano, salt, and pepper to taste then mix/shake properly to coat. Place the pork into the refrigerator and let sit for about twenty-four hours.
2. Next, set a Wood Pellet Smoker and Grill to 180 degrees F then preheat for about fifteen minutes. Make sure the lid is closed. Place the pork on the grill and smoke for about one hour thirty minutes. Wrap the pork with aluminum foil then cook for another forty-five minutes.
3. Remove the foil, place the pork back on the grill and cook for another fifteen minutes until cooked through, tender, and attains an internal temperature of 160 degrees F. Once cooked, let the pork cool for a few minutes, slice, and serve.

Nutrition: Calories 384 Fat 27g Carbohydrates 3g Protein 29g

197. Delicious Parmesan Roast Pork

Preparation Time: 10 minutes
Cooking Time: 3 hours 45 minutes
Servings: 10

Ingredients:

- 4 chopped garlic cloves.
- 2 tablespoons of olive oil.
- 1 tablespoon of minced dried basil.
- 1 tablespoon of dried and crushed oregano.
- 1 pound of boneless pork loin.
- 1 cup of bread crumbs.
- 1/4 cup of grated Parmesan cheese.

Directions:

1. Using a small mixing bowl, add in the garlic, olive oil, basil, and oregano then mix properly to combine. Rub the mixture on the pork loin, coating all sides then place in the large bowl. Cover the bowl with a plastic wrap then place in the refrigerator for about two hours to overnight.
2. In another mixing bowl, add in the bread crumbs and cheese then mix properly to combine. Dredge the pork in the cheese mixture then set aside. Preheat a Wood Pellet Smoker and Grill to 225 degrees F, place the pork on the grill, cover the lid and smoke the pork for about three to four hours until an inserted thermometer reads 155 degrees F.
3. Next, wrap the pork in aluminum foil and let stand for about ten minutes. Slice and serve.

Nutrition: Calories 250 Fat 10g Carbohydrates 3g Protein 34g

198. Bacon Wrapped Pork Tenderloin

Preparation Time: 15 minutes
Cooking Time: 40 minutes
Servings: 4

Ingredients:

- 1 pork tenderloin.
- 4 strips of bacon.
- Rub:
- 8 tablespoons of brown sugar.
- 3 tablespoons of kosher salt to taste.
- 1 tablespoon of chili powder.
- 1 teaspoon of black pepper to taste.
- 1 teaspoon of onion powder.
- 1 teaspoon of garlic powder.

Directions:

1. Using a small mixing bowl, add in the sugar, chili powder, onion powder, garlic powder, salt, and pepper to taste then mix properly to combine, set aside. Use a sharp knife to trim off fats present on the pork then coat with 1/4 of the prepared rub. Make sure you coat all sides.
2. Next, roll each pork tenderloin with a piece of bacon, lay the meat on a cutting board then pound with a meat mallet to give an even thickness, secure the ends of the bacon with toothpicks to hold still. Coat the meat again with just a little more of the rub spice then set aside.
3. Preheat a Wood Pellet Smoker and Grill to 350 degrees F, place the pork tenderloin on the grill, and grill for about fifteen minutes. Increase the temperature of the grill to 400 degrees F and cook for another fifteen minutes until it is cooked through and reads an internal temperature of 145 degrees F.
4. Once cooked, let the pork rest for a few minutes, slice, and serve.

Nutrition: Calories 236 Fat 8g Carbohydrates 10g Protein 29g

199. Naked St. Louis Ribs

Preparation Time: 30 Minutes
Cooking Time: 5-6 Hours
Servings: 6

Ingredients:

- 3 St. Louis-style pork blacks
- 1 cup and 1 tablespoon of Yang's original dry lab or your favorite pork club

Directions:

1. Insert the spoon handle between the membrane and the rib bone and remove the membrane under the rib bone rack. Pat the membrane using a paper towel and pull it down slowly from the rack to remove it.
2. Rub both sides of the rib with a sufficient amount of friction.
3. Configure a wood pellet smoker grill for indirect cooking and preheat to 225 ° F using hickory or apple pellets.
4. If using reblack, place the ribs on the grill grid rack. Otherwise, you can use a Teflon-coated fiberglass mat or place the ribs directly on the grill.
5. Slice rib bone at 225 ° F for 5-6 hours with hickory pellets until the internal

temperature of the thickest part of the ick bone reaches 185 ° F to 190 ° F.
6. Place ribs under loose foil tent for 10 minutes before carving and serving.

Nutrition: Calories: 200 Carbs: 13g Fat: 5g Protein: 23g

200. Bacon and Sausage Bites

Servings: 10
Calories: 620
Cooking Time: 45 minutes

Ingredients:

- Smoked sausages - 1 pack
- Thick-cut bacon - 1 lb.
- Brown sugar - 2 cups

Directions:

- Slice ⅓ of the sausages and wrap them around small pieces of sausage. Use a toothpick to secure them.
- Line a baking tray with baking paper and place the small pieces of wrapped sausage on it.
- Sprinkle brown sugar on top.
- Preheat the wood pellet to 300 degrees.
- Keep the baking tray with the wrapped sausages inside for 30 minutes.
- Remove and let it stay outside for 15 minutes.
- Serve warm with a dip of your choice.

Nutrition: Carbohydrates: 61 g; Protein: 27 g; Fat: 30 g; Sodium: 1384 mg

201. Grilled Pork Chops

Servings: 6
Calories: 300
Cooking Time: 20 minutes

Ingredients:

- Pork chops - 6, thickly cut
- Barbeque mix

Directions:

- Preheat your wood pellet grill to 450 degrees.
- Place the seasoned chops on the grill. Close the lid.
- Cook for 6 minutes. The temperature should be around 145 degrees when you remove the lid.
- Remove the pork chops.
- Let it remain open for 5-10 minutes.
- Serve with your choice of side dish.

Nutrition: Carbohydrates: 0 g; Protein: 23 g; Fat: 7.8 g

202. Pigs in a Blanket

Servings: 12
Calories: 267
Cooking Time: 30 minutes

Ingredients:

- Pork sausages - 1 pack
- Biscuit dough - 1 pack

Directions:

- Preheat your wood pellet grill to 350 degrees.

- Cut the sausages and the dough into thirds.
- Wrap the dough around the sausages. Place them on a baking sheet.
- Grill with a closed lid for 20-25 minutes or until they look cooked.
- Take them out when they are golden brown.
- Serve with a dip of your choice.

Nutrition: Protein: 9 g; Fat: 22 g; Sodium: 732 mg; Cholesterol: 44 mg

203. Smoked Bacon

Servings: 6
Calories: 315
Cooking Time: 30 minutes
Ingredients:

- Thick cut bacon - 1 lb.

Directions:

- Preheat your wood pellet grill to 375 degrees.
- Line a huge baking sheet. Place a single layer of thick-cut bacon on it.
- Bake for 20 minutes and then flip it to the other side.
- Cook for another 10 minutes or until the bacon is crispy.
- Take it out and enjoy your tasty grilled bacon.

Nutrition: Protein: 9 g; Fat: 10 g; Sodium: 500 mg; Cholesterol: 49 mg

204. Smoked, Candied, and Spicy Bacon

Servings: 2
Calories: 458
Cooking Time: 40 minutes

Ingredients:

- Center-cut bacon - 1 lb.
- Brown sugar - ½ cup
- Maple syrup - ½ cup
- Hot sauce - 1 tbsp
- Pepper - ½ tbsp

Directions:

- Mix the maple syrup, brown sugar, hot sauce, and pepper in a bowl.
- Preheat your wood pellet grill to 300 degrees.
- Line a baking sheet and place the bacon slices on it.
- Generously spread the brown sugar mix on both sides of the bacon slices.
- Place the pan on the wood pellet grill for 20 minutes. Flip the bacon pieces.
- Leave them for another 15 minutes until the bacon looks cooked and the sugar is melted.
- Remove from the grill and let it stay for 10-15 minutes.
- Voila! Your bacon candy is ready!

Nutrition: Carbohydrates: 37 g; Protein: 9 g; Sodium: 565 mg; Cholesterol: 49 mg

205. Stuffed Pork Crown Roast

Servings: 2-4
Calories: 1010.8
Cooking Time: 3 hours 30 minutes

Ingredients:

- 12-14 ribs or 1 Snake River Pork Crown Roast
- Apple cider vinegar - 2 tbsp
- Apple juice - 1 cup
- Dijon mustard - 2 tbsp
- Salt - 1 tsp
- Brown sugar - 1 tbsp
- Freshly chopped thyme or rosemary - 2 tbsp
- Cloves of minced garlic - 2
- Olive oil - ½ cup
- Coarsely ground pepper - 1 tsp
- Your favorite stuffing - 8 cups

Directions:

- Set the pork properly in a shallow roasting pan on a flat rack. Cover both ends of the bone with a piece of foil.
- To make the marinade, boil the apple cider or apple juice on high heat until it reduces to about half its quantity. Remove the content from the heat and whisk in the mustard, vinegar, thyme, garlic, brown sugar, pepper, and salt. Once all that is properly blended, whisk in the oil slowly.
- Use a pastry brush to apply the marinade to the roast. Ensure that you coat all the surfaces evenly. Cover it on all sides using plastic wrap. Allow it to sit for about 60 minutes, until the meat has reached room temperature.
- At this time, feel free to brush the marinade on the roast again. Cover it and return it to the refrigerator until it is time to cook it. When you are ready to cook it, allow the meat to reach room temperature, then put in on the pellet grill. Ensure that the grill is preheated for about 15 minutes before you do.
- Roast the meat for 30 minutes, then reduce the temperature of the grill. Fill the crown loosely with the stuffing and mound it at the top. Cover the stuffing properly with foil. You can also bake the stuffing separately alongside the roast in a pan.
- Roast the pork thoroughly for 90 more minutes. Get rid of the foil and continue to roast the stuffing for 30-90 minutes until the pork reaches an internal temperature of 150 degrees Fahrenheit. Ensure that you do not touch the bone of the meat with the temperature probe or you will get a false reading.
- Remove the roast from the grill. Allow it to rest for around 15 minutes so that the meat soaks in all the juices. Remove the foil covering the bones. Leave the butcher's string on until you are ready to carve it. Now, transfer it to a warm platter, carve between the bones, and enjoy!

Nutrition: Carbohydrates: 5.5 g; Protein: 107.9 g; Fat: 58.9 g; Sodium: 702 mg; Cholesterol: 325.3 mg

206. Smoked Pork Chops Marinated with Tarragon

Preparation Time: 3 hours and 20 minutes
Servings: 4

Ingredients:

- 1/2 cup olive oil
- 4 Tbsp of fresh tarragon chopped
- 2 tsp fresh thyme, chopped
- salt and grated black pepper
- 2 tsp apple-cider vinegar
- 4 pork chops or fillets

Directions:

- Whisk the olive oil, tarragon, thyme, salt, pepper, apple cider and stir well. Place the pork chops in a container and pour with tarragon mixture.
- Refrigerate for 2 hours. Start pellet grill on, lid open, until the fire is established (4-5 minutes).
- Increase the temperature to 225°F and allow to pre-heat, lid closed, for 10 - 15 minutes. Remove chops from marinade and pat dry on kitchen towel.
- Arrange pork chops on the grill rack and smoke for about 3 to 4 hours.
- Transfer chops on a serving platter and let rest 15 minutes before serving.

Nutrition: Calories: 528.8; Carbs: 0.6g; Fat: 35g; Fiber: 0.14g; Protein: 51g

207. Smoked Pork Cutlets in Citrus-Herbs Marinade

Preparation Time: 1 hour and 45 minutes
Servings: 4

Ingredients:

- 4 pork cutlets
- 1 fresh orange juice
- 2 large lemons freshly squeezed
- 10 twigs of coriander chopped
- 2 Tbs of fresh parsley finely chopped
- 3 cloves of garlic minced
- 2 Tbs of olive oil
- salt and ground black pepper

Directions:

- Place the pork cutlets in a large container along with all remaining ingredients; toss to cover well.
- Refrigerate at least 4 hours, or overnight. When ready, remove the pork cutlets from marinade and pat dry on kitchen towel.
- Start pellet grill on, lid open, until the fire is established (4-5 minutes).
- Increase the temperature to 250°F and allow to pre-heat, lid closed, for 10 - 15 minutes. Place pork cutlets on grill grate and smoke for 1 1/2 hours.

Nutrition: Calories: 260; Carbs: 5g; Fat: 12g; Fiber: 0.25g; Protein: 32.2g

208. Smoked Pork Cutlets with Caraway and Dill

Preparation Time: 1 hour and 45 minutes
Servings: 4

Ingredients:

- 4 pork cutlets
- 2 lemons freshly squeezed
- 2 Tbs fresh parsley finely chopped
- 1 Tbsp of ground caraway
- 3 Tbsp of fresh dill finely chopped
- 1/4 cup of olive oil
- salt and ground black pepper

Directions:

- Place the pork cutlets in a large resealable bag along with all remaining ingredients; shake to combine well. Refrigerate for at least 4 hours.
- Remove the pork cutlets from marinade and pat dry on kitchen towel.
- Start the pellet grill (recommended maple pellet) on SMOKE with the lid open until the fire is established.
- Set the temperature to 250 °F and preheat, lid closed, for 10 to 15 minutes.
- Arrange pork cutlets on the grill rack and smoke for about 1 1/2 hours. Allow cooling on room temperature before serving.

Nutrition: Calories: 308; Carbs: 2.4g; Fat: 18.5g; Fiber: 0.36g; Protein: 32g

209. Smoked Pork Loin in Sweet-Beer Marinade

Preparation Time: 3 hours
Servings: 6

Ingredients:

- Marinade
- 1 onion finely diced
- 1/4 cup honey (preferably a darker honey)
- 1 1/2 cups of dark beer
- 4 Tbs of mustard
- 1 Tbs fresh thyme finely chopped
- Salt and pepper
- Pork 3 1/2 lbs of pork loin

Directions:

- Combine all ingredients for the marinade in a bowl.
- Place the pork along with marinade mixture in a container, and refrigerate overnight. Remove the pork from marinade and dry on kitchen towel.
- Prepare the grill on Smoke with the lid open until the fire is established (4 to 5 minutes).
- Set the temperature to 250F and preheat, lid closed, for 10 to 15 minutes.
- Place the pork on the grill rack and smoke until the internal temperature of the pork is at least 145-150 °F (medium-rare), 2-1/2 to 3 hours.
- Remove meat from the smoker and let rest for 15 minutes before slicing.
- Serve hot or cold.

Nutrition: Calories: 444.6; Carbs: 17g; Fat: 12.7g; Fiber: 0.8g; Protein: 60.5g

210. Smoked Pork Ribs with Fresh Herbs

Preparation Time: 3 hours and 20 minutes
Servings: 6

Ingredients:

- 1/4 cup olive oil
- 1 Tbs garlic minced
- 1 Tbs crushed fennel seeds
- 1 tsp of fresh basil leaves finely chopped
- 1 tsp fresh parsley finely chopped
- 1 tsp fresh rosemary finely chopped
- 1 tsp fresh sage finely chopped
- Salt and ground black pepper to taste
- 3 lbs. pork rib roast bone-in

Directions:

- Combine the olive oil, garlic, fennel seeds, parsley, sage, rosemary, salt, and pepper in a bowl; stir well.
- Coat each chop on both sides with the herb mixture.
- Start the pellet grill (recommended hickory pellet) on SMOKE with the lid open until the fire is established.
- Set the temperature to 225 °F and preheat, lid closed, for 10 to 15 minutes. Smoke the ribs for 3 hours.
- Transfer the ribs to a serving platter and serve hot.

Nutrition: Calories: 459.2; Carbs: 0.6g; Fat: 31.3g; Fiber: 0.03g; Protein: 41g

211. Smoked Pork Side Ribs with Chives

Preparation Time: 3 hours and 20 minutes
Servings: 6

Ingredients:

- 1/3 cup of olive oil (or garlic-infused olive oil)
- 3 Tbsp of ketchup
- 3 Tbsp chives finely chopped
- 3 lbs of pork side ribs
- Salt and black pepper to taste

Directions:

- In a bowl, stir together olive oil, finely chopped chives, ketchup, and the salt and pepper.
- Cut pork into individual ribs and generously coat with chives mixture. Start the pellet grill on SMOKE with the lid open until the fire is established.
- Set the temperature to 250 °F and preheat, lid closed, for 10 to 15 minutes.
- Arrange pork chops on the grill rack and smoke for about 3 to 4 hours.
- Allow resting 15 minutes before serving.

Nutrition: Calories: 689.7; Carbs: 2g; Fat: 65g; Fiber: 0.1g; Protein: 35.2g

212. Smoked Spicy Pork Medallions

Preparation Time: 1 hour and 45 minutes
Servings: 6

Ingredients:

- 2 lbs pork medallions
- 3/4 cup chicken stock
- 1/2 cup tomato sauce (organic)
- 2 Tbs of smoked hot paprika (or to taste)
- 2 Tbsp of fresh basil finely chopped
- 1 Tbsp oregano
- Salt and pepper to taste

Directions:

- In a bowl, combine together the chicken stock, tomato sauce, paprika, oregano, salt, and pepper.
- Brush generously over the outside of the tenderloin. Start the pellet grill on Smoke with the lid open until the fire is established (4 to 5 minutes).
- Set the temperature to 250°F and preheat, lid closed, for 10 to 15 minutes.
- Place the pork on the grill grate and smoke until the internal temperature of the pork is at least medium-rare (about 145°F), for 1 1/2 hours.

Nutrition: Calories: 364.2; Carbs: 4g; Fat: 14.4g; Fiber: 2g; Protein: 52.4g

213. Pork Shoulder

Servings: 12

Preparation time: 10 minutes

Cooking time: 14 hours and 50 minutes

Ingredients:

- 5 pounds boneless pork shoulder, fat trimmed
- 1 cup pork seasoning

For Vinegar Sauce:

- 1-gallon apple cider vinegar
- 1 cup paprika
- 1 cup brown sugar
- 1¼ cup white sugar
- ⅓ cup salt
- 3 tablespoons cayenne pepper
- 3 tablespoons black pepper
- 1 cup water
- 2/3 tablespoon xanthan gum

For Injection:

- 67.6 fluid ounce peach nectar
- 24-ounce prepared vinegar sauce

For Mop:

- 3 cups prepared vinegar sauce
- 3 cups apple juice
- 3 tablespoons grenadine

Directions:

- Open hopper of the smoker, add dry pallets, make sure ash-can is in place, then open the ash damper, power on the smoker and close the ash damper.
- Set the temperature of the smoker to 350 degrees F, let preheat for 30

minutes, then set it to 225 degrees F and continue preheating for 20 minutes or until the green light on the dial blinks that indicate smoker has reached to set temperature.
- Place pork on the smoker grill, shut with lid and smoke for 5 hours.
- Meanwhile, prepare vinegar sauce and for this, place all the ingredients for the sauce in a large pot, stir until mix, then place pot over medium-high heat and bring the sauce to boil.
- Remove pot from heat, let cool for 10 minutes, then pour 24 ounces of the sauce in another pot, pour in peach nectar and stir until combined.
- Take a meat injector, fill it with sauce-nectar mixture, then inject into the pork after 5 hours of smoking; make sure injection, and pork temperature is same, and continue smoking the pork for 9 hours or until thoroughly cooked.
- Whisk together all the ingredients for mop and brush pork with it during the last hour of smoking.
- When done, transfer pork to a cutting board or large dish, let rest for 10 minutes, then shred the meat using two forks and serve.

Nutrition: Calories: 343.4; Total Fat: 23.9 g; Saturated Fat: 7 g; Protein: 33.3 g; Carbs: 0.8 g; Fiber: 0 g; Sugar: 0.4 g

214. Pork Loin Roulade

Servings: 2

Preparation time: 4 hours and 45 minutes

Cooking time: 155 minutes

Ingredients:

- 1 pork loin, fat trimmed
- 2 cups basil leaves, fresh
- 3 tablespoons garlic powder
- 1 tablespoon salt
- 2 tablespoons ground black pepper
- 3 tablespoons dried thyme
- 3 tablespoons dried oregano
- 1 cup grated mozzarella cheese
- 24 ounces ricotta cheese
- ½ cup grated parmesan cheese
- 24 ounces marinara sauce

Directions:

- Place all the cheeses in a large bowl, add garlic powder, salt, black pepper, thyme, and oregano, stir until well combined, then cover the bowl and let mixture cool in the refrigerator for 30 minutes.
- Then place pork loin on working space, make some shallow cuts on it and roll it using a rolling pin until flat.
- Spread prepared cheese mixture on the pork loin, leaving 1-inch of edge, then lay basil leaves to cover cheese spread completely and roll the pork loin very gently and secure with toothpicks or kitchen twine.
- Place pork loin on a dish and then place it in refrigerator to chill for 4 hours.

- When ready to smoke, open hopper of the smoker, add dry pallets, make sure ash-can is in place, then open the ash damper, power on the smoker and close the ash damper.
- Set the temperature of the smoker to 350 degrees F, let preheat for 30 minutes, then set it to 275 degrees F and continue preheating for 20 minutes or until the green light on the dial blinks that indicate smoker has reached to set temperature.
- Place pork loin on the smoker grill, shut with lid and smoke for 1 hour and 45 minutes or until thoroughly cooked and the internal temperature reach to 140 degrees F.
- When done, transfer pork loin to a dish and let it rest for 10 minutes.
- Meanwhile, place a pot over medium heat, pour in marinara sauce, bring it to boil and then remove pot from heat.
- Cut pork loin into slices, remove toothpicks or kitchen twine and top pork slices with marinara sauce.
- Serve straight away.

Nutrition: Calories: 390; Total Fat: 24 g; Saturated Fat: 10 g; Protein: 21 g; Carbs: 22 g; Fiber: 2 g; Sugar: 2 g

215. Pork Butt

Servings: 8

Preparation time: 10 minutes

Cooking time: 10 hours and 50 minutes

Ingredients:

- 10 pounds pork butt
- Salt as needed

Directions:

- Open hopper of the smoker, add dry pallets, make sure ash-can is in place, then open the ash damper, power on the smoker and close the ash damper.
- Set the temperature of the smoker to 350 degrees F, let preheat for 30 minutes, then set it to 250 degrees F and continue preheating for 20 minutes or until the green light on the dial blinks that indicate smoker has reached to set temperature.
- Meanwhile, score pork butts and then season with salt until well coated.
- Place pork butts on the smoker grill, shut with lid and smoke for 10 hours or until thoroughly cooked and the internal temperature of pork reach to 195 degrees F.
- When done, transfer pork butts to a cutting board, let rest for 10 minutes, then shred with two forks and serve straight away.

Nutrition: Calories: 256; Total Fat: 12 g; Saturated Fat: 4 g; Protein: 32 g; Carbs: 0 g; Fiber: 0 g; Sugar: 0 g

216. Pineapple Bourbon Glazed Ham

Servings: 8

Preparation time: 10 minutes

Cooking time: 4 hours and 50 minutes

Ingredients:

- 4 pounds spiral cut ham, precooked
- ½ cup ham rub
- 1/2 cup brown sugar
- 1 tablespoon ground mustard
- 1/3 cup molasses
- 1 cup honey
- 18-ounce pineapple preserves
- 1 cup bourbon

Directions:

- Open hopper of the smoker, add dry pallets, make sure ash-can is in place, then open the ash damper, power on the smoker and close the ash damper.
- Set the temperature of the smoker to 350 degrees F, let preheat for 30 minutes, then set it to 225 degrees F and continue preheating for 20 minutes or until the green light on the dial blinks that indicate smoker has reached to set temperature.
- Meanwhile, prepare glaze and for this, place a pot over low heat, add all the ingredients except for ham, whisk well and cook for 20 minutes or until glaze thickens.
- Then remove pot from heat and let the glaze cool until required.
- Take an aluminum foil tray, take a wire rack on top of it, place ham on it and then place on the smoker grill.
- Shut smoker with lid and smoke for 4 hours or until thoroughly cooked, brush with prepared glaze every 15 minutes during the last hour.
- When done, transfer ham to a cutting board, let rest for 15 minutes and then slice to serve.

Nutrition: Calories: 146.3; Total Fat: 0.7 g; Saturated Fat: 0.3 g; Protein: 18 g; Carbs: 15 g; Fiber: 0.6 g; Sugar: 17 g

217. Pork Tacos

Servings: 6

Preparation time: 25 minutes

Cooking time: 4 hours and 30 minutes

Ingredients:

- 5-pounds country-style pork ribs
- ½ cup all-purpose rub
- ½ cup chopped cilantro
- Chopped mixed greens as needed
- Barbecue sauce as needed
- 6 shell tortillas, soft, warmed
- For Pico De Gallo:
- 1/4 of white onion, peeled and sliced
- 3 Roma tomatoes
- 1 serrano pepper
- ½ teaspoon garlic salt
- 1 lime, juiced

Directions:

- Open hopper of the smoker, add dry pallets, make sure ash-can is in place, then open the ash damper, power on the smoker and close the ash damper.

- Set the temperature of the smoker to 350 degrees F, let preheat for 30 minutes or until the green light on the dial blinks that indicate smoker has reached to set temperature.
- Meanwhile, season pork ribs with the rub until evenly coated on all sides.
- Place pork ribs on the smoker grill, shut with lid and smoke for 3 to 4 hours or until thoroughly cooked and fall part tender.
- In the meantime, prepare Pico De Gallo and for this, place all its ingredients in a bowl and stir until well combined, set aside until required.
- When done, transfer ribs to a cutting board, let rest for 10 minutes and then cut into slices.
- Place pork in warmed taco shells, top with mixed greens and cilantro, drizzle with barbecue sauce and serve.

Nutrition: Calories: 260; Total Fat: 10 g; Saturated Fat: 4 g; Protein: 16 g; Carbs: 28 g; Fiber: 3 g; Sugar: 1 g

218. Cider Pork Steak

Servings: 4
Preparation time: 2 hours and 40 minutes
Cooking time: 3 hours

Ingredients:

- 4 pork steaks
- 1/3 cup sea salt
- ¼ cup pork rub
- 2 teaspoons dried thyme
- 1 cup maple syrup
- 2 teaspoons hot sauce
- ¼ cup BBQ sauce
- 1½ cup apple cider
- 1½ cup ice water
- 1 cup water

Directions:

- Prepare brine and for this, place a small saucepan over medium heat, pour in 1 cup water, salt, thyme and 1/3 cup maple syrup and cook for 5 to 10 minutes or until salt dissolves completely and brine is hot.
- Then remove pan from the heat, pour in apple cider, 1 teaspoon hot sauce and ice water, stir well until ice dissolves and let brine chill for 30 minutes or until temperature of the brine reach to 45 degrees.
- Then place pork steaks in a large plastic bag, pour in brine, seal the bag, turn it upside down to coat steaks with brine and marinate in refrigerator for 2 hours.
- Meanwhile, place remaining maple syrup in a small bowl, add hot sauce and barbecue sauce, whisk until combined and set aside until required.
- When ready to smoke, open hopper of the smoker, add dry pallets, make sure ash-can is in place, then open the ash damper, power on the smoker and close the ash damper.
- Set the temperature of the smoker to 300 degrees F, let preheat for 30 minutes or until the green light on the dial blinks that indicate smoker has reached to set temperature.
- Then remove pork steaks from the brine, pat dry with paper towels, place

- pork steaks on the smoker grill, shut with lid and smoke for 1 hour 30 minutes to 2 hours or until thoroughly cooked, brushing pork steaks with maple syrup every 3 minutes during the last 10 minutes, flipping steaks halfway through.
- When done, transfer steaks to a cutting board, let rest for 5 minutes and serve straight away.

Nutrition: Calories: 150; Total Fat: 9 g; Saturated Fat: 3 g; Protein: 14 g; Carbs: 3 g; Fiber: 0 g; Sugar: 6 g

219. Raspberry Chipotle Pork Kebabs

Servings: 8

Preparation time: 10 minutes

Cooking time: 50 minutes

Ingredients:

- 1-pound pork loin, boneless, cut into cubes
- 3 medium green bell peppers, cored and sliced
- 1 large red onion, peeled and cut into cubes
- 2 tablespoons raspberry chipotle spice rub
- 1 tablespoon honey
- 1/8 cup vinegar apple cider
- 1 tablespoon olive oil

Directions:

- Whisk together chipotle seasoning, vinegar, honey and olive oil in a large bowl until combined, then add cubed pork, toss until well coated, then cover the bowl with a plastic wrap and marinate in refrigerator for 1 hour.
- Meanwhile, place eight wooden skewers in a shallow dish, cover with water and let soak for 1 hour.
- When ready to smoke, Open hopper of the smoker, add dry pallets, make sure ash-can is in place, then open the ash damper, power on the smoker and close the ash damper.
- Set the temperature of the smoker to 400 degrees F, let preheat for 30 minutes or until the green light on the dial blinks that indicate smoker has reached to set temperature.
- Meanwhile, remove marinated pork pieces from the marinade and thread evenly in wooden skewers, alternating with onion and pepper pieces.
- Place pork skewers on the smoker grill, shut with lid and smoke for 20 minutes or pork is nicely browned and vegetables are tender, turning halfway through
- Serve straight away.

Nutrition: Calories: 193.2; Total Fat: 10.4 g; Saturated Fat: 3 g; Protein: 15 g; Carbs: 9.4 g; Fiber: 1.4 g; Sugar: 2.5 g

220. Bacon Cheese Fries

Servings: 2

Preparation time: 10 minutes

Cooking time: 100 minutes

Ingredients:

- ½-pound bacon slices
- 2 large potatoes
- ¼ cup olive oil
- 3 teaspoons minced garlic
- ½ teaspoon salt
- ¼ teaspoon ground black pepper
- 2 sprigs of rosemary
- ½ cup grated mozzarella cheese

Directions:

- Open hopper of the smoker, add dry pallets, make sure ash-can is in place, then open the ash damper, power on the smoker and close the ash damper.
- Set the temperature of the smoker to 375 degrees F, let preheat for 30 minutes or until the green light on the dial blinks that indicate smoker has reached to set temperature.
- Meanwhile, take a large baking sheet, line it with parchment paper and place bacon slices on it in a single layer.
- Place baking sheet on the smoker grill, shut with lid, smoke for 20 minutes, then flip the bacon and continue smoking for 5 minutes or until bacon is crispy.
- When done, transfer bacon to a dish lined with paper towels to soak excess fat, then cut bacon into small pieces and set aside until required.
- Set the temperature of the smoker to 325 degrees F, let preheat for 15 minutes or until the green light on the dial blinks that indicate smoker has reached to set temperature.
- Prepare fries and for this, slice each potato into eight wedges, then spread potato wedges on a rimmed baking sheet in a single layer, drizzle with oil, sprinkle with garlic, salt, black pepper and rosemary and toss until well coat.
- Place baking sheet on the smoker grill, shut with lid, smoke for 20 to 30 minutes or until potatoes are nicely golden brown and tender.
- Then remove baking sheet from the smoker, sprinkle bacon and cheese on top of fries and continue smoking for 1 minute or until cheese melt.
- Serve straight away.

Nutrition: Calories: 388; Total Fat: 22 g; Saturated Fat: 6.8 g; Protein: 9.9 g; Carbs: 38 g; Fiber: 3.5 g; Sugar: 0.5 g

221. Candied Bacon

Servings: 8

Preparation time: 10 minutes

Cooking time: 2 hours and 30 minutes

Ingredients:

- 1-pound bacon, thick-cut
- 1 cup brown sugar
- 4 tablespoons chipotle spice

Directions:

- Open hopper of the smoker, add dry pallets, make sure ash-can is in place, then open the ash damper, power on the smoker and close the ash damper.
- Set the temperature of the smoker to 200 degrees F, let preheat for 30 minutes or until the green light on the dial blinks that indicate smoker has reached to set temperature.
- Meanwhile, take an aluminum foil tray or a cookie sheet, place a wire rack on it, place bacon slices on the rack and sprinkle its one side with sugar and chipotle spice.
- Place cookie sheet on the smoker grill, shut with lid, smoke for 1 hour, then flip the bacon slices, sprinkle with remaining sugar and chipotle spice and continue smoking for 1 hour until bacon is crispy and nicely browned.
- Serve straight away.

Nutrition: Calories: 77; Total Fat: 4 g; Saturated Fat: 1.4 g; Protein: 4 g; Carbs: 5.2 g; Fiber: 0 g; Sugar: 6 g

222. Bacon Wrapped Onion Rings

Servings: 8

Preparation time: 10 minutes

Cooking time: 2 hours and 30 minutes

Ingredients:

- 1 pack bacon, thick cut
- 2 white onions

Directions:

- Open hopper of the smoker, add dry pallets, make sure ash-can is in place, then open the ash damper, power on the smoker and close the ash damper.
- Set the temperature of the smoker to 250 degrees F, let preheat for 30 minutes or until the green light on the dial blinks that indicate smoker has reached to set temperature.
- Meanwhile, peel the onions, cut into thirds, separate the onion slices into rings and wrap each onion rings with two bacon slices, securing with a toothpick.
- Prepare more bacon wrapped onion rings until all the bacon is used up.
- Place onion rings on the smoker grill, shut with lid and smoke for 2 hours or until bacon is cooked, turning halfway through.
- Serve straight away.

Nutrition: Calories: 86; Total Fat: 6 g; Saturated Fat: 1.7 g; Protein: 6 g; Carbs: 2 g; Fiber: 0 g; Sugar: 0.4 g

223. Braised Pork Carnitas

Servings: 6

Preparation time: 15 minutes

Cooking time: 3 hours and 50 minutes

Ingredients:

- 4 pounds pork shoulder, boneless, scored
- 2 teaspoons salt
- 1/2 teaspoon ground cumin
- 2 tablespoons vegetable shortening
- 12 ounces beer
- Water as needed
- For Serving:
- Corn tortillas as required
- Diced onions as needed
- Shredded lettuce
- Sliced radishes
- Fresh cilantro
- Salsa Verde
- Pico de Gallo
- Guacamole

Directions:

- Open hopper of the smoker, add dry pallets, make sure ash-can is in place, then open the ash damper, power on the smoker and close the ash damper.
- Set the temperature of the smoker to 300 degrees F, let preheat for 30 minutes or until the green light on the dial blinks that indicate smoker has reached to set temperature.
- Meanwhile, cut pork into 2-inch pieces, then place in a roasting pan, then pour in beer and enough water to cover the pork pieces and then stir in salt and cumin.
- Place pan on the smoker grill, uncover the pan, shut the smoker with lid and smoke for 3 hours or until pork is tender, stirring occasionally.
- When done, remove the pan from the grill and break the pork into bite-size pieces using a fork.
- Add lard to the pork, then return the pan on the smoker grill and cook for 20 minutes or more until pork is nicely browned.
- Serve pork in corn tortilla along with onion, lettuce, radish, cilantro, salsa, Pico and guacamole.

Nutrition: Calories: 470; Total Fat: 8 1 g; Saturated Fat: 2 g; Protein: 44 g; Carbs: 55 g; Fiber: 8 g; Sugar: 3 g

224. Cuban Pork

Servings: 10

Preparation time: 20 minutes

Cooking time: 12 hours and 30 minutes

Ingredients:

- 8 pounds pork shoulder, boneless
- 2 medium onions, peeled and cut into rings
- 2 heads of garlic, peeled and chopped
- 3 tablespoons salt
- 1 tablespoon ground black pepper
- 1 tablespoon cumin
- 2 tablespoons oregano
- 4 cups orange juice
- 2 2/3 cups lime juice

Directions:

- Place all the ingredients in a bowl except for salt, black pepper, and pork and stir until well mixed.
- Season pork with salt and black, then place in a large plastic bag, pour in prepared orange juice mixture, seal the bag, turn it upside down and let marinate in the refrigerator for a minimum of 1 hour.
- When ready to smoke, open hopper of the smoker, add dry pallets, make sure ash-can is in place, then open the ash damper, power on the smoker and close the ash damper.
- Set the temperature of the smoker to 205 degrees F, let preheat for 30 minutes or until the green light on the dial blinks that indicate smoker has reached to set temperature.
- Meanwhile, remove pork from marinade, place it in a high sided baking pan and strain marinade on it.
- Place baking pan containing pork on the smoker grill, shut with lid and smoke for 12 hours or until thoroughly cooked and the internal temperature of pork reach to 205 degrees F.
- When done, remove pork from the grill, then shred with two forks, toss until mixed with its liquid and serve.

Nutrition: Calories: 280.1; Total Fat: 9.4 g; Saturated Fat: 3.3 g; Protein: 42.5 g; Carbs: 4.6 g; Fiber: 0.9 g; Sugar: 0.3 g

225. Seared Garlic Scallops

Cooking Time:5 Minutes
Servings: 3

The Heat: Hardwood Alder

Ingredients:

- 1 dozen scallops
- Salt and black pepper, to taste
- 1 tablespoon of butter
- 1 tablespoon of olive oil
- 1 tablespoon of parsley, chopped for garnishing
- Ingredients for Garlic Butter
- 4 tablespoons of butter
- 1 lemon, zest
- 2 garlic cloves

Directions:

- Preheat the smoker grill to 400 degrees F, by closing the lid for 1 minute.
- Season the cleaned scallops with salt and black pepper.
- Once the grill is preheated, take a skillet and heat butter and olive oil by putting on the grate.
- Once, butter melts, add scallops.
- In a small bowl, combine all garlic butter ingredients and mix well.
- Flip the scallop and spoon it with a few tablespoons of garlic butter.
- Cook it for one minute.
- Remove it from the grill and then add more garlic butter if liked.
- Then serve with parsley.
- Enjoy.

226. Pineapple Maple Glaze Fish

Cooking Time: 15 Minutes
Servings: 6

The Heat: Hardwood Apple

Ingredients:

- 3 pounds of fresh salmon
- 1/4 cup maple syrup
- 1/2 cup pineapple juice
- Brine Ingredients:
- 3 cups of water
- Sea salt, to taste
- 2 cups of pineapple juice
- ½ cup of brown sugar
- 5 tablespoons of Worcestershire sauce
- 1 tablespoon of garlic salt

Directions:

- Combine all the brine ingredients in a large cooking pan.
- Place the fish into the brine and let it sit for 2 hours for marinating.
- After 2 hours take out the fish and pat dry with a paper towel and set aside.
- Preheat the smoker grill to 250 degrees Fahrenheit, until the smoke started to appear.
- Put salmon on the grill and cook for 15 minutes.
- Meanwhile, mix pineapple and maple syrup in a bowl and baste fish every 5 minutes.
- Once the salmon is done, serve and enjoy.

227. Classic Smoked Trout

Cooking Time: 1 Hour
Servings: 3

The Heat: Hardwood Apple

Ingredients for The Brine:

- 4 cups of water
- 1-2 cups dark-brown sugar
- 1 cup of sea salt

Ingredients for The Trout's:

- 3 pounds of trout, backbone and pin bones removed
- 4 tablespoons of olive oil

Directions:

- Preheat the electrical smoker grill, by setting the temperature to 250 degrees F, for 15 minutes by closing the lid.
- Take a cooking pot, and combine all the brine ingredients, including water, sugar, and salt.
- Submerged the fish in the brine mixture for a few hours.
- Afterward, take out the fish, and pat dry with the paper towel.
- Drizzle olive oil over the fish, and then place it over the grill grate for cooking.
- Smoke the fish, until the internal temperature reaches 140 degrees Fahrenheit, for 1 hour.
- Then serve.

228. Smoked Sea Bass

Preparation Time: 10 Minutes

Cooking Time: 40 Minutes

Servings: 4

Ingredients:

Marinade:

- 1 tsp. Blackened Saskatchewan
- 1 tbsp. Thyme, fresh
- 1 tbsp. Oregano, fresh
- 8 cloves of Garlic, crushed
- 1 lemon, the juice
- ¼ cup oil

Sea Bass:

- 4 Sea bass fillets, skin off
- Chicken Rub Seasoning
- Seafood Seasoning (like Old Bay)
- 8 tbsp. Gold Butter

For garnish:

- Thyme
- Lemon

Directions:

- Make the marinade: In a ziplock bag combine the ingredients and mix. Add the fillets and marinate for 30 min in the fridge. Turn once.
- Preheat the grill to 325F with closed lid.
- In a dish for baking add the butter. Remove the fish from marinade and pour it in the baking dish. Season the fish with chicken and seafood rub. Place it in the baking dish and on the

grill. Cook 30 minutes. Baste 1 - 2 times.
- Remove from the grill when the internal temperature is 160F.
- Garnish with lemon slices and thyme. Enjoy!

Nutrition: Calories: 220; Protein: 32g; Carbs: 1g; Fat: 8g

229. Grilled Clams with Garlic Butter

Preparation Time: 10 Minutes

Cooking Time: 8 Minutes

Servings: 6 - 8

Ingredients:

- 1 Lemon, cut wedges
- 1 - 2 tsp. Anise
- flavored Liqueur
- 2 tbsp. Parsley, minced
- 2 - 3 Garlic cloves, minced
- 8 tbsp. butter, chunks
- 24 of Littleneck Clams

Directions:

- Clean the clams with cold water. Discard those who are with broken shells or don't close.
- Preheat the grill to 450F with closed lid.
- In a casserole dish squeeze juice from 2 wedges, and add parsley, garlic, butter, and liqueur. Arrange the littleneck clams on the grate. Grill 8 minutes, until open. Discard those that won't open.
- Transfer the clams in the baking dish.
- Serve in a shallow dish with lemon wedges. Enjoy!

Nutrition: Calories: 273; Protein: 4g; Carbs: 0.5g; Fat: 10g

230. Simple but Delicious Fish Recipe

Preparation Time: 45 Minutes

Cooking Time: 10 Minutes

Servings: 4 - 6

Ingredients:

- 4 lbs. fish, cut it into pieces (portion size)
- 1 tbsp. minced Garlic
- 1/3 cup of Olive oil
- 1 cup of Soy Sauce
- Basil, chopped

- 2 Lemons, the juice

Directions:

- Preheat the grill to 350F with closed lid.
- Combine the ingredients in a bowl. Stir to combine. Marinade the fish for 45 min.
- Grill the fish until it reaches 145F internal temperature.
- Serve with your favorite side dish and enjoy!

Nutrition: Calories: 153; Protein: 25g; Carbs:1g; Fat: 4g

231. Crab Legs on the Grill

Preparation Time: 15 Minutes

Cooking Time: 30 Minutes

Servings: 4 - 6

Ingredients:

- 1 cup melted Butter
- 3 lb. Halved Crab Legs
- 2 tbsp. Lemon juice, fresh
- 1 tbsp. Old Bay
- 2 Garlic cloves, minced
- For garnish: chopped parsley
- For serving: Lemon wedges

Directions:

- Place the crab legs in a roasting pan.
- In a bowl combine the lemon juice, butter, and garlic. Mix. Pour over the legs. Coat well. Sprinkle with old bay.
- Preheat the grill to 350F with closed lid.
- Place the roasting pan on the grill and cook 20 - 30 minutes busting two times with the sauce in the pan.
- Place the legs on a plate. Divide the crab sauce among 4 bowls for dipping.
- Serve and enjoy!

Nutrition: Calories: 170; Proteins: 20g; Carbs: 0; Fat: 8g

232. Seared Tuna Steaks

Preparation Time: 5 Minutes

Cooking Time: 5 Minutes

Servings: 2 - 4

Ingredients:

- 3 -inch Tuna
- Black pepper
- Sea Salt
- Olive oil
- Sriracha
- Soy Sauce

Directions:

- Baste the tuna steaks with oil and sprinkle with black pepper and salt.
- Preheat the grill to high with closed lid.
- Grill the tuna for 2 ½ minutes per side.
- Remove from the grill. Let it rest for 5 minutes.
- Cut into thin pieces and serve with Sriracha and Soy Sauce. Enjoy.

Nutrition: Calories: 120; Proteins: 34g; Carbs: 0; Fat: 1.5g

233. Roasted Shrimp Mix

Preparation Time: 30 Minutes

Cooking Time: 1h 30min Total: 2h

Servings: 8 - 12

Ingredients:

- 3 lb. Shrimp (large), with tails, divided
- 2 lb. Kielbasa Smoked Sausage
- 6 corns, cut into 3 pieces
- 2 lb. Potatoes, red
- Old Bay

Directions:

- Preheat the grill to 275F with closed lid.
- First, cook the sausage on the grill. Cook 1 hour.
- Increase the temperature to high. Season the corn and potatoes with Old Bay. Now roast until they become tender.
- Season the shrimp with the Old Bay and cook on the grill for 20 minutes.
- In a bowl combine the cooked ingredients. Toss.
- Adjust seasoning with Old Bay and serve. Enjoy!

Nutrition: Calories: 530; Proteins: 20g; Carbs: 32g; Fat: 35g

234. Finnan Haddie Recipe

Preparation Time: 5 minutes

Cooking Time: 35 minutes

Hardwood: Alder

Ingredients:

- 2 pounds smoked haddock fillets
- 2 tablespoons all-purpose flour
- 1/4 cup melted butter
- 2 cups warm milk

Directions:

- Preheat the oven to 325 degrees F (165 degrees C).
- Place smoked haddock into a glass baking dish.
- Whisk the flour into the melted butter until smooth, then whisk in milk, and pour over the haddock.
- Bake in the preheated oven until the sauce has thickened and the fish flakes easily with a fork, about 35 minutes.

235. Smoked Fish Pie

Preparation Time: 20 minutes
Cooking time: 45 minutes
Yield servings: 8 individuals

Ingredients for the Crust:

- 1/2 cup (1 stick) butter (cold)
- 1 1/4 cup all-purpose flour
- 1/2 teaspoon salt
- 3 tablespoons Emmenthal cheese, grated
- 3 tablespoons water

For the Filling:

- 1 cup smoked salmon, chopped
- 1/2 cup leeks white part only, chopped
- 3 eggs
- 1/2 cup whole milk
- 3/4 cup heavy whipping cream
- 1/3 cup dill chopped, fresh
- 1 teaspoon lemon pepper
- 1/2 teaspoon salt
- 3/4 cup cheese Emmenthal cheese, grated

Directions:

- Preheat the oven to 400°F. In a medium bowl, cut the butter into the flour and salt until it resembles coarse crumbs.
- Add the grated cheese and water and combine until the crust comes together. Roll out the crust on a lightly floured surface.
- Press the crust into a 9-inch pie pan, folding and crimping the top edge. Prick the crust all over with a fork. Then bake at 400°F for 15 minutes. Prepare the Pie Gather the ingredients.
- Place the chopped salmon and leek into the prepared pie crust. In a large bowl, whisk together the eggs, milk, cream, dill, lemon pepper, and salt. Pour the egg mixture over the salmon and leek in the crust. Sprinkle with grated Emmenthal cheese.
- Return the pie to the oven and bake for 27 to 30 minutes.
- Cool at least 15 minutes before cutting and serving. Enjoy!

236. Garlic Shrimp Pesto Bruschetta

Servings: 12

Preparation time: 10 minutes

Cooking time: 45 minutes

Ingredients:

- 12 slices of baguette bread
- 12 jumbo shrimp, smoked
- ½ teaspoon ground black pepper
- 1 ½ teaspoon salt
- 1/2 teaspoon garlic powder
- 2 teaspoons minced garlic
- 1/2 teaspoon red chili pepper flakes
- 1/4 teaspoon parsley, leaves
- 1/2 teaspoon paprika, smoked
- 2 tablespoons olive oil
- 12 tablespoons basil pesto

Directions:

- Open hopper of the smoker, add dry pallets, make sure ash-can is in place,

then open the ash damper, power on the smoker and close the ash damper.
- Set the temperature of the smoker to 350 degrees F, let preheat for 30 minutes or until the green light on the dial blinks that indicate smoker has reached to set temperature.
- Meanwhile, take a large baking sheet, line it with aluminum foil and place baguette slices on it.
- Whisk together minced garlic and olives, brush the mixture on both sides of bread slices, then place the baking sheet on smoker grill, shut with lid and smoke for 15 minutes.
- Take a skillet pan, add shrimps, season with salt, black pepper, paprika, garlic powder, and red chili powder, drizzle with olive oil and then place the pan on the grill grate to cook for 5 minutes or more until shrimp is pink, stirring often; set aside until required.
- When bread slices are toasted, let them cool for 5 minutes, then spread 1 tablespoon pesto on each slice and evenly top with shrimp.
- Serve straight away.

Nutrition: Calories: 168; Total Fat: 7 g; Saturated Fat: 1 g; Protein: 5 g; Carbs: 19 g; Fiber: 1 g; Sugar: 1 g

237. Blackened Mahi-Mahi Tacos

Servings: 6

Preparation time: 15 minutes

Cooking time: 45 minutes

Ingredients:

- 4 mahi-mahi fillets, each about 4 ounces
- 4 teaspoons habanero spice
- 2 tablespoons olive oil
- 12 small tortilla, toasted

For Topping:

- 1/3 cup diced red onion
- 2 tablespoons lime juice
- 2 cups shredded red cabbage
- 1 cup mango salsa
- 1 cup sour cream

Directions:

- Open hopper of the smoker, add dry pallets, make sure ash-can is in place, then open the ash damper, power on the smoker and close the ash damper.
- Set the temperature of the smoker to 400 degrees F, switch smoker to open flame cooking mode, press the open flame 3, remove the grill grates and the batch, replace batch with direct flame insert, then return grates on the grill in the lower position and let preheat for 30 minutes or until the green light on the dial blinks that indicate smoker has reached to set temperature.
- Meanwhile, rinse the fillets with water, pat dry with paper towels, then rub with oil and season with habanero

- seasoning on both sides, set aside until required.
- Place fillets on the smoker grill, shut with lid and smoke for 7 minutes per side or until cooked.
- When done, transfer fillets to a cutting board, let rest for 5 minutes and then cut into bite-size pieces.
- Assemble the tacos and for this, stack the tortillas in two, then evenly place fillet pieces in them, top with onion, cabbage, salsa and cream and drizzle with lime juice.
- Serve straight away.

Nutrition: Calories: 230; Total Fat: 8 g; Saturated Fat: 1.5 g; Protein: 13 g; Carbs: 26 g; Fiber: 4 g; Sugar: 2 g

238. Shrimp Tacos

Servings: 4

Preparation time: 10 minutes

Cooking time: 40 minutes

Ingredients:

- 1-pound Shrimp, peeled, deveined
- 3 tablespoons taco seasoning
- 1/2 teaspoon salt
- 2 tablespoons olive oil

Directions:

- Open hopper of the smoker, add dry pallets, make sure ash-can is in place, then open the ash damper, power on the smoker and close the ash damper.
- Set the temperature of the smoker to 400 degrees F, switch smoker to open flame cooking mode, press the open flame 3, remove the grill grates and the batch, replace batch with direct flame insert, then return grates on the grill in the lower position and let preheat for 30 minutes or until the green light on the dial blinks that indicate smoker has reached to set temperature.
- Meanwhile, place shrimps in a large bowl, add oil, toss until well coated, then thread 4 shrimps per skewer and sprinkle with taco seasoning.
- Place the shrimp skewers on the smoker grill, shut with lid and smoke for 5 minutes per side or until shrimp is pink.
- Serve straight away.

Nutrition: Calories: 139; Total Fat: 6 g; Saturated Fat: 1.2 g; Protein: 19 g; Carbs: 1 g; Fiber: 0 g; Sugar: 0 g

239. Buffalo Shrimp

Servings: 6

Preparation time: 10 minutes

Cooking time: 36 minutes

Ingredients:

- 1-pound shrimp, peeled, deveined
- 1/4 teaspoon garlic powder
- 1/2 teaspoon salt
- 1/4 teaspoon onion powder
- 1/2 cup buffalo sauce

Directions:

- Open hopper of the smoker, add dry pallets, make sure ash-can is in place,

then open the ash damper, power on the smoker and close the ash damper.
- Set the temperature of the smoker to 450 degrees F, switch smoker to open flame cooking mode, press the open flame 3, remove the grill grates and the batch, replace batch with direct flame insert, then return grates on the grill in the lower position and let preheat for 30 minutes or until the green light on the dial blinks that indicate smoker has reached to set temperature.
- Meanwhile, place the shrimps in a large bowl, add garlic powder, salt and onion powder and toss until well coated.
- Place shrimps on the smoker grill, shut with lid and smoke for 3 minutes per side or until pink.
- When done, transfer the shrimps in a bowl, add buffalo sauce and toss until well coated.
- Serve straight away.

Nutrition: Calories: 293; Total Fat: 8 g; Saturated Fat: 4 g; Protein: 46 g; Carbs: 2 g; Fiber: 1 g; Sugar: 0 g

240. Salmon Cakes

Servings: 4

Preparation time: 45 minutes

Cooking time: 7 minutes

Ingredients:

- For Salmon Cakes:
- 1 cup cooked salmon, flaked
- 1/2 of red pepper, diced
- 1 tablespoon mustard
- 1/2 tablespoon rib rub
- 1 1/2 cups breadcrumb
- 2 eggs
- 1/2 tablespoon olive oil
- 1/4 cup mayonnaise

For the Sauce:

- 1 cup mayonnaise, divided
- 1/2 tablespoon capers, diced
- 1/4 cup dill pickle relish

Directions:

- Place all the ingredients for the salmon cakes in a bowl, except for oil, stir until well mixed and then let rest for 15 minutes.
- Open hopper of the smoker, add dry pallets, make sure ash-can is in place, then open the ash damper, power on the smoker and close the ash damper.
- Set the temperature of the smoker to 350 degrees F, switch smoker to open flame cooking mode, press the open flame 3, remove the grill grates and the batch, replace batch with direct flame insert, then return grates on the grill in the lower position, place the sheet pan and let preheat for 30 minutes or until the green light on the dial blinks that indicate smoker has reached to set temperature.
- Meanwhile, prepare the sauce and for this, place all the ingredients for the sauce in a bowl and whisk until combined, set aside until required.
- Then place the salmon mixture on the heated sheet pan, about 2 tablespoons per patty, press with a spatula to form a patty, grill for 5 minutes, then flip

the patties, continue smoking for 2 minutes.
- When done, transfer salmon cakes to a dish and serve with prepared sauce.

Nutrition: Calories: 229.7; Total Fat: 9 g; Saturated Fat: 2 g; Protein: 22.4 g; Carbs: 13 g; Fiber: 1.1 g; Sugar: 2.8 g

241. Crab Stuffed Mushroom Caps

Servings: 4

Preparation time: 10 minutes

Cooking time: 35 minutes

Ingredients:

- 12 porcini mushroom caps, destemmed, cleaned
- 2 tablespoon parsley
- 2 tablespoons imitation crab, chopped
- 1 teaspoon steak rub
- 2 teaspoon lemon, zest
- 8 ounces cream cheese, softened
- 1 tablespoon lemon, juice
- 3/4 cup panko bread crumbs
- 1/2 cup shredded parmesan cheese, divided

Directions:

- Open hopper of the smoker, add dry pallets, make sure ash-can is in place, then open the ash damper, power on the smoker and close the ash damper.
- Set the temperature of the smoker to 450 degrees F, switch smoker to open flame cooking mode, press the open flame 3, remove the grill grates and the batch, replace batch with direct flame insert, then return grates on the grill in the lower position and let preheat for 30 minutes or until the green light on the dial blinks that indicate smoker has reached to set temperature.
- Meanwhile, place all the ingredients in a bowl, except for mushroom caps and ¼ cup cheese, stir well until combined and then use a spoon to stuff this mixture into each mushroom cap until packed.
- Top mushrooms with remaining cheese, then place them on the smoker grill, shut with lid and smoke for 5 minutes or cheese bubbles and mushrooms are tender.
- Serve straight away.

Nutrition: Calories: 289.1; Total Fat: 13.5 g; Saturated Fat: 6 g; Protein: 23.7 g; Carbs: 18.2 g; Fiber: 1.2 g; Sugar: 7.7 g

242. Grilled Red Snapper

Servings: 4

Preparation time: 10 minutes

Cooking time: 1 hour and 15 minutes

Ingredients:

- 4 fillets of red snapper, large
- 1 lime, juiced, zested, sliced
- 2 medium onions, thinly sliced
- For the Baste:
- 1 teaspoon minced garlic
- 3 teaspoon chopped cilantro
- 1 teaspoon ground black pepper
- 1 teaspoon ancho chili powder
- ½ teaspoon lime zest

- 1 teaspoon salt
- ½ teaspoon cumin
- 1/3 cup olive oil
- ¼ cup ponzu sauce

Directions:

- Open hopper of the smoker, add dry pallets, make sure ash-can is in place, then open the ash damper, power on the smoker and close the ash damper.
- Set the temperature of the smoker to 350 degrees F, switch smoker to open flame cooking mode, press the open flame 3, remove the grill grates and the batch, replace batch with direct flame insert, then return grates on the grill in the lower position and let preheat for 30 minutes or until the green light on the dial blinks that indicate smoker has reached to set temperature.
- Meanwhile, prepare the baste and for this, place all its ingredients in a bowl and stir until mixed.
- Take a heatproof cooking basket, then breakfast onion slices and line them on the bottom of the basket along with lime slices.
- Brush both sides of fillets with the baste mixture, place them in the basket, then the basket on the smoker grill, shut with lid and smoke for 30 to 45 minutes or more until cooked, brushing with the baste every 15 minutes.
- Serve straight away.

Nutrition: Calories: 311; Total Fat: 8.8 g; Saturated Fat: 1.2 g; Protein: 45 g; Carbs: 11 g; Fiber: 3 g; Sugar: 1.3 g

243. Sugar Cured Salmon

Servings: 4

Preparation time: 6 hours and 10 minutes

Cooking time: 1 hour

Ingredients:

- 3 pounds salmon, skinned
- 2 tablespoons seafood Rub
- 3 cups brown sugar
- 1 ½ cup sea salt
- 1/3 cup BBQ Sauce

Directions:

- Prepare the cure and for this, place salt, sugar, and rub in a medium bowl and stir until mixed.
- Take a large baking dish, spread ¼-inch of the cure in its bottom, top with salmon, then top evenly with the remaining cure, cover the dish and let it rest in the refrigerator for 6 hours.
- When ready to cook, open hopper of the smoker, add dry pallets, make sure ash-can is in place, then open the ash damper, power on the smoker and close the ash damper.
- Set the temperature of the smoker to 300 degrees F, let preheat for 30 minutes or until the green light on the dial blinks that indicate smoker has reached to set temperature.
- Meanwhile, remove the baking dish from the refrigerator, uncover it, remove the salmon from the cure, rinse well and pat dry with paper towels.

- Place salmon on the smoker grill, shut with lid, smoke it for 20 minutes, then brush salmons with BBQ sauce and continue smoking for 10 minutes or until glazed.
- Serve straight away.

Nutrition: Calories: 70; Total Fat: 1 g; Saturated Fat: 0 g; Protein: 14 g; Carbs: 2 g; Fiber: 0 g; Sugar: 2 g

244. Shrimp Scampi

Servings: 4

Preparation time: 10 minutes

Cooking time: 45 minutes

Ingredients:

- 1 pound shrimp, peeled, deveined
- 1/2 teaspoon salt
- 1/2 teaspoon garlic powder
- 1 tablespoon lemon juice
- 1/2 cup butter, salted, melted
- 1/4 cup dry white wine
- 1/2 teaspoon minced garlic

Directions:

- Open hopper of the smoker, add dry pallets, make sure ash-can is in place, then open the ash damper, power on the smoker and close the ash damper.
- Set the temperature of the smoker to 400 degrees F, switch smoker to open flame cooking mode, press the open flame 3, remove the grill grates and the batch, replace batch with direct flame insert, then return grates on the grill in the lower position and let preheat for 30 minutes or until the green light on the dial blinks that indicate smoker has reached to set temperature.
- Meanwhile, stir together minced garlic, lemon juice, wine, and melted butter until combined, then pour the mixture in a cast iron pan, place it over medium heat and cook for 3 to 4 minutes or until heated through; remove the pan from heat.
- Season shrimps with salt and garlic powder, then place them carefully in the pan, place it on the smoker grill, shut with lid and smoke shrimps for 10 minutes or until pink.
- Serve straight away.

Nutrition: Calories: 137.6; Total Fat: 8.5 g; Saturated Fat: 1 g; Protein: 13.7 g; Carbs: 1.3 g; Fiber: 0.1 g; Sugar: 0.2 g

245. Spicy Lime Shrimp

Servings: 4

Preparation time: 40 minutes

Cooking time: 36 minutes

Ingredients:

- 1-pound shrimps, peeled, deveined
- For the Marinade:
- 1 tablespoon minced garlic
- 1/2 teaspoon salt
- 1/2 teaspoon ground cumin
- 1/4 teaspoon paprika
- 1/4 teaspoon red flakes pepper
- 2 teaspoons red chili paste
- 1 large lime, juiced

Directions:

- Prepare the marinade and for this, place all the ingredients of marinade in a bowl and stir until mixed.
- Place shrimps in a large plastic bag, pour in marinade, seal the bag, turn it upside down to coat shrimps with the marinade and let marinate for 30 minutes in the refrigerator.
- Meanwhile, open hopper of the smoker, add dry pallets, make sure ash-can is in place, then open the ash damper, power on the smoker and close the ash damper.
- Set the temperature of the smoker to 400 degrees F, switch smoker to open flame cooking mode, press the open flame 3, remove the grill grates and the batch, replace batch with direct flame insert, then return grates on the grill in the lower position and let preheat for 30 minutes or until the green light on the dial blinks that indicate smoker has reached to set temperature.
- Then thread the marinated shrimps on soaked skewers, place them on the smoker grill, shut with lid and smoke for 2 to 3 minutes or until done.
- Serve straight away.

Nutrition: Calories: 84.4; Total Fat: 1 g; Saturated Fat: 0.2 g; Protein: 16.3 g; Carbs: 2.2 g; Fiber: 0.4 g; Sugar: 0.2 g

246. Jerk Shrimp

Preparation Time: 15 Minutes
Cooking Time: 6
Servings: 12

Ingredients:

- 2 pounds shrimp, peeled, deveined
- 3 tablespoons olive oil
- For the Spice Mix:
- 1 teaspoon garlic powder
- 1 teaspoon of sea salt
- ¼ teaspoon ground cayenne
- 1 tablespoon brown sugar
- 1/8 teaspoon smoked paprika
- 1 tablespoon smoked paprika
- ¼ teaspoon ground thyme
- 1 lime, zested

Directions:

1. Switch on the Traeger grill, fill the grill hopper with flavored wood pellets, power the grill on by using the control panel, select 'smoke' on the temperature dial, or set the temperature to 450 degrees F and let it preheat for a minimum of 5 minutes.
2. Meanwhile, prepare the spice mix and for this, take a small bowl, place all of its ingredients in it and stir until mixed.
3. Take a large bowl, place shrimps in it, sprinkle with prepared spice mix, drizzle with oil and toss until well coated.
4. When the grill has preheated, open the lid, place shrimps on the grill grate, shut the grill and smoke for 3 minutes per side until firm and thoroughly cooked.
5. When done, transfer shrimps to a dish and then serve.

247. Spicy Shrimps Skewers

Preparation Time: 10 Minutes
Cooking Time: 6 Minutes
Servings: 4

Ingredients:

- 2 pounds shrimp, peeled, and deveined
- For the Marinade:
- 6 ounces Thai chilies
- 6 cloves of garlic, peeled
- 1 ½ teaspoon sugar
- 2 tablespoons Napa Valley rub
- 1 ½ tablespoon white vinegar
- 3 tablespoons olive oil

Directions:

1. Prepare the marinade and for this, place all of its ingredients in a food processor and then pulse for 1 minute until smooth.
2. Take a large bowl, place shrimps on it, add prepared marinade, toss until well coated, and let marinate for a minimum of 30 minutes in the refrigerator.
3. When ready to cook, switch on the Traeger grill, fill the grill hopper with apple-flavored wood pellets, power the grill on by using the control panel, select 'smoke' on the temperature dial, or set the temperature to 450 degrees F and let it preheat for a minimum of 5 minutes.
4. Meanwhile, remove shrimps from the marinade and then thread onto skewers.
5. When the grill has preheated, open the lid, place shrimps' skewers on the grill grate, shut the grill and smoke for 3 minutes per side until firm.
6. When done, transfer shrimps' skewers to a dish and then serve.

Nutrition: Calories: 187.2 Fat: 2.7g Carbs: 2.7g Protein: 23.2g

248. Lobster Tails

Preparation Time: 10 Minutes
Cooking Time: 35 Minutes
Servings: 4

Ingredients:

- 2 lobster tails, each about 10 ounces
- For the Sauce:
- 2 tablespoons chopped parsley
- ¼ teaspoon garlic salt
- 1 teaspoon paprika
- ¼ teaspoon ground black pepper
- ¼ teaspoon old bay seasoning
- 8 tablespoons butter, unsalted
- 2 tablespoons lemon juice

Directions:

1. Switch on the Traeger grill, fill the grill hopper with flavored wood pellets, power the grill on by using the control panel, select 'smoke' on the temperature dial, or set the temperature to 450 degrees F and let it preheat for a minimum of 15 minutes.
2. Meanwhile, prepare the sauce and for this, take a small saucepan, place it over medium-low heat, add butter in it and when it melts, add remaining ingredients for the sauce and stir until combined, set aside until required.
3. Prepare the lobster and for this, cut the shell from the middle to the tail by

using kitchen shears and then take the meat from the shell, keeping it attached at the base of the crab tail.

4. Then butterfly the crab meat by making a slit down the middle, then place lobster tails on a baking sheet and pour 1 tablespoon of sauce over each lobster tail, reserve the remaining sauce.
5. When the grill has preheated, open the lid, place crab tails on the grill grate, shut the grill and smoke for 30 minutes until opaque.
6. When done, transfer lobster tails to a dish and then serve with the remaining sauce.

Nutrition: Calories: 290 Fat: 22g Carbs: 1g Protein: 20g

249. Lemon Garlic Scallops

Preparation Time: 10 Minutes
Cooking Time: 5 Minutes
Servings: 6

Ingredients:

- 1 dozen scallops
- 2 tablespoons chopped parsley
- Salt as needed
- 1 tablespoon olive oil
- 1 tablespoon butter, unsalted
- 1 teaspoon lemon zest
- For the Garlic Butter:
- ½ teaspoon minced garlic
- 1 lemon, juiced
- 4 tablespoons butter, unsalted, melted

Directions:

1. Switch on the Traeger grill, fill the grill hopper with alder flavored wood pellets, power the grill on by using the control panel, select 'smoke' on the temperature dial, or set the temperature to 400 degrees F and let it preheat for a minimum of 15 minutes.
2. Meanwhile, remove frill from scallops, pat dry with paper towels and then season with salt and black pepper.
3. When the grill has preheated, open the lid, place a skillet on the grill grate, add butter and oil, and when the butter melts, place seasoned scallops on it and then cook for 2 minutes until seared.
4. Meanwhile, prepare the garlic butter and for this, take a small bowl, place all of its ingredients in it and then whisk until combined.
5. Flip the scallops, top with some of the prepared garlic butter, and cook for another minute.
6. When done, transfer scallops to a dish, top with remaining garlic butter, sprinkle with parsley and lemon zest, and then serve.

Nutrition: Calories: 184 Fat: 10g Carbs: 1g Protein: 22g

250. Halibut in Parchment

Preparation Time: 15 Minutes
Cooking Time: 15 Minutes
Servings: 4

Ingredients:

- 16 asparagus spears, trimmed, sliced into 1/2-inch pieces
- 2 ears of corn kernels
- 4 ounces halibut fillets, pin bones removed
- 2 lemons, cut into 12 slices
- Salt as needed
- Ground black pepper as needed
- 2 tablespoons olive oil
- 2 tablespoons chopped parsley

Directions:

1. Switch on the Traeger grill, fill the grill hopper with flavored wood pellets, power the grill on by using the control panel, select 'smoke' on the temperature dial, or set the temperature to 450 degrees F and let it preheat for a minimum of 5 minutes.
2. Meanwhile, cut out 18-inch long parchment paper, place a fillet in the center of each parchment, season with salt and black pepper, and then drizzle with oil.
3. Cover each fillet with three lemon slices, overlapping slightly, sprinkle one-fourth of asparagus and corn on each fillet, season with some salt and black pepper, and seal the fillets and vegetables tightly to prevent steam from escaping the packet. When the grill has preheated, open the lid, place fillet packets on the grill grate, shut the grill and smoke for 15 minutes until packets have turned slightly brown and puffed up.
4. When done, transfer packets to a dish, let them stand for 5 minutes, then cut 'X' in the center of each packet, carefully uncover the fillets and vegetables, sprinkle with parsley, and then serve.

Nutrition: Calories: 186.6 Fat: 2.8 g Carbs: 14.2 g Protein: 25.7 g

251. Chilean Sea Bass

Preparation Time: 30 Minutes
Cooking Time: 40 Minutes
Servings: 6

Ingredients:

- 4 sea bass fillets, skinless, each about 6 ounces
- Chicken rub as needed
- 8 tablespoons butter, unsalted
- 2 tablespoons chopped thyme leaves
- Lemon slices for serving
- For the Marinade:
- 1 lemon, juiced
- 4 teaspoons minced garlic
- 1 tablespoon chopped thyme
- 1 teaspoon blackened rub
- 1 tablespoon chopped oregano
- ¼ cup oil

Directions:

1. Prepare the marinade and for this, take a small bowl, place all of its ingredients in it, stir until well combined, and then pour the mixture into a large plastic bag.

2. Add fillets in the bag, seal it, turn it upside down to coat fillets with the marinade and let it marinate for a minimum of 30 minutes in the refrigerator.
3. When ready to cook, switch on the Traeger grill, fill the grill hopper with apple-flavored wood pellets, power the grill on by using the control panel, select 'smoke' on the temperature dial, or set the temperature to 325 degrees F and let it preheat for a minimum of 15 minutes.
4. Meanwhile, take a large baking pan and place butter on it.
5. When the grill has preheated, open the lid, place baking pan on the grill grate, and wait until butter melts.
6. Remove fillets from the marinade, pour marinade into the pan with melted butter, then season fillets with chicken rubs until coated on all sides, then place them into the pan, shut the grill and cook for 30 minutes until internal temperature reaches 160 degrees F, frequently basting with the butter sauce.
7. When done, transfer fillets to a dish, sprinkle with thyme and then serve with lemon slices.

Nutrition: Calories: 232 Fat: 12.2 g Carbs: 0.8 g Protein: 28.2 g

252. Sriracha Salmon

Preparation Time: 2 Hours And 10 Minutes
Cooking Time: 25 Minutes
Servings: 4

Ingredients:

- 3-pound salmon, skin on
- For the Marinade:
- 1 teaspoon lime zest
- 1 tablespoon minced garlic
- 1 tablespoon grated ginger
- Sea salt as needed
- Ground black pepper as needed
- ¼ cup maple syrup
- 2 tablespoons soy sauce
- 2 tablespoons Sriracha sauce
- 1 tablespoon toasted sesame oil
- 1 tablespoon rice vinegar
- 1 teaspoon toasted sesame seeds

Directions:

1. Prepare the marinade and for this, take a small bowl, place all of its ingredients in it, stir until well combined, and then pour the mixture into a large plastic bag.
2. Add salmon in the bag, seal it, turn it upside down to coat salmon with the marinade and let it marinate for a minimum of 2 hours in the refrigerator.
3. When ready to cook, switch on the Traeger grill, fill the grill hopper with flavored wood pellets, power the grill on by using the control panel, select 'smoke' on the temperature dial, or set the temperature to 450 degrees F and let it preheat for a minimum of 5 minutes.

4. Meanwhile, take a large baking sheet, line it with parchment paper, place salmon on it skin-side down and then brush with the marinade.
5. When the grill has preheated, open the lid, place baking sheet containing salmon on the grill grate, shut the grill and smoke for 25 minutes until thoroughly cooked.
6. When done, transfer salmon to a dish and then serve.

Nutrition: Calories: 360 Fat: 21 g Carbs: 28 g Protein: 16 g

253. Grilled Rainbow Trout

Preparation Time: 1 Hour
Cooking Time: 2 Hours
Servings: 6

Ingredients:

- 6 rainbow trout, cleaned, butterfly
- For the Brine:
- ¼ cup salt
- 1 tablespoon ground black pepper
- ½ cup brown sugar
- 2 tablespoons soy sauce
- 16 cups water

Directions:

1. Prepare the brine and for this, take a large container, add all of its ingredients in it, stir until sugar has dissolved, then add trout and let soak for 1 hour in the refrigerator.
2. When ready to cook, switch on the Traeger grill, fill the grill hopper with oak flavored wood pellets, power the grill on by using the control panel, select 'smoke' on the temperature dial, or set the temperature to 225 degrees F and let it preheat for a minimum of 15 minutes.
3. Meanwhile, remove trout from the brine and pat dry with paper towels.
4. When the grill has preheated, open the lid, place trout on the grill grate, shut the grill and smoke for 2 hours until thoroughly cooked and tender.
5. When done, transfer trout to a dish and then serve.

Nutrition: Calories: 250 Fat: 12 g Carbs: 1.4 g Protein: 33 g

254. Cider Salmon

Preparation Time: 9 Hours
Cooking Time: 1 Hour
Servings: 4

Ingredients:

- 1 ½ pound salmon fillet, skin-on, center-cut, pin bone removed
- For the Brine:
- 4 juniper berries, crushed
- 1 bay leaf, crumbled
- 1 piece star anise, broken
- 1 ½ cups apple cider
- For the Cure:
- ½ cup salt
- 1 teaspoon ground black pepper
- ¼ cup brown sugar
- 2 teaspoons barbecue rub

Directions:

1. Prepare the brine and for this, take a large container, add all of its ingredients in it, stir until mixed, then add salmon and let soak for a minimum of 8 hours in the refrigerator.
2. Meanwhile, prepare the cure and for this, take a small bowl, place all of its ingredients in it and stir until combined.
3. After 8 hours, remove salmon from the brine, then take a baking dish, place half of the cure in it, top with salmon skin-side down, sprinkle remaining cure on top, cover with plastic wrap and let it rest for 1 hour in the refrigerator.
4. When ready to cook, switch on the Traeger grill, fill the grill hopper with oak flavored wood pellets, power the grill on by using the control panel, select 'smoke' on the temperature dial, or set the temperature to 200 degrees F and let it preheat for a minimum of 5 minutes.
5. Meanwhile, remove salmon from the cure, pat dry with paper towels, and then sprinkle with black pepper.
6. When the grill has preheated, open the lid, place salmon on the grill grate, shut the grill, and smoke for 1 hour until the internal temperature reaches 150 degrees F.
7. When done, transfer salmon to a cutting board, let it rest for 5 minutes, then remove the skin and serve.

Nutrition: Calories: 233 Fat: 14 g Carbs: 0 g Protein: 25 g

255. Cajun Shrimp

Preparation Time: 4 Hours
Cooking Time: 8 Minutes
Servings: 6

Ingredients:

- 2 pounds shrimp, peeled, deveined
- For the Marinade:
- 1 teaspoon minced garlic
- 1 lemon, juiced
- 1 teaspoon salt
- 1 tablespoon Cajun shake
- 4 tablespoons olive oil

Directions:

1. Prepare the marinade and for this, take a small bowl, place all of its ingredients in it, stir until well combined, and then pour the mixture into a large plastic bag.
2. Add shrimps in the bag, seal it, turn it upside down to coat salmon with the marinade and let it marinate for a minimum of 4 hours in the refrigerator.
3. When ready to cook, Switch on the Traeger grill, fill the grill hopper with flavored wood pellets, power the grill on by using the control panel, select 'smoke' on the temperature dial, or set the temperature to 450 degrees F and let it preheat for a minimum of 5 minutes.
4. Meanwhile, when the grill has preheated, open the lid, place shrimps on the grill grate, shut the grill and smoke for 4 minutes per side until firm. When done, transfer shrimps to a dish and then serve.

Nutrition: Calories: 92 Fat: 7.6 g Carbs: 2.2 g Protein: 4.6 g

256. Grilled Lobster Tail

Preparation Time: 10 minutes
Cooking Time: 15 minutes
Servings: 4

Ingredients:

- 2 (8 ounces each) lobster tails
- 1/4 teaspoon old bay seasoning
- ½ teaspoon oregano
- 1 teaspoon paprika
- Juice from one lemon
- 1/4 teaspoon Himalayan salt
- 1/4 teaspoon freshly ground black pepper
- 1/4 teaspoon onion powder
- 2 tablespoons freshly chopped parsley
- ¼ cup melted butter

Directions:

1. Slice the tail in the middle with a kitchen shear. Pull the shell apart slightly and run your hand through the meat to separate the meat partially, keeping it attached to the base of the tail partially.
2. Combine the old bay seasoning, paprika, oregano, salt, pepper and onion powder in a mixing bowl.
3. Drizzle lobster tail with lemon juice and season generously with the seasoning mixture.
4. Preheat your wood pellet smoker to 450°F, using apple wood pellets.
5. Place the lobster tail directly on the grill grate, meat side down. Cook for about 15 minutes or until the internal temperature of the tails reaches 140°F.
6. Remove the tail from the grill and let them rest for a few minutes to cool.
7. Drizzle melted butter over the tails.
8. Serve and garnish with fresh chopped parsley.

Nutrition: Calories 146 Fat 11.7g Carbs: 2.1g Protein 9.3g

257. Halibut

Preparation Time: 10 minutes
Cooking Time: 30 minutes
Servings: 4

Ingredients:

- 1 pound fresh halibut filet (cut into 4 equal sizes)
- 1 tablespoon fresh lemon juice
- 2 garlic cloves (minced)
- 2 teaspoons soy sauce
- ½ teaspoon ground black pepper
- ½ teaspoon onion powder
- 2 tablespoons honey
- ½ teaspoon oregano
- 1 teaspoon dried basil
- 2 tablespoons butter (melted)
- Maple syrup for serving

Directions:

1. In a mixing bowl, combine the lemon juice, honey, soy sauce, onion powder, oregano, dried basil, pepper and garlic.
2. Brush the halibut filets generously with the filet the mixture. Wrap the filets with aluminum foil and refrigerate for 4 hours.
3. Remove the filets from the refrigerator and let them sit for about 2 hours, or until they are at room temperature.

4. Activate your wood pellet grill on smoke, leaving the lid opened for 5 minutes or until fire starts.
5. Close the lid and allow to preheat your grill to 275°F 15 minutes, using fruit wood pellets.
6. Place the halibut filets directly on the grill grate and smoke for 30 minutes or until the internal temperature of the fish reaches 135°F.
7. Remove the filets from the grill and let them rest for 10 minutes.
8. Serve and top with maple syrup to taste

Nutrition: Calories 180 Fat: 6.3g Carbs: 10g Protein 20.6g

258. Grilled Salmon

Preparation Time: 10 minutes
Cooking Time: 30 minutes
Servings: 6

Ingredients:

- 2 pounds salmon (cut into fillets)
- 1/2 cup low sodium soy sauce
- 2 garlic cloves (grated)
- 4 tablespoons olive oil
- 2 tablespoons honey
- 1 teaspoon ground black pepper
- ½ teaspoon smoked paprika
- ½ teaspoon Italian seasoning
- Garnish:
- 2 tablespoons chopped green onion

Directions:

1. In a huge container, combine the honey, pepper, paprika, Italian seasoning, garlic, soy sauce and olive oil. Add the salmon fillets and toss to combine. Cover the bowl and refrigerate for 1 hour.
2. Remove the fillets from the marinade and let it sit for about 2 hours, or until it is at room temperature.
3. Start the wood pellet on smoke, leaving the lid opened for 5 minutes, or until fire starts.
4. Close the lid and preheat grill to 350°F for 15 minutes.
5. Grease the grill grate with oil and arrange the fillets on the grill grate, skin side up. Close the grill lid and cook for 4 minutes.
6. Flip the fillets and cook for additional 25 minutes or until the fish is flaky.
7. Remove the fillets from heat and let it sit for a few minutes.
8. Serve warm and garnish with chopped green onion.

Nutrition: Calories 317 Fat 18.8g Carbs: 8.3g Protein 30.6g

259. BBQ Shrimp

Preparation Time: 20 Minutes
Cooking Time: 8 Minutes
Servings: 6

Ingredients:

- 2 pound raw shrimp (peeled and deveined)
- ¼ cup extra virgin olive oil
- ½ teaspoon paprika
- ½ teaspoon red pepper flakes
- 2 garlic cloves (minced)
- 1teaspoon cumin
- 1 lemon (juiced)
- 1 teaspoon kosher salt
- 1 tablespoon chili paste
- Bamboo or wooden skewers (soaked for 30 minutes, at least)

Directions:

1. In a large mixing bowl, combine the pepper flakes, cumin, lemon, salt, chili, paprika, garlic and olive oil. Add the shrimp and toss to combine.
2. Transfer the shrimp and marinade into a zip-lock bag and refrigerate for 4 hours.
3. Take off the shrimp meat from the marinade and let it rest until it is a room temperature.
4. Start your grill on smoke, leaving the lid opened for 5 minutes, or until fire starts. Use hickory wood pellet.
5. Close the lid and preheat the grill too HIGH for 15 minutes.
6. Thread shrimps onto skewers and arrange the skewers on the grill grate.
7. Smoke shrimps for 8 minutes, 4 minutes per side. Serve and enjoy.

260. Grilled Tuna

Preparation Time: 5 Minutes
Cooking Time: 4 Minutes
Servings: 4

Ingredients:

- 4 (6 ounce each) tuna steaks (1 inch thick)
- 1 lemon (juiced)
- 1 clove garlic (minced)
- 1 teaspoon chili
- 2 tablespoons extra virgin olive oil
- 1 cup white wine
- 3 tablespoons brown sugar
- 1 teaspoon rosemary

Directions:

1. In a huge container, combine the chili, lemon, white wine, sugar, rosemary, olive oil and garlic. Add the tuna steaks and toss to combine.
2. Transfer the tuna and marinade to a zip lock bag. Refrigerate for 3 hours.
3. Remove the tuna steaks from the marinade and let them rest for about 1 hour, or until the steaks are at room temperature.
4. Start your grill on smoke, leaving the lid opened for 5 minutes, or until fire starts. Use hickory or mesquite wood pellet.
5. Close the grill lid and preheat the grill on HIGH for 15 minutes.
6. Grease the grill grate with oil and place the tuna on the grill grate. Grill tuna steaks for 4 minutes, 2 minutes per side.
7. Remove the tuna from the grill and let them rest for a few minutes. Serve and enjoy.

261. Oyster in Shell

Preparation Time: 25 Minutes
Cooking Time: 8 Minutes
Servings: 4

Ingredients:

- 12 medium oysters * note that all oysters should be completely closed. Opened/dead oysters are of no use here.
- 1 teaspoon oregano
- 1 lemon (juiced)
- 1 teaspoon freshly ground black pepper
- 6 tablespoons unsalted butter (melted)
- 1 teaspoon salt or more to taste
- 2 garlic cloves (minced)
- Garnish:
- 2 ½ tablespoons grated parmesan cheese
- 2 tablespoons freshly chopped parsley

Directions:

1. Start by scrubbing the outside of the shell with a scrub brush under cold running water to remove dirt.
2. Hold an oyster in a towel, flat side up. Insert an oyster knife in the hinge of the oyster.
3. Twist the knife with pressure to pop open the oyster. Run the knife along the oyster hinge to open the shell completely. Discard the top shell.
4. Gently run the knife under the oyster to loosen the oyster foot from the bottom shell.
5. Repeat step 2 and 3 for the remaining oysters.
6. Combine melted butter, lemon, pepper, salt, garlic and oregano in a mixing bowl. 6. Pour ½ to 1 teaspoon of the butter mixture on each oyster.
7. Start your wood pellet grill on smoke, leaving the lid opened for 5 minutes, or until fire starts. Close the lid and preheat the grill to HIGH with lid closed for 15 minutes.
8. Gently arrange the oysters onto the grill grate. Grill oyster for 6 to 8 minutes or until the oyster juice is bubbling and the oyster is plump.
9. Remove oysters from heat. Serve and top with grated parmesan and chopped parsley.

Nutrition: Calories 200 Fat 19.2g Carbs: 3.9g Protein 4.6g

262. Grilled King Crab Legs

Preparation Time: 10 Minutes
Cooking Time: 25 Minutes
Servings: 4

Ingredients:

- 4 pounds king crab legs (split)
- 4 tablespoons lemon juice
- 2 tablespoons garlic powder
- 1 cup butter (melted)
- 2 teaspoons brown sugar
- 2 teaspoons paprika
- 2 teaspoons powdered black pepper for taste

Directions:

1. In a mixing bowl, combine the lemon juice, butter, sugar, garlic, paprika and pepper.

2. Arrange the split crab on a baking sheet, split side up.
3. Drizzle ¾ of the butter mixture over the crab legs.
4. Configure your pellet grill for indirect cooking and preheat it to 225°F, using mesquite wood pellets.
5. Arrange the crab legs onto the grill grate, shell side down.
6. Cover the grill and cook 25 minutes.
7. Remove the crab legs from the grill.
8. Serve and top with the remaining butter mixture.

Nutrition: Calories 894 Fat 53.2g Carbs: 6.1g Protein 88.6g

263. Smoked Salmon

Preparation Time: 12 hours
Cooking Time: 4 hours
Servings: 6

Ingredients:

- For The Brine
- 4 Cups of water
- 1 Cup of brown sugar
- 1/3 Cup of kosher salt
- For The Salmon
- 1 Large skin-on salmon filet
- Real maple syrup

Directions:

1. Combine the ingredients of the brine until the sugar is completely dissolved; then place it into a large zip lock bag or a large covered container; then Place the cleaned salmon into that brine, and refrigerate for about 10 to 12 hours.
2. Once the fish is perfectly brined, remove it from the liquid; then rinse and pat dry with clean paper towels.
3. Let the fish sit out at the room temperature for about 1 to 2 hours to let the pellicle form.
4. Turn the smoker on to get the fire started; then place the salmon over a baking rack sprayed with cooking spray
5. Place the rack over the smoker; then close the lid
6. Baste the salmon with the pure syrup generously
7. Smoke for about 3 to 4 hours; then serve and enjoy.

Nutrition: Calories: 101 Fat: 2g Carbs: 16g Protein: 4g

264. Smoked Sardines

Preparation Time: 12 hours
Cooking Time: 5 hours
Servings: 5

Ingredients:

- 20 to 30 fresh gutted sardines
- 4 Cups of water
- ¼ Cup of Kosher salt
- ¼ Cup of honey
- 4 to 5 Bay leaves
- 1 Chopped or finely grated onion
- 2 Smashed garlic cloves
- ½ Cup of chopped parsley or cilantro
- 3 to 4 crushed dried or hot chilies
- 2 Tablespoons of cracked black peppercorns

Directions:

1. Start by gutting and washing the sardines; then remove the backbone and the ribs
2. In making a brine, put all the ingredients above except for the sardines in a pot; then bring the mixture to a boil and turn off the heat after that
3. Stir your ingredients to combine; then cover and let come to the room temperature
4. When the brine is perfectly cool, submerge the sardines in it in a large, covered and non-reactive container.
5. Let the sardines soak in the refrigerator for about 12 hours or for an overnight
6. Take the sardines out of the brine; then rinse under the cold water quickly and pat dry.
7. Let dry over a rack in a cool place for about 30 to 60 minutes
8. Be sure to turn the fish over once; then once the sardines look dry to you; place them in a smoker as far away from the heat as possible.
9. Smoke the sardines for about 4 to 5 hours over almond wood

Nutrition: Calories: 180 Fat: 10g Carbs: 0g Protein: 13g

265. Smoked Catfish

Preparation Time: 2 hours
Cooking Time: 2 hours 30 minutes
Servings: 3

Ingredients:

- 4 to 5 catfish fillets
- 1 Cup of oil
- ½ Cup of red wine vinegar
- 1 Juiced lemon
- 1Minced garlic clove
- 2 Tablespoons of oregano
- 1 Tablespoon of thyme
- 1 Tablespoon of basil
- 1 Teaspoon of black pepper
- 1 Teaspoon of cayenne pepper
- 1 Tablespoon of salt
- 3 Tablespoons of sugar
- Wood of your choice

Directions:

1. Mix the ingredients of the marinade ingredients all together.
2. Place the catfish in a shallow baking bowl or dish
3. Pour the marinade on top of the fish and turn them to ensure that the catfish are evenly coated.
4. Cover your dish with a plastic wrap and place it in the fridge for about 1 hour.
5. Start the smoker with wood of your choice and set it aside for a temperature of about 225°F
6. Place the catfish over racks and place it in the smoker.
7. Smoke the cat fish for about 2 and 1/3 hours.

Nutrition: Calories: 139 Fat: 5.37g Carbohydrates: 0g Protein: 21.5g

266. Spicy Smoked Shrimp

Preparation Time: 10 minutes
Cooking Time: 30 minutes
Servings: 4

Ingredients:

- 2 pounds of peeled and deveined shrimp
- 6 ounces of Thai chilies
- 6 Garlic cloves
- 2 Tablespoons of chicken rub of your choice
- 1 and ½ teaspoons of sugar
- 1 and ½ tablespoons of white vinegar
- 3 Tablespoons of olive oil

Directions:

1. Place all your ingredients besides the shrimp in a blender; then blend until you get a paste
2. Put the shrimp in a shallow container; then add in the chili garlic mixture; then place in the refrigerator and let marinate for about 30 minutes
3. Remove from the fridge and thread the shrimp on metal or bamboo skewers for about 30 minutes
4. Start your Wood pellet smoker to about 225° F and preheat with the lid closed for about 10 to 15 minutes
5. Place the shrimp on a grill and cook for about 2 to 3 minutes per side or until the shrimp are pink

Nutrition: Calories: 206 Fat: 10g Carbs: 10g Protein: 16g

267. Smoked Ahi Tuna

Preparation Time: 40 minutes
Cooking Time: 4 hours
Servings: 6

Ingredients:

- 6 Albacore Tuna Filets of about 8 OZ
- 1 Cup of Kosher salt
- 1 Cup of brown sugar
- The zest of 1 orange
- The zest of 1 lemon

Directions:

1. In a small bowl; combine altogether the salt with the sugar and the citrus zest.

2. Layer the fish and the brine in a container making sure that there is enough brine between each of the filet; then let the brine sit in the refrigerator for about 6 hours
3. Start your wood pellet smoker grill on smoke with the lid open for about 4 to 5 minutes
4. Leave the temperature setting on smoke and preheat for about 10 to 15 minutes
5. Remove the filets from the brine and rinse any excess; then pat dry and place over a rack in the refrigerator for about 30 to 40 minutes
6. Remove the filets from the fridge and place it on the grill grate cooking for about 3 hours
7. Increase the temperature to about 225°F and cook for about 1 additional hour
8. Remove from the grill; then serve and enjoy your dish.

Nutrition: Calories: 150 Fat: 14.1g Carbs: 0.9g Protein: 4.8g

268. Smoked Crab Legs

Preparation Time: 5 minutes
Cooking Time: 15 minutes
Servings: 5

Ingredients:

- 1 Pinch of black pepper
- ¾ Stick of butter to the room temperature
- 2 Tablespoons of chopped chives
- 1 Minced garlic clove
- 1 Sliced lemon
- 3 Lobster, tail, about 7 ounces
- 1 Pinch of kosher salt

Directions:

1. Start your Wood Pellet grill on smoke with the lid open for about 3 to 7 minutes
2. Preheat to about 350°F; then blend the butter, the chives, the minced garlic and the black pepper in a bowl
3. Sealed it with a plastic wrapper and set aside
4. Blend the butter, the chives, the minced garlic, and the black pepper in a bowl; then cover with a plastic wrap and set it aside.
5. Butterfly the lobster tails into the middle of the soft part of the underside of the shell and don't cut completely through the center of the meat
6. Rub some olive oil and season it with 1 pinch of salt
7. Smoke Grill the lobsters with the cut side down for about 5 minutes
8. Flip the tails and top with 1 tablespoon of herbed butter; then grill for an additional 4 minutes
9. Remove from the smoker grill and serve with more quantity of herb butter
10. Top with lemon wedges; then serve and enjoy your dish.

Nutrition: Calories: 90 Fat: 1g Carbohydrates: 0g Protein: 20g

269. Cedar Plank Salmon

Preparation Time: 2 Hours And 10 Minutes
Cooking Time: 2 Hours And 45 Minutes
Servings: 4

Ingredients:

- 2 pounds salmon fillets
- ¼ cup brown sugar
- ¼ cup soy sauce
- 3 tablespoons apple cider vinegar
- ¼ cup red wine
- ¼ cup sweet Thai chili sauce
- Cedar plank as needed

Directions:

1. Place sugar in a small bowl, add soy sauce, vinegar, and red wine and stir well until combined.
2. Place salmon fillets in a large plastic bag, pour in brown sugar mixture, then seal the bag, turn it upside down to coat salmon with the mixture and marinate in the refrigerator for 2 hours.
3. In the meantime, soak the cider plank in the water.
4. When ready to cook, open hopper of the smoker, add dry pallets, make sure ash-can is in place, then open the ash damper, power on the smoker and close the ash damper.
5. Set the temperature of the smoker to20 degrees F, let preheat for 30 minutes or until the green light on the dial blinks that indicate smoker has reached to set temperature.
6. Meanwhile, pat dries the plant, then arrange marinated salmon fillets on it and place on the smoker grill.
7. Shut the smoker with lid, smoke for 1 hour, then baste salmon fillets with half of the chili sauce and continue smoking for 1 hour.
8. Set to increase the temperature of the smoker to 350 degrees F, baste fillets with remaining chili sauce, flip the fillets and continue smoking the fillets until the internal temperature of the salmon reaches to 130 degrees F. Serve straight away.

Nutrition: Calories: 370 Fat: 13.1 g Protein: 38.3 g Carbs: 20.4 g

270. Garlic Salmon

Preparation Time: 10 Minutes
Cooking Time: 55 Minutes
Servings: 4

Ingredients:

- 3 pounds salmon fillets, skin on
- 2 tablespoons minced garlic
- 1/2 tablespoons minced parsley
- 4 tablespoons seafood seasoning
- 1/4 cup olive oil

Directions:

1. Open hopper of the smoker, add dry pallets, make sure ash-can is in place, then open the ash damper, power on the smoker and close the ash damper.
2. Set the temperature of the smoker to 450 degrees F, switch smoker to open flame cooking mode, press the open flame 3, remove the grill grates and the batch, replace batch with direct flame insert, then return grates on the grill in the lower position and let preheat for

30 minutes or until the green light on the dial blinks that indicate smoker has reached to set temperature.

3. Meanwhile, take a baking sheet, line it with parchment sheet and place salmon on it, skin-side down, and then season salmon with seafood seasoning on both sides.
4. Stir together garlic, parsley, and oil until combined and then brush this mixture on the salmon fillets.
5. Place baking sheet containing salmon fillets on the smoker grill, shut with lid and smoke for 25 minutes or until the internal temperature of salmon reach to 140 degrees F.
6. When done, transfer salmon fillets to a dish, brush with more garlic-oil mixture and serve with lemon wedges.

Nutrition: Calories: 130 Fat: 6 g Protein: 13 g Carbs: 6 g

DESSERT RECIPES

271. Spicy Sausage & Cheese Balls

Preparation Time: 20 minutes
Cooking Time: 40 minutes
Servings: 4

Ingredients:

- 1lb Hot Breakfast Sausage
- 2 cups Bisquick Baking Mix
- 8 ounces Cream Cheese
- 8 ounces Extra Sharp Cheddar Cheese
- 1/4 cup Fresno Peppers
- 1 tablespoon Dried Parsley
- 1 teaspoon Killer Hogs AP Rub
- 1/2 teaspoon Onion Powder

Directions:

1. Get ready smoker or flame broil for roundabout cooking at 400⁰.
2. Blend Sausage, Baking Mix, destroyed cheddar, cream cheddar, and remaining fixings in a huge bowl until all-around fused.
3. Utilize a little scoop to parcel blend into chomp to estimate balls and roll tenderly fit as a fiddle.
4. Spot wiener and cheddar balls on a cast-iron container and cook for 15mins.
5. Present with your most loved plunging sauces.

Nutrition: Calories: 95 Carbs: 4g Fat: 7g Protein: 5g

272. White Chocolate Bread Pudding

Preparation Time: 20 minutes
Cooking Time: 1hr 15 minutes
Servings: 12

Ingredients:

- 1 loaf French bread
- 4 cups Heavy Cream
- 3 Large Eggs
- 2 cups White Sugar
- 1 package White Chocolate morsels
- ¼ cup Melted Butter
- 2 teaspoons Vanilla
- 1 teaspoon Ground Nutmeg
- 1 teaspoon Salt
- Bourbon White Chocolate Sauce
- 1 package White Chocolate morsels
- 1 cup Heavy Cream
- 2 tablespoons Melted Butter
- 2 tablespoons Bourbon
- ½ teaspoon Salt

Directions:

1. Get ready pellet smoker or any flame broil/smoker for backhanded cooking at 350⁰.
2. Tear French bread into little portions and spot in a massive bowl. Pour four cups of Heavy Cream over Bread and douse for 30mins.
3. Join eggs, sugar, softened spread, and vanilla in a medium to estimate bowl. Include a package of white chocolate pieces and a delicate blend. Season with Nutmeg and Salt.
4. Pour egg combo over the splashed French bread and blend to sign up for.
5. Pour the combination right into a properly to buttered nine X 13 to inchmeal dish and spot it at the smoker.
6. Cook for 60Secs or until bread pudding has set and the top is darker.
7. For the sauce: Melt margarine in a saucepot over medium warm temperature. Add whiskey and hold on cooking for three to 4mins until liquor vanished and margarine begins to darkish-colored.
8. Include vast cream and heat till a mild stew. Take from the warmth and consist of white chocolate pieces a bit at a time continuously blending until the complete percent has softened. Season with a hint of salt and serve over bread pudding.

Nutrition: Calories: 372 Carbs: 31g Fat: 25g Protein: 5g

273. Cheesy Jalapeño Skillet Dip

Preparation Time: 10 minutes
Cooking Time: 15 minutes
Serving: 8

Ingredients:

- 8 ounces cream cheese
- 16 ounces shredded cheese
- 1/3 cup mayonnaise
- 4 ounces diced green chilies
- 3 fresh jalapeños
- 2 teaspoons Killer Hogs AP Rub
- 2 teaspoons Mexican Style Seasoning

For the topping:

- ¼ cup Mexican Blend Shredded Cheese
- Sliced jalapeños
- Mexican Style Seasoning
- 3 tablespoons Killer Hogs AP Rub
- 2 tablespoons Chili Powder
- 2 tablespoons Paprika
- 2 teaspoons Cumin
- ½ teaspoon Granulated Onion
- ¼ teaspoon Cayenne Pepper
- ¼ teaspoon Chipotle Chili Pepper ground
- ¼ teaspoon Oregano

Directions:

1. Preheat smoker or flame broil for roundabout cooking at 350^0
2. Join fixings in a big bowl and spot in a cast to press skillet
3. Top with Mexican Blend destroyed cheddar and cuts of jalapeno's
4. Spot iron skillet on flame broil mesh and cook until cheddar hot and bubbly and the top has seared
5. Marginally about 25mins.
6. Serve warm with enormous corn chips (scoops), tortilla chips, or your preferred vegetables for plunging.

Nutrition: Calories: 150 Carbs: 22g Fat: 6g Protein: 3g

274. Cajun Turkey Club

Preparation Time: 5 Minutes
Cooking Time: 10 Minutes
Servings: 3

Ingredients:

- 1 3lbs Turkey Breast
- 1 stick Butter (melted)
- 8 ounces Chicken Broth
- 1 tablespoon Killer Hogs Hot Sauce
- 1/4 cup Malcolm's King Craw Seasoning
- 8 Pieces to Thick Sliced Bacon
- 1 cup Brown Sugar
- 1 head Green Leaf Lettuce
- 1 Tomato (sliced)
- 6 slices Toasted Bread
- ½ cup Cajun Mayo
- 1 cup Mayo
- 1 tablespoon Dijon Mustard
- 1 tablespoon Killer Hogs Sweet Fire Pickles (chopped)
- 1 tablespoon Horseradish
- ½ teaspoon Malcolm's King Craw Seasoning
- 1 teaspoon Killer Hogs Hot Sauce
- Pinch of Salt & Black Pepper to taste

Directions:

1. Get ready pellet smoker for backhanded cooking at 325⁰ utilizing your preferred wood pellets for enhancing.
2. Join dissolved margarine, chicken stock, hot sauce, and 1 Tbl-Spn of Cajun Seasoning in a blending bowl. Infuse the blend into the turkey bosom scattering the infusion destinations for even inclusion.
3. Shower the outside of the turkey bosom with a Vegetable cooking splash and season with Malcolm's King Craw Seasoning.
4. Spot the turkey bosom on the smoker and cook until the inside temperature arrives at 165⁰. Utilize a moment read thermometer to screen temp during the cooking procedure.
5. Consolidate darker sugar and 1 teaspoon of King Craw in a little bowl. Spread the bacon with the sugar blend and spot on a cooling rack.
6. Cook the bacon for 12 to 15mins or until darker. Make certain to turn the bacon part of the way through for cooking.
7. Toast the bread, cut the tomatoes dainty, and wash/dry the lettuce leaves.
8. At the point when the turkey bosom arrives at 165 take it from the flame broil and rest for 15mins. Take the netting out from around the bosom and cut into slender cuts.
9. To cause the sandwich: To slather Cajun Mayo* on the toast, stack on a few cuts of turkey bosom, lettuce, tomato, and bacon. Include another bit of toast and rehash a similar procedure. Include the top bit of toast slathered with more Cajun mayo, cut the sandwich into equal parts and appreciate.

Nutrition: Calories: 130 Carbs: 1g Fat: 4g Protein: 21g

275. Juicy Loosey Cheeseburger

Preparation Time: 10 minutes
Cooking Time: 10 minutes
Servings: 6

Ingredients:

- 2 lbs. ground beef
- 1 egg beaten
- 1 Cup dry bread crumbs
- 3 tablespoons evaporated milk
- 2 tablespoons Worcestershire sauce
- 1 tablespoons Grilla Grills All Purpose Rub
- 4 slices of cheddar cheese
- 4 buns

Directions:

1. Start by consolidating the hamburger, egg, dissipated milk, Worcestershire and focus on a bowl. Utilize your hands to blend well. Partition this blend into 4 equivalent parts. At that point take every one of the 4 sections and partition them into equal parts. Take every one of these little parts and smooth them. The objective is to have 8 equivalent level patties that you will at that point join into 4 burgers.
2. When you have your patties smoothed, place your cheddar in the center and afterward place the other patty over

this and firmly squeeze the sides to seal. You may even need to push the meat back towards the inside a piece to shape a marginally thicker patty. The patties ought to be marginally bigger than a standard burger bun as they will recoil a bit of during cooking.

3. Preheat your Kong to 300^0.
4. Keep in mind during flame broiling that you fundamentally have two meager patties, one on each side, so the cooking time ought not to have a place. You will cook these for 5 to 8mins per side—closer to 5mins on the off chance that you favor an uncommon burger or more towards 8mins in the event that you like a well to done burger.
5. At the point when you flip the burgers, take a toothpick and penetrate the focal point of the burger to permit steam to getaway. This will shield you from having a hit to out or having a visitor who gets a jaw consume from liquid cheddar as they take their first nibble.
6. Toss these on a pleasant roll and top with fixings that supplement whatever your burgers are loaded down with.

Nutrition: Calories: 300 Carbs: 33g Fat: 12g Protein: 15g

276. No Flip Burgers

Preparation Time: 30 minutes
Cooking Time: 30 minutes
Servings: 2

Ingredients:

- Ground Beef Patties
- Grilla Grills Beef Rub
- Choice of Cheese
- Choice of Toppings
- Pretzel Buns

Directions:

1. To start, you'll need to begin with freezing yet not solidified meat patties. This will help guarantee that you don't overcook your burgers. Liberally sprinkle on our Beef Rub or All to Purpose Rub and delicately knead into the two sides of the patty. As another option, you can likewise season with salt and pepper and some garlic salt.
2. Preheat your Silverbac to 250^0 Fahrenheit and cook for about 45mins. Contingent upon the thickness of your burgers you will need to keep an eye on them after around 30 to 45mins, yet there's no compelling reason to flip. For a medium to uncommon burger, we recommend cooking to about 155^0.
3. After the initial 30 to 40mins, in the event that you like liquefied cheddar on your burger feel free to mix it up. Close your barbecue back up and let them wrap up for another 10mins before evacuating. For an additional punch of flavor, finish your burger off with a sprinkle of Grilla Grill's Gold 'N Bold sauce. Appreciate.

Nutrition: Calories: 190 Carbs: 17g Fat: 9g Protein: 13g

277. Juicy Loosey Smokey Burger

Preparation Time: 30 minutes
Cooking Time: 30 minutes
Servings: 2

Ingredients:

- 1 pound Beef
- 1/3 pound per burger
- Cheddar cheese
- Grilla AP Rub
- Salt
- Freshly Ground Black Pepper
- Hamburger Bun
- BBQ Sauce

Directions:

1. Split every 1/3 pound of meat, which is 2.66 ounces per half.
2. Level out one half to roughly six inches plate. Put wrecked of American cheddar, leaving 1/2 inch clear.
3. Put another portion of the meat on top, and seal edges. Rehash for all burgers.
4. Sprinkle with Grilla AP rub, salt, and pepper flame broil seasonings.
5. Smoke at 250 for 50mins. No compelling reason to turn.
6. Apply Smokey Dokey BBQ sauce, ideally a mustard-based sauce like Grilla Gold and Bold, or Sticky Fingers Carolina Classic. Cook for an extra 10 minutes, or to favored doneness.

Nutrition: Calories: 264 Carbs: 57g Fat: 2g Protein: 4g

278. Chipotle Turkey Burgers

Preparation Time: 5 minutes
Cooking Time: 35 minutes
Servings: 4

Ingredients:

- 2 lbs. Ground Turkey
- ½ Cup Onion Chopped and Sauteed
- 3 tablespoons Fresh Chopped Cilantro
- 2 chipotle chili in adobo sauce
- Finely chopped 2 teaspoons Garlic Powder
- 2 teaspoons onion powder
- 3 tablespoons Grilla Beef Rub
- 8 slices pepper jack cheese
- 8 hamburger buns

Directions:

1. Finely hack the onion. Spot turkey, onion, cilantro, chile pepper, garlic powder, onion powder, and Grilla Beef Rub in a massive bowl. Utilize your arms to combination nicely and structure into 8 patties.
2. Preheat either your Silverbac, Kong, or Grilla to 375⁰. Cook the turkey burgers till they attain in any event 165⁰. Cook time may be directed by using the thickness of burgers, but, keep in mind approximately 45min of prepare dinner time.
3. Utilize a little, profoundly precise thermometer, as an example, a Thermopop to test the inward temp. Try not to confide inside the vibe of the patties, you have to prepare dinner them till they arrive at one hundred sixty five to keep a strategic distance from any troubles. Just before the patties are Takeled from the flame

broil, pinnacle with the cheddar and serve. On the off danger which you are hoping to keep all of the carbs out of the dish, avoid the bun and consume it like burger steak with sautéed onions and mushrooms.

4. In the occasion which you want fewer carbs but something nonetheless hand held, those are high-quality in a wrap with some farm dressing and tomato and lettuce. Our personal considered one of a kind Gold and Bold is an excellent mixing with these burgers too.

Nutrition: Calories: 376 Carbs: 26g Fat: 15g Protein: 0g

279. Bread Pudding

Preparation Time: 15 minutes

Cooking Time: 45 minutes

Servings: 4

Ingredients:

- 8 stale donuts
- 3 eggs
- 1 cup milk
- 1 cup heavy cream
- ½ cup brown sugar
- 1 teaspoon vanilla
- 1 pinch salt
- Blueberry Compote
- 1 pint blueberries
- 2/3 cup granulated sugar
- ¼ cup water
- 1 lemon
- Oat Topping
- 1 cup quick oats
- ½ cup brown sugar
- 1 teaspoon flour
- 2 to 3 tablespoons room temperature butter

Directions:

1. Warmth your Grilla Grill to 350⁰.
2. Cut your doughnuts into 6 pieces for every doughnut and put it in a safe spot. Blend your eggs, milk, cream, darker sugar, vanilla, and salt in a bowl until it's everything fused. Spot your doughnuts in a lubed 9 by 13 container at that point pour your custard blend over the doughnuts. Press down on the doughnuts to guarantee they get covered well and absorb the juices.
3. In another bowl, consolidate your oats, dark colored sugar, flour and gradually join the spread with your hand until the blend begins to cluster up like sand. When that is prepared, sprinkle it over the highest point of the bread pudding and toss it on the barbecue around 40 to 45mins until it gets decent and brilliant dark-colored.
4. While the bread pudding is preparing, place your blueberries into a skillet over medium-high warmth and begin to cook them down so the juices begin to stream. When that occurs, include your sugar and water and blend well. Diminish the warmth to drug low and let it cook down until it begins to thicken up. Right when the blend begins to thicken, pizzazz your lemon and add the get-up-and-go to the blueberry compote and afterward cut your lemon down the middle and squeeze it into the blend. What you're left with is a tasty, splendid compote

that is ideal for the sweetness of the bread pudding.

5. Watch out for your bread pudding around the 40 to 50mins mark. The blend will, in any case, shake a piece in the middle however will solidify as it stands once you pull it off. You can pull it early on the off chance that you like your bread pudding more sodden however to me, the ideal bread pudding will be more dim with some caramelization yet will at present have dampness too!
6. Presently this is the point at which I'd snatch an attractive bowl, toss a pleasant aiding of bread pudding in there then top it off with the compote and a stacking scoop of vanilla bean frozen yogurt at that point watch faces light up. In addition to the fact that this is an amazingly beautiful dish, the flavor will take you out. Destined to be an enormous hit in your family unit. Give it a shot and express gratitude toward me.
7. What's more, as usual, ensure you snap a photo of your manifestations and label us in your dishes! We'd love to include your work.

Nutrition: Calories: 290 Carbs: 62g Fat: 4g Protein: 5g

280. Smoked Chocolate Bacon Pecan Pie

Preparation Time: 1hr 45 minutes

Cooking Time: 45 minutes

Servings: 8

Ingredients:

- 4 eggs
- 1 cup chopped pecans
- 1 tablespoon of vanilla
- ½ cup semi to sweet chocolate chips
- ½ cup dark corn syrup
- ½ cup light corn syrup
- ¾ cup bacon (crumbled)
- ¼ cup bourbon
- 4 tablespoons or ¼ cup of butter
- ½ cup brown sugar
- ½ cup white sugar
- 1 tablespoon cornstarch
- 1 package refrigerated pie dough
- 16 ounces heavy cream
- ¾ cup white sugar
- ¼ cup bacon
- 1 tablespoon vanilla

Directions:

1. Pie:
2. Carry Smoker to 350⁰.
3. Blend 4 tablespoons spread, ½ cup darker sugar, and ½ cup white sugar in blending bowl.
4. In a different bowl, blend 4 eggs and 1 tablespoon cornstarch together and add to blender.
5. Include ½ cup dull corn syrup, ½ cup light corn syrup, ¼ cup whiskey, 1 cup

slashed walnuts, 1 cup bacon, and 1 tablespoon vanilla to blend.

6. Spot pie batter in 9-inch pie skillet.
7. Daintily flour mixture.
8. Uniformly place ½ cup chocolate contributes pie dish.
9. Take blend into the pie dish.
10. Smoke at 350⁰ for 40mins or until the focus is firm.
11. Cool and top with bacon whipped cream.
12. Bacon whipped Cream:
13. Consolidate fixings (16 ounces substantial cream, ¾ cup white sugar, ¼ cup bacon to finely cleaved, and 1 tablespoon vanilla) and mix at rapid until blend thickens. This formula can be separated into 6mins pie container or custard dishes or filled in as one entire pie.

Nutrition: Calories: 200 Carbs: 18g Fat: 0g Protein: 3g

281. Bacon Sweet Potato Pie

Preparation Time: 15 minutes

Cooking Time: 50 minutes

Servings: 8

Ingredients:

- 1 pound 3 ounces sweet potatoes
- 1 ¼ cups plain yogurt
- ¾ cup packed, dark brown sugar
- ½ teaspoon of cinnamon
- ¼ teaspoon of nutmeg
- 5 egg yolks
- ¼ teaspoon of salt
- 1 (up to 9 inch) deep dish, frozen pie shell
- 1 cup chopped pecans, toasted
- 4 strips of bacon, cooked and diced
- 1 tablespoon maple syrup
- Optional: Whipped topping

Directions:

1. In the first region, 3D shapes the potatoes right into a steamer crate and sees into a good-sized pot of stew water. Ensure the water is not any nearer than creeps from the base of the bushel. When steamed for 20mins, pound with a potato masher and installed a safe spot.
2. While your flame broil is preheating, location the sweet potatoes within the bowl of a stand blender and beat with the oar connection.
3. Include yogurt, dark colored sugar, cinnamon, nutmeg, yolks, and salt, to flavor, and beat until very a whole lot joined. Take this hitter into the pie shell and see onto a sheet dish. Sprinkle walnuts and bacon on pinnacle and bathe with maple syrup.
4. Heat for 45 to 60mins or until the custard arrives at 165 to 180⁰. Take out from fish fry and funky. Keep refrigerated within the wake of cooling.

Nutrition: Calories: 270 Carbs: 39g Fat: 12g Protein: 4g

282. Grilled Fruit with Cream

Preparation Time: 15 minutes

Cooking Time: 10 minutes

Servings: 6

Ingredients:

- 2 halved Apricot
- 1 halved Nectarine
- 2 halved peaches
- ¼ cup of Blueberries
- ½ cup of Raspberries
- 2 tablespoons of Honey
- 1 orange, the peel
- 2 cups of Cream
- ½ cup of Balsamic Vinegar

Directions:

1. Preheat the grill to 400F with closed lid.
2. Grill the peaches, nectarines and apricots for 4 minutes on each side.
3. Place a pan over the stove and turn on medium heat. Add 2 tablespoons of honey, vinegar, and orange peel. Simmer until medium thick.
4. In the meantime add honey and cream in a bowl. Whip until it reaches a soft form.
5. Place the fruits on a serving plate. Sprinkle with berries. Drizzle with balsamic reduction. Serve with cream and enjoy!

Nutrition: Calories: 230 Protein: 3g Carbs: 35g Fat: 3g

283. Apple Pie on the Grill

Preparation Time: 15 minutes

Cooking Time: 30 minutes

Servings: 6

Ingredients:

- ¼ cup of Sugar
- 4 Apples, sliced
- 1 tablespoon of Cornstarch
- 1 teaspoon Cinnamon, ground
- 1 Pie Crust, refrigerated, soften in according to the directions on the box
- ½ cup of Peach preserves

Directions:

1. Preheat the grill to 375F with closed lid.
2. In a bowl combine the cinnamon, cornstarch, sugar, and apples. Set aside.
3. Place the piecrust in a pie pan. Spread the preserves and then place the apples. Fold the crust slightly.
4. Place a pan on the grill (upside - down) so that you don't brill/bake the pie directly on the heat.
5. Cook 30 - 40 minutes. Once done, set aside to rest. Serve and enjoy

Nutrition: Calories: 160 Protein: 0.5g Carbs: 35g Fat: 1g

284. Grilled Layered Cake

Preparation Time: 10 minutes

Cooking Time: 20 minutes

Servings: 6

Ingredients:

- 2 x pound cake
- 3 cups of whipped cream
- ¼ cup melted butter
- 1 cup of blueberries
- 1 cup of raspberries
- 1 cup sliced strawberries

Directions:

1. Preheat the grill to high with closed lid.
2. Slice the cake loaf (3/4 inch), about 10 per loaf. Brush both sides with butter.
3. Grill for 7 minutes on each side. Set aside.
4. Once cooled completely start layering your cake. Place cake, berries then cream.
5. Sprinkle with berries and serve.

Nutrition: Calories: 160 Protein: 2.3g Carbs: 22g Fat: 6g

285. Coconut Chocolate Simple Brownies

Preparation Time: 15 minutes
Cooking Time: 25 minutes
Servings: 6

Ingredients:

- 4 eggs
- 1 cup Cane Sugar
- ¾ cup of Coconut oil
- 4 ounces chocolate, chopped
- ½ teaspoon of Sea salt
- ¼ cup cocoa powder, unsweetened
- ½ cup flour
- 4 ounces Chocolate chips
- 1 teaspoon of Vanilla

Directions:

1. Preheat the grill to 350F with closed lid.
2. Take a baking pan (9x9), grease it and line a parchment paper.
3. In a bowl combine the salt, cocoa powder and flour. Stir and set aside.
4. In the microwave or double boiler melt the coconut oil and chopped chocolate. Let it cool a bit.
5. Add the vanilla, eggs, and sugar. Whisk to combine.
6. Add into the flour, and add chocolate chips. Pour the mixture into a pan.
7. Place the pan on the grate. Bake for 20 minutes. If you want dryer brownies to bake for 5 - 10 minutes more.
8. Let them cool before cutting.
9. Cut the brownies into squares and serve.

Nutrition: Calories: 135 Protein: 2g Carbs: 16g Fat: 3g

286. Seasonal Fruit on the Grill

Preparation Time: 5 minutes
Cooking Time: 10 minutes
Servings: 4

Ingredients:

- 2 plums, peaches apricots, etc. (choose seasonally)
- 3 tablespoons Sugar, turbinate
- ¼ cup of Honey
- Gelato, as desired

Directions:

1. Preheat the grill to 450F with closed lid.
2. Slice each fruit in halves and remove pits. Brush with honey. Sprinkle with some sugar.
3. Grill on the grate until you see that there are grill marks. Set aside.
4. Serve each with a scoop of gelato. Enjoy.

Nutrition: Calories: 120 Protein: 1g Carbs: 15g Fat: 3g

287. Bacon Chocolate Chip Cookies

Preparation Time: 30 minutes

Cooking Time: 30 minutes

Servings: 6

Ingredients:

- 8 slices cooked and crumbled bacon
- 2 ½ teaspoon apple cider vinegar
- 1 teaspoon vanilla
- 2 cup semisweet chocolate chips
- 2 room temp eggs
- 1 ½ teaspoon baking soda
- 1 cup granulated sugar
- ½ teaspoon salt
- 2 ¾ cup all-purpose flour
- 1 cup light brown sugar
- 1 ½ stick softened butter

Directions:

1. Mix salt, baking soda and flour.
2. Cream the sugar and the butter together. Lower the speed. Add in the eggs, vinegar, and vanilla.
3. Put it on low fire, slowly add in the flour mixture, bacon pieces, and chocolate chips.
4. Add Preferred Wood Pellet pellets to your smoker and follow your cooker's startup procedure. Preheat your smoker, with your lid closed, until it reaches 375.
5. Put a parchment paper on a baking sheet you are using and drop a teaspoonful of cookie batter on the baking sheet. Let them cook on the grill, covered, for approximately 12 minutes or until they are browned.

Nutrition: Calories: 167 Carbs: 21g Fat: 9g Protein: 2g

288. Chocolate Chip Cookies

Preparation Time: 30 minutes
Cooking Time: 30 minutes
Servings: 8

Ingredients:

- 1 ½ cup chopped walnuts
- 1 teaspoon vanilla
- 2 cup chocolate chips
- 1 teaspoon baking soda
- 2 ½ cup plain flour
- ½ teaspoon salt
- 1 ½ stick softened butter
- 2 eggs
- 1 cup brown sugar
- ½ cup sugar

Directions:

1. Add Preferred Wood Pellet pellets to your smoker and follow your cooker's startup procedure. Preheat your smoker, with your lid closed, until it reaches 350.
2. Mix the baking soda, salt, and flour.
3. Cream the brown sugar, sugar, and butter. Mix in the vanilla and eggs until it comes together.
4. Slowly add in the flour while continuing to beat. Once all flour has been incorporated, add in the chocolate chips and walnuts. Using a spoon, fold into batter.
5. Place an aluminum foil onto grill. In an aluminum foil, drop spoonful of dough and bake for 17 minutes.

Nutrition: Calories: 150 Carbs: 18g Fat: 5g Protein: 10g

289. Apple Cobbler

Preparation Time: 30 minutes
Cooking Time: 1 hour 50 minutes
Servings: 8

Ingredients:

- 8 Granny Smith apples
- 1 cup sugar
- 1 stick melted butter
- 1 teaspoon cinnamon
- Pinch salt
- ½ cup brown sugar
- 2 eggs
- 2 teaspoons baking powder
- 2 cup plain flour
- 1 ½ cup sugar

Directions:

1. Peel and quarter apples, place into a bowl. Add in the cinnamon and one c. sugar. Stir well to coat and let it set for one hour.
2. Add Preferred Wood Pellet pellets to your smoker and follow your cooker's startup procedure. Preheat your smoker, with your lid closed, until it reaches 350.
3. In a large bowl add the salt, baking powder, eggs, brown sugar, sugar, and flour. Mix until it forms crumbles.
4. Place apples into a Dutch oven. Add the crumble mixture on top and drizzle with melted butter.
5. Place on the grill and cook for 50 minutes.

Nutrition: Calories: 152 Carbs: 26g Fat: 5g Protein: 1g

290. Caramel Bananas

Preparation Time: 15 minutes.
Cooking Time: 15 minutes.
Servings: 4

Ingredients:

- 1/3 cup chopped pecans
- ½ cup sweetened condensed milk
- 4 slightly green bananas
- ½ cup brown sugar
- 2 tablespoons corn syrup
- ½ cup butter

Directions:

1. Add pellet to your smoker and follow your cooker's startup procedure. Preheat your smoker, with the lid closed, until it reaches 350.
2. Place the milk, corn syrup, butter, and brown sugar into a heavy saucepan and bring to boil. For five minutes simmer the mixture in low heat. Stir frequently.
3. Place the bananas with their peels on, on the grill and let them grill for five minutes. Flip and cook for five minutes more. Peels will be dark and might split.
4. Place on serving platter. Cut the ends off the bananas and split peel down the middle. Take the peel off the bananas and spoon caramel on top. Sprinkle with pecans.

Nutrition: Calories: 152 Carbs: 36g Fat: 1g Protein: 1g

291. Exotic Apple Pie

Preparation Time: 20 minutes
Cooking Time: 1 hour 30 minutes
Servings: 4

Ingredients:

- 3 Apples (large, thinly sliced)
- 1/3 Cup of Sugar
- 1 Tablespoon of Flour
- 1/4 Teaspoon of Cinnamon (ground)
- 1 Tablespoon of Lemon juice
- Pinch of Nutmeg (ground)
- Pinch of salt
- Homemade pie or box of pie dough

Directions:

1. Pepping for the Grill
2. Mix apples, flour, and sugar, cinnamon, nutmeg, salt, and lemon juice in a bowl thoroughly
3. Cut the dough into two
4. Put one half of the dough into the 10" pie plate and press firmly with your hand
5. Pour the apple mix into the dough and cover it with the other half
6. Use your hand again to crimp the edges of the pie together
7. Use knife cut the top of the dough
8. Pepping on the Pellet Smoker
9. Set the Smoker grill to indirect cooking and preheat to 425°F
10. Transfer to the smoker and bake, and then cover the edges of the pie with foil to avoid burn
11. Bake the dough until it turns golden brown approximately 45 minutes
12. Remove and allow cooling for 1 hour
13. Slice, serve and enjoy

292. Cinnamon Sugar Pumpkin Seeds

Preparation Time: 15 minutes
Cooking Time: 30 minutes
Servings: 8

Ingredients:

- 2 tablespoons sugar
- Seeds from a pumpkin
- 1 teaspoon cinnamon
- 2 tablespoons melted butter

Directions:

1. Add wood pellets to your smoker and follow your cooker's startup procedure. Preheat your smoker, with your lid closed, until it reaches 350.
2. Clean the seeds and toss them in the melted butter. Add them to the sugar and cinnamon. Spread them out on a baking sheet, place on the grill, and smoke for 25 minutes. Serve.

Nutrition: Calories: 127 Protein: 5g Carbs: 15g Fat: 21g

293. Blackberry Pie

Preparation Time: 15 minutes
Cooking Time: 40 minutes
Servings: 8

Ingredients:

- Butter, for greasing
- ½ cup all-purpose flour
- ½ cup milk
- 2 pints blackberries
- 2 cup sugar, divided
- 1 box refrigerated piecrusts
- 1 stick melted butter
- 1 stick of butter
- Vanilla ice cream

Directions:

1. Add wood pellets to your smoker and follow your cooker's startup procedure. Preheat your smoker, with your lid closed, until it reaches 375.
2. Butter a cast iron skillet.
3. Unroll a piecrust and lay it in the bottom and up the sides of the skillet. Use a fork to poke holes in the crust.
4. Lay the skillet on the grill and smoke for five mins, or until the crust is browned. Set off the grill.
5. Mix together 1 ½ c. of sugar, the flour, and the melted butter together. Add in the blackberries and toss everything together.
6. The berry mixture should be added to the skillet. The milk should be added on the top afterward. Sprinkle on half of the diced butter.
7. Unroll the second pie crust and lay it over the skillet. You can also slice it into strips and weave it on top to make it look like a lattice. Place the rest of the diced butter over the top. Sprinkle the rest of the sugar over the crust and place it skillet back on the grill.
8. Lower the lid and smoke for 15 to 20 minutes or until it is browned and bubbly. You may want to cover with some foil to keep it from burning during the last few minutes of cooking. Serve the hot pie with some vanilla ice cream.

Nutrition: Calories: 393 Protein: 4.25g Carbs: 53.67g Fat: 18.75g

294. S'mores Dip

Preparation Time: 10 minutes
Cooking Time: 25 minutes
Servings: 8

Ingredients:

- 12 ounces semisweet chocolate chips
- 1/4 cup milk
- 2 tablespoons melted salted butter
- 16 ounces marshmallows
- Apple wedges
- Graham crackers

Directions:

1. Add wood pellets to your smoker and follow your cooker's startup procedure. Preheat your smoker, with your lid closed, until it reaches 450.
2. Put a cast iron skillet on your grill and add in the milk and melted butter. Stir together for a minute.
3. Once it has heated up, top with the chocolate chips, making sure it makes a single layer. Place the marshmallows on top, standing them on their end and covering the chocolate.
4. Cover, and let it smoke for five to seven minutes. The marshmallows should be toasted lightly.
5. Take the skillet off the heat and serve with apple wedges and graham crackers.

Nutrition: Calories: 216.7 Protein: 2.7g Carbs: 41g Fat: 4.7g

295. Ice Cream Bread

Preparation Time: 10 minutes

Cooking Time: 1 hour

Servings: 6

Ingredients:

- 1 1/2 quart full-fat butter pecan ice cream, softened
- 1 teaspoon salt
- 2 cups semisweet chocolate chips
- 1 cup sugar
- 1 stick melted butter
- Butter, for greasing
- 4 cups self-rising flour

Directions:

1. Add wood pellets to your smoker and follow your cooker's startup procedure. Preheat your smoker, with your lid closed, until it reaches 350.
2. Mix together the salt, sugar, flour, and ice cream with an electric mixer set to medium for two minutes.
3. As the mixer is still running, add in the chocolate chips, beating until everything is blended.
4. Spray a Bundt pan or tube pan with cooking spray. If you choose to use a pan that is solid, the center will take too long to cook. That's why a tube or Bundt pan works best.
5. Add the batter to your prepared pan.
6. Set the cake on the grill, cover, and smoke for 50 minutes to an hour. A toothpick should come out clean.
7. Take the pan off of the grill. For 10 minutes cool the bread. Remove

carefully the bread from the pan and then drizzle it with some melted butter.

Nutrition: Calories: 148.7 Protein: 3.5g Carbs: 27g Fat: 3g

RUBS, SAUCES, MARINADES, AND GLAZES

296. North Carolina Barbecue Sauce

Ingredients:

- 1 qt cider vinegar
- 12 oz ketchup
- 2/3 C packed brown sugar
- 2 Tbsp. salt
- ¼ C lemon juice
- 1 Tbsp. red pepper flakes
- 1 Tbsp. smoked paprika
- 1 Tbsp. onion powder
- 1 tsp each: black pepper, dry mustard

Directions:

- Bring all ingredients to the boil, and then simmer for 30-45 minutes, stirring frequently.
- Allow to cool and serve or bottle.

297. Memphis-Style Barbecue Sauce

Ingredients:

- 1 Tbsp. butter
- ¼ C finely chopped onion
- 1 ½ C ketchup
- ¼ C chili sauce
- 4 Tbsp. brown sugar
- 4 Tbsp. molasses
- 2 Tbsp. yellow mustard
- 1 Tbsp. fresh lemon juice
- 1 Tbsp. Worcestershire sauce
- 1 Tbsp. liquid hickory smoke
- ½ tsp garlic powder
- ½ tsp salt
- ½ tsp ground black pepper
- 1 tsp chili powder
- dash cayenne pepper

Directions:

- Bring all ingredients to the boil, and then simmer for 30-45 minutes, stirring frequently.
- Allow to cool and serve or bottle.

298. Texas Brisket Sauce

Ingredients:

- ½ C brisket drippings (defatted)
- ½ C vinegar
- 1 Tbsp. Worcestershire sauce
- ½ C ketchup
- ½ tsp hot pepper sauce (Franks)
- 1 lg onion, diced
- 2 cloves of garlic, pressed
- 1 Tbsp. salt
- ½ tsp chili powder
- Juice of one lemon

Directions:

- Combine all ingredients.
- Simmer, stirring occasionally, for 15 minutes.
- Allow to cool and refrigerate 24-48 hours before using.

299. Sweet Hawaiian Pork Sauce

Ingredients:

- 15oz peaches & juice
- 15oz pineapple & juice
- 16oz peach preserves
- 1 cup brown sugar
- 2 Tbsp. liquid smoke
- 2 Tbsp. minced garlic
- 1 Tbsp. red pepper flakes

Directions:

- Combine all and bring to boil.
- Lower heat and simmer on low until sauce has begun to thicken.
- Keep warm until serving. Drizzle over pulled pork.

300. Spicy Thai Peanut Sauce

Ingredients:

- 3 C creamy peanut butter
- 3/4 C coconut milk
- 1/3 C fresh lime juice
- 1/3 C soy sauce
- 1 Tbsp. fish sauce
- 1 Tbsp. hot sauce
- 1 Tbsp. minced fresh ginger root
- 5 cloves garlic. minced

Directions:

- In a bowl, mix the peanut butter, coconut milk, lime juice, soy sauce, fish sauce, hot sauce, ginger, and garlic.
- Simmer 10 minutes, cool and serve.

301. Garlic Mojo

Ingredients:

- 8 garlic cloves
- 1 tsp salt
- 1/4 C sweet orange juice
- 1/8 C of fresh lime or lemon juice.
- 1 Habanero pepper, diced (optional)

Directions:

- Chop garlic fine with salt, or crush using a mortar and pestle or food processor with salt to form a thick paste.
- Wearing gloves, carefully core Habanero pepper and wash out all seeds and membranes. Dice pepper, set aside. Wash prep area, dispose of gloves and wash your hands with dish soap.
- In a mixing bowl, combine the garlic paste, pepper, and juice, and let the mixture sit at room temperature for 30 minutes or longer.

302. Gorgonzola Dipping Sauce

Ingredients:

- 1 C crumbled blue cheese
- 2/3 C sour cream
- ½ C mayonnaise
- 1 clove garlic, minced
- 1 oz white wine
- 2 tsp Worcestershire sauce

- 1 tsp salt
- 1 tsp fresh ground black pepper

Directions:

- In a glass or plastic bowl, combine all ingredients, using the salt and pepper to finalize the taste and the white wine to set the consistency.

303. "Burning' Love" Rub

Ingredients:

- ¼ C coarse sea salt
- ¼ C light brown sugar
- 2 Tbsp. garlic powder
- 2 Tbsp. onion powder
- 2 Tbsp. Italian seasonings
- 4 Tbsp. smoked paprika
- 2 Tbsp. course black pepper
- 1 Tbsp. hickory salt
- 1 tsp cayenne powder

Directions:

- Combine and mix well.
- Good for 6-8lbs. of pork.

304. Brisket Rub

Ingredients:

- (For 4 full briskets 7-8lbs. each)
- 1 C fine sea salt
- 1 C coarse pepper
- 1 C granulated garlic
- 1/4 C smoked paprika

Directions:

- Rub briskets and refrigerate 12-24 hours.
- Allow briskets to come to room temp before smoking.
- Smoke brisket(s) with a combination of oak and pecan wood pellets, at a temp between 225-25
- The difference between good brisket and amazing brisket is patience.
- Double wrap the finishing brisket in foil, wrap that in a towel, and let the whole thing rest in a closed cooler for 1-2 hours.
- Then, once you've unwrapped it, allow it to sit and cool slightly for 15-20 minutes for slicing or pulling.

305. Smokey Beef Rib Rub

Ingredients:

- 2 Tbsp. brown sugar
- 2 Tbsp. black pepper
- 2 Tbsp. smoked paprika
- 2 Tbsp. chili powder
- 2 tsp onion salt
- 2 tsp garlic powder
- 2 tsp celery salt
- 2 tsp seasoning salt

Directions:

- Mix well and rub both sides of ribs, wrap tightly in plastic wrap, and refrigerate overnight.
- Bring ribs to room temperature before cooking.

306. Hellfire Cajun Rub

Ingredients:

- 8 Tbsp. smoked paprika
- 4 Tbsp. cayenne powder
- 4 Tbsp. dried parsley
- 4 Tbsp. black pepper
- 2 Tbsp. garlic powder
- 6 Tbsp. fine sea salt
- 2 Tbsp. ground cumin
- 4 Tbsp. dried oregano
- 1 tsp ghost chili powder (to taste)

Directions:

- Combine all the ingredients, mix well and store 24-48 hours, in an airtight container, before using.
- Note: Wear gloves, and use extreme caution, when handling ghost chili powder, even breathing the tiniest amount will be painful.
- This chili has been measured at over 1 million Scoville units (by comparison, Jalapeno peppers are about 4500 Scoville units.)
- This is the hottest Chili Powder available anywhere.
- Start with just a teaspoon...trust me. ;)

307. Carolina Basting Mop

Ingredients:

- 2 qtrs. Water
- 2 qtrs. Apple Cider Vinegar
- 2 qtrs. vegetable oil
- 1 C liquid smoke
- ½ C salt
- ¼ C cayenne pepper
- ¼ C black pepper
- 1 sweet onion, diced fine

Directions:

- Combine all ingredients and bring to a simmer.
- Allow to cool overnight, and warm before using.
- Use as a rib/chicken baste, or sprinkle on pulled or chopped pork before serving.

308. Basic Vinegar Mop

Ingredients:

- 2 C cider vinegar
- ½ C vegetable oil
- 5 tsp salt
- 4 tsp red pepper flakes or powder

Directions:

- Combine all ingredients and bring to a simmer, allow to cool overnight to help the flavors marry.
- Keep warm and apply to meat before you close your pellet grill/smoker, when you flip the meat, and again when the meat is done cooking.
- Allow the meat to rest at least 30 minutes to soak up the mop.

309. Perry's Pig Picking' Mop

Ingredients:

- 1 qt. apple juice
- 1 qt. apple cider vinegar
- ¼ C fine sea salt
- ¼ C garlic powder
- ¼ C smoked paprika
- 1 C light oil
- 1 tsp black pepper
- 1 tsp cayenne pepper

Directions:

- Simmer for 15-20 minutes.
- Keep warm and apply to pig before you close your pellet smoker, when you flip the pig, and again when the pig is done cooking.
- For a more traditional "Eastern" North Carolina mop, use only the apple juice, vinegar, salt, and cayenne. For South Carolina, add 1 cup prepared mustard to that.

310. Beef Rib Mop

Ingredients:

- 3/4 C brown sugar
- 1/2 C bottled barbecue sauce
- 1/2 C ketchup
- 1/2 C cider vinegar
- 1/2 C Worcestershire sauce
- 1 C water
- 1 Tbsp. salt
- 1 Tbsp. chili powder
- 1 Tbsp. paprika

Directions:

- Combine all ingredients in a quart jar. Shake to blend thoroughly.
- Best if made ahead of time; will keep indefinitely in the refrigerator.
- This mop is great for brisket, as well. Keep warm and apply to ribs before you close your pellet smoker, when you flip the ribs, and again when the ribs are done cooking.

311. Traditional Cuban Mojo

Ingredients:

- 1 C sour orange juice
- 1 Tbsp. oregano
- 1 Tbsp. bay leaves
- 1 garlic bulb
- 1 tsp cumin
- 3 tsp salt
- 4 oz of water

Directions:

- Peel and mash the garlic cloves. Mix all the ingredients and let it sit for a minimum of one hour.
- For marinade, add the above recipe to 1 ½ gallons of water, and 13 oz. of table salt.
- Blend all ingredients and let it sit for a minimum of one hour, strain and inject, or place meat in a cooler and pour marinade to cover overnight.
- You can replace the sour orange juice with the following mix: 6 oz. orange juice, 2 oz. lemon juice.

312. Hawaiian Mojo

Ingredients:

- 1 C orange juice
- 1 C pineapple juice
- ½ C mesquite liquid smoke
- 1 Tbsp. oregano
- 1 Tbsp. minced garlic
- 1 tsp cumin
- 3 tsp salt
- 4 oz. of water

Directions:

- Mix all the ingredients and let it sit for a minimum of one hour.
- For marinade/injection, add the above recipe to 1 ½ gallons of water, and 13 oz. of table salt.
- Blend all ingredients and let it sit for a minimum of one hour, strain and inject, or place meat in a cooler and pour marinade to cover.
- Allow to marinate overnight.
- After injecting/soaking the pig or shoulder, pat dry with paper towels and apply a salt rub all over the meat, use Kosher salt or coarse sea salt.

313. BBQ Smoke Rubs

Preparation Time: 10 minutes
Cooking Time: 2 Hours
Servings: 4

Ingredients:

- 6 tablespoons brown sugar
- 2 tablespoons paprika
- 1 tablespoon salt
- 1 tablespoon ground black pepper
- 2 teaspoons garlic powder

Directions:

1. First, preheat the smoker grill at 225 degrees Fahrenheit, by closing the lid.
2. Mild wood chips can be used to create the smoke until the temperature is 100 degrees Fahrenheit.
3. Take a bowl and combine all the listed ingredients.
4. Transfer the bowl spices into an aluminum pie pan and place the pie pan directly onto the smoker grill grate and close the lid.
5. Smoke the spices for 2 hours.
6. Once done store in an air-tight glass jars for further use.

Nutrition: Calories: 70 Carbs: 16g Fat: 0g Protein: 0g

314. Queen Spice Rub

Preparation Time: 10 minutes
Cooking Time: 60 Minutes
Servings: 4

Ingredients:

- 1 tablespoon salt
- 6 teaspoons ground cayenne pepper
- 6 teaspoons ground white pepper
- 2 teaspoons ground black pepper
- 4 teaspoons paprika
- 5 teaspoons onion powder
- 2 teaspoons garlic powder

Directions:

1. Preheat the smoker grill at 220 degrees Fahrenheit by closing the lid.
2. You can use apple wood chip to create the smoke.
3. The internal temperature should be 100 degrees Fahrenheit, to smoke the spices.
4. Take a bowl and combine all the listed ingredients.
5. Transfer all the spices into aluminum pipe and place it on the grate.
6. Close the lid of the smoker and let it smoke for 1 hour.
7. Afterward, remove the foil tin from the grill and store in a tight jar for further use.

Nutrition: Calories: 24 Carbs: 5g Fat: 1g Protein: 1g

315. Seafood Seasoning

Preparation Time: 15 minutes
Cooking Time: 120 Minutes
Servings: 4

Ingredients:

- 4 teaspoons paprika
- 2 teaspoons cinnamon
- 2 teaspoons ground ginger
- 3 teaspoons ground cumin
- 4 teaspoons ground coriander
- 3 teaspoons dried lemon peel
- 4 teaspoons onion powder
- 1 teaspoon lemon pepper
- 3 teaspoons dried parsley
- 3 teaspoons dried cilantro
- 3 teaspoons garlic powder

Directions:

1. Take a bowl, and combine all the listed ingredients in it.
2. Preheat the smoker grill for few minutes at 220 degrees F.
3. The smoker will be preheated until the internal temperature reaches 100 degrees F.
4. Transfer the bowl spices to the aluminum pie pan.
5. Put the pie pan on grill grate and smoke for 2 hours.
6. Once done, let it get cool and then store in the tight jar.

Nutrition: Calories: 50 Carbs: 12g Fat: 0g Protein: 1g

316. Three Pepper Rub

Preparation Time: 10 minutes
Cooking Time: 3 Hours
Servings: 3

Ingredients:

- 2 tablespoons of black pepper
- 2 tablespoons of white pepper
- 2 tablespoons of red pepper
- 1 tablespoon of onion powder
- 2 teaspoons of garlic powder
- 2 tablespoons of dried thyme
- 4 tablespoons of paprika
- 2 tablespoons of dried oregano

Directions:

1. Mix all the spices in the bowl and transfer to aluminum foil tin.
2. Preheat the smoker grill at 220 degrees F for 20 minutes.
3. Put the aluminum foil tin onto the grill grate and smoke for 3 hours by closing the lid.
4. Once done, store it in the tight jar for further use.

Nutrition: Calories: 67 Carbs: 16g Fat: 2g Protein: 2g

317. Jerky Seasoning

Preparation Time: 10 minutes
Cooking Time: 2 Hours
Servings: 4

Ingredients:

- 8 tablespoons dried minced onion
- 6 teaspoons dried thyme
- 4 teaspoons ground allspice
- 2 teaspoons ground black pepper
- 4 teaspoons ground cinnamon
- 4 teaspoons cayenne pepper
- 2 teaspoons sea salt

Directions:

1. Mix all the spices in the bowl, and transfer to aluminum foil tin.
2. Preheat the smoker grill to 220 degrees F, by closing the lid for 22 minutes.
3. Place the aluminum foil pan onto the smoker grill grate, and let it smoke for 2 hours.
4. The spices will be smoked to perfection until now.
5. Let the spices get cool down before storing in the airtight jars.

Nutrition: Calories: 63 Carbs: 0g Fat: 1g Protein: 13g

318. Not-Just-For-Pork Rub

Preparation Time: 5 minutes
Cooking Time: 0 minute
Servings: 4

Ingredients:

- ½ teaspoon ground thyme
- ½ teaspoon paprika
- ½ teaspoon coarse kosher salt
- ½ teaspoon garlic powder
- ½ teaspoon onion powder
- ½ teaspoon chili powder
- ¼ teaspoon dried oregano leaves
- ¼ teaspoon freshly ground black pepper
- ¼ teaspoon ground chipotle chili pepper
- ¼ teaspoon celery seed

Directions:

1. Using an airtight bag, combine the thyme, paprika, salt, garlic powder, onion powder, chili powder, oregano, black pepper, chipotle pepper, and celery seed. Close the container and shake to mix. Unused rub will keep in an airtight container for months.

Nutrition: Calories: 64 Carbs: 10g Fat: 1g Protein: 1g

319. Chicken Rub

Preparation Time: 5 minutes
Cooking Time: 0 minute
Servings: 4

Ingredients:

- 2 tablespoons packed light brown sugar
- 1½ teaspoons coarse kosher salt
- 1¼ teaspoons garlic powder
- ½ teaspoon onion powder
- ½ teaspoon freshly ground black pepper
- ½ teaspoon ground chipotle chili pepper
- ½ teaspoon smoked paprika
- ¼ teaspoon dried oregano leaves
- ¼ teaspoon mustard powder
- ¼ teaspoon cayenne pepper

Directions:

1. Using an airtight bag, combine the brown sugar, salt, garlic powder, onion powder, black pepper, chipotle pepper, paprika, oregano, mustard, and cayenne. Close the container and shake to mix. Unused rub will keep in an airtight container for months.

Nutrition: Calories: 15 Carbs: 3g Fat: 0g Protein: 0g

320. Dill Seafood Rub

Preparation Time: 5 minutes
Cooking Time: 0 minute
Servings: 2

Ingredients:

- 2 tablespoons coarse kosher salt
- 2 tablespoons dried dill weed
- 1 tablespoon garlic powder
- 1½ teaspoons lemon pepper

Directions:

1. Using an airtight bag combine the salt, dill, garlic powder, and lemon pepper. Close the container and shake to mix. Unused rub will keep in an airtight container for months.

Nutrition: Calories: 15 Carbs: 3g Fat: 0g Protein: 0g

321. Cajun Rub

Preparation Time: 5 minutes
Cooking Time: 0 minute
Servings: 2

Ingredients:

- 1 teaspoon freshly ground black pepper
- 1 teaspoon onion powder
- 1 teaspoon coarse kosher salt
- 1 teaspoon garlic powder
- 1 teaspoon sweet paprika
- ½ teaspoon cayenne pepper
- ½ teaspoon red pepper flakes
- ½ teaspoon dried oregano leaves
- ½ teaspoon dried thyme
- ½ teaspoon smoked paprika

Directions:

1. Using an airtight bag combine the black pepper, onion powder, salt, garlic powder, sweet paprika, cayenne, red pepper flakes, oregano, thyme, and smoked paprika. Close the container and shake to mix. Unused rub will keep in an airtight container for months.

Nutrition: Calories: 23 Carbs: 2g Fat: 1g Protein: 2g

322. Espresso Brisket Rub

Preparation Time: 5 minutes
Cooking Time: 0 minute
Servings: 2

Ingredients:

- 3 tablespoons coarse kosher salt
- 2 tablespoons ground espresso coffee
- 2 tablespoons freshly ground black pepper
- 1 tablespoon garlic powder
- 1 tablespoon light brown sugar
- 1½ teaspoons dried minced onion
- 1 teaspoon ground cumin

Directions:

1. In a small airtight container or zip-top bag, combine the salt, espresso, black pepper, garlic powder, brown sugar, minced onion, and cumin. Close the container and shake to mix. Unused rub will keep in an airtight container for months.

Nutrition: Calories: 56 Carbs: 13g Fat: 1g Protein: 2g

323. Sweet Brown Sugar Rub

Preparation Time: 5 minutes
Cooking Time: 0 minute
Servings: 4

Ingredients:

- 2 tablespoons light brown sugar
- 1 teaspoon coarse kosher salt
- 1 teaspoon garlic powder
- 1 teaspoon onion powder
- 1 teaspoon sweet paprika
- ½ teaspoon freshly ground black pepper
- ½ teaspoon cayenne pepper
- ½ teaspoon dried oregano leaves
- ¼ teaspoon smoked paprika

Directions:

1. In a small airtight container or zip-top bag, combine the brown sugar, salt, garlic powder, onion powder, sweet paprika, black pepper, cayenne, oregano, and smoked paprika. Close the container and shake to mix. Unused rub will keep in an airtight container for months.

Nutrition: Calories: 17 Carbs: 5g Fat: 0g Protein: 0g

324. Sweet and Spicy Cinnamon Rub

Preparation Time: 10 minutes
Cooking Time: 0 minute
Servings: 3

Ingredients:

- 2 tablespoons light brown sugar
- 1 teaspoon coarse kosher salt
- 1 teaspoon garlic powder
- 1 teaspoon onion powder
- 1 teaspoon sweet paprika
- ½ teaspoon freshly ground black pepper
- ½ teaspoon cayenne pepper
- ½ teaspoon dried oregano leaves
- ½ teaspoon ground ginger
- ½ teaspoon ground cumin
- ¼ teaspoon smoked paprika
- ¼ teaspoon ground cinnamon
- ¼ teaspoon ground coriander
- ¼ teaspoon chili powder

Directions:

1. Using an airtight bag, combine the brown sugar, salt, garlic powder, onion powder, sweet paprika, black pepper, cayenne, oregano, ginger, cumin, smoked paprika, cinnamon, coriander, and chili powder. Close the container and shake to mix. Unused rub will keep in an airtight container for months.

Nutrition: Calories: 25 Carbs: 6g Fat: 0g Protein: 1g

325. Easy Teriyaki Marinade

Preparation Time: 5 minutes
Cooking Time: 0 minute
Servings: 2

Ingredients:

- ¼ cup water
- ¼ cup soy sauce
- ¼ cup packed light brown sugar
- ¼ cup Worcestershire sauce
- 2 garlic cloves, sliced

Directions:

1. In a container, whisk the water, soy sauce, brown sugar, Worcestershire sauce, and garlic until combined. Refrigerate any unused marinade in an airtight container for 2 or 3 days.

Nutrition: Calories: 40 Carbs: 7g Fat: 1g Protein: 1g

326. Thanksgiving Turkey Brine

Preparation Time: 5 minutes
Cooking Time: 0 minute
Servings: 4

Ingredients:

- 2 gallons water
- 2 cups coarse kosher salt
- 2 cups packed light brown sugar

Directions:

1. In a clean 5-gallon bucket, stir together the water, salt, and brown sugar until the salt and sugar dissolve completely.

Nutrition: Calories: 52 Carbs: 0g Fat: 2g Protein: 8g

327. Bill's Best BBQ Sauce

Preparation Time: 10 minutes
Cooking Time: 30 minutes
Servings: 3

Ingredients:

- 1 small onion, finely chopped
- 2 garlic cloves, finely minced
- 2 cups ketchup
- 1 cup water
- ½ cup molasses
- ½ cup apple cider vinegar
- 5 tablespoons granulated sugar
- 5 tablespoons light brown sugar
- 1 tablespoon Worcestershire sauce
- 1 tablespoon freshly squeezed lemon juice
- 2 teaspoons liquid smoke
- 1½ teaspoons freshly ground black pepper
- 1 tablespoon yellow mustard

Directions:

1. On the stovetop, in a saucepan over medium heat, combine the onion, garlic, ketchup, water, molasses, apple cider vinegar, granulated sugar, brown sugar, Worcestershire sauce, lemon juice, liquid smoke, black pepper, and mustard. Wait to boil, then reduce the heat to low and simmer for 30 minutes, straining out any bigger chunks, if desired.
2. If the sauce is cool completely then you can transfer to an airtight container and refrigerate for up to 2 weeks, or use a canning process to store for longer.

Nutrition: Calories: 60 Carbs: 13g Fat: 1g Protein: 0g

328. Chimichurri Sauce

Preparation Time: 5 minutes
Cooking Time: 0 minute
Servings: 2

Ingredients:

- ½ cup extra-virgin olive oil
- 1 bunch fresh parsley, stems removed
- 1 bunch fresh cilantro, stems removed
- 1 small red onion, chopped
- 3 tablespoons dried oregano
- 1 tablespoon minced garlic
- Juice of 1 lemon
- 2 tablespoons red wine vinegar
- 1 teaspoon salt
- 1 teaspoon freshly ground black pepper
- 1 teaspoon cayenne pepper

Directions:

1. Using a blender or processor, combine all of the ingredients and pulse several times until finely chopped.
2. The chimichurri sauce will keep in an airtight container in the refrigerator for up to 5 days.

Nutrition: Calories: 51 Carbs: 1g Fat: 5g Protein: 1g

329. Chipotle Butter

Preparation Time: 10 minutes
Cooking Time: 5 minutes
Servings: 1

Ingredients:

- 1 cup (2 sticks) salted butter
- 2 chipotle chilies in adobo sauce, finely chopped
- 2 teaspoons adobo sauce
- 2 teaspoons salt
- Juice of 1 lime

Directions:

1. On the stove top, in a small saucepan over medium heat, melt the butter. Stir in the chopped chilies, adobo sauce, salt, and lime juice, continuing to stir until the salt is dissolved, about 5 minutes. Remove from the heat.
2. Serve the chipotle butter hot or cold. It will give shelf life in an airtight container in the refrigerator for up to 2 weeks.

Nutrition: Calories: 60 Carbs: 1g Fat: 6g Protein: 0g

330. Cilantro-Balsamic Drizzle

Preparation Time: 5 minutes
Cooking Time: 0 minutes
Servings: 2

Ingredients:

- ½ cup balsamic vinegar
- ½ cup dry white wine
- ¼ cup extra-virgin olive oil
- ½ cup chopped fresh cilantro
- 2 teaspoons garlic powder
- 1 teaspoon salt
- 1 teaspoon freshly ground black pepper
- 1 teaspoon red pepper flakes
- Splash of Sriracha

Directions:

1. In a medium container, whisk together the balsamic vinegar, wine, olive oil, cilantro, garlic powder, salt, pepper, and red pepper flakes until well combined.
2. Add a dash of Sriracha and stir.
3. The best storage is an airtight container put in the refrigerator for up to 2 weeks.

Nutrition: Calories: 21 Carbs: 6g Fat: 0g Protein: 2g

331. Sweet Potato Mustard

Preparation Time: 25 minutes
Cooking Time: 20 minutes
Servings: 1

Ingredients:

- ½ cup apple cider vinegar
- ⅓ cup yellow mustard seeds
- 1 bay leaf
- 1 cup water
- 1 tablespoon molasses
- 1 tablespoon bourbon
- ⅔ Cup sweet potato purée
- ¼ cup packed brown sugar
- 2 tablespoons ground mustard
- ½ teaspoon smoked paprika
- 1 teaspoon salt
- ½ teaspoon ground cinnamon
- ½ teaspoon ground allspice
- ½ teaspoon cayenne pepper

Directions:

1. Put your saucepan on top the stove over a medium-high heat, bring the apple cider vinegar to a boil.
2. Take it from the heat and stir the mustard seeds and bay leaf, and let steep, uncovered, for 1 hour. Discard the bay leaf after steeping.
3. Pour the liquid into a food processor or blender, making sure to scrape in the mustard seeds as well. Add the water, molasses, and bourbon, and pulse until smooth.
4. Put all the mixture back into the saucepan over medium heat and stir in the sweet potato purée. Wait to boil, then reduce the heat to low and cook, stirring occasionally, for 5 minutes.
5. Whisk in the brown sugar, ground mustard, smoked paprika, salt, cinnamon, allspice, and cayenne, and simmer until thickened, about 10 minutes.
6. Take it from the heat and allow to cool completely before refrigerating.
7. The sweet potato mustard is best served cold. It will keep in an airtight container in the refrigerator for up to 2 weeks.

332. Mandarin Glaze

Preparation Time: 5 minutes
Cooking Time: 25 minutes
Servings: 2

Ingredients:

- 1 (11-ounce) can mandarin oranges, with their juices
- ½ cup ketchup
- 3 tablespoons brown sugar
- 1 tablespoon apple cider vinegar
- 1 tablespoon yellow mustard
- 1 teaspoon ground cloves
- 1 teaspoon ground cinnamon
- 1 teaspoon garlic powder
- 1 teaspoon onion powder
- 1 teaspoon salt
- 1 teaspoon freshly ground black pepper

Directions:

1. Using a blender, combine the mandarin oranges and juice, the ketchup, brown sugar, apple cider vinegar, mustard, cloves, cinnamon, garlic powder, onion powder, salt, and pepper, and pulse until the oranges are in tiny pieces.
2. Transfer all the mixture to a small saucepan on the stove top and bring to a boil over medium heat, stirring occasionally.
3. Reduce the heat to low and simmer for 15 minutes.
4. Remove from the heat and strain out the orange pieces if desired. Serve the glaze hot.
5. Keeping the glaze in an airtight container in the refrigerator for up to 5 days.

Nutrition: Calories: 35 Carbs: 9g Fat: 0g Protein: 0g

333. Jamaican Jerk Paste

Preparation Time: 10 minutes
Cooking Time:
Servings: 2

Ingredients:

- ¼ cup cane syrup
- 8 whole cloves
- 6 Scotch bonnet or habanero chilies, stemmed and seeded
- ¼ cup chopped scallions
- 2 tablespoons whole allspice (pimento) berries
- 2 tablespoons salt
- 2 teaspoons freshly ground black pepper
- 2 teaspoons ground cinnamon
- 1 teaspoon cayenne pepper
- 1 teaspoon dried thyme
- 1 teaspoon ground cumin

Directions:

1. Using a blender combine the cane syrup, cloves, chilies, scallions, allspice, salt, pepper, cinnamon, cayenne pepper, thyme, and cumin until smooth and sticky.
2. The paste will keep in an airtight container in the refrigerator for up to 1 week.

Nutrition: Calories: 65 Carbs: 8g Fat: 4g Protein: 1g

334. Our House Dry Rub

Preparation Time: 10 minutes
Cooking Time: 0 minute
Servings: 2

Ingredients:

- ¼ cup paprika
- ¼ cup turbinate sugar
- 3 tablespoons Cajun seasoning
- 1 tablespoon packed brown sugar
- 1½ teaspoons chili powder
- 1½ teaspoons cayenne pepper
- 1½ teaspoons ground cumin

Directions:

1. In a small bowl, combine the paprika, turbinate sugar, Cajun seasoning, brown sugar, chili powder, cayenne pepper, and cumin.
2. Store the rub in an airtight container at room temperature for up to a month.

Nutrition: Calories: 63 Carbs: 0g Fat: 0g Protein: 0g

335. Blueberry BBQ Sauce

Preparation Time: 5 minutes
Cooking Time: 10 minutes
Servings: 2

Ingredients:

- 2 cups water
- ½ cup minced fresh blueberries
- 1 tablespoon balsamic vinegar
- ½ cup ketchup
- 1 tablespoon Worcestershire sauce
- 1 teaspoon Sriracha
- 1 teaspoon liquid smoke
- 1 teaspoon Dijon mustard
- Salt
- Freshly ground black pepper

Directions:

1. Put your saucepan top of the stove then low heat, simmer the water, blueberries, and balsamic vinegar for 5 minutes.
2. Stir in the ketchup, Worcestershire sauce, Sriracha, liquid smoke, and Dijon mustard, season with salt and pepper, and continue simmering for 5 minutes.
3. Remove from the heat and strain out most of the blueberry pulp.
4. The barbecue sauce will keep in an airtight container in the refrigerator for up to 1 week.

Nutrition: Calories: 40 Carbs: 11g Fat: 0g Protein: 0g

336. Sweet and Spicy Jalapeño Relish

Preparation Time: 10 minutes
Cooking Time: 0 minute
Servings: 2

Ingredients:

- 6 jalapeño peppers, stemmed, seeded, and cut into pieces
- 1 Serrano chili, stemmed, seeded, and cut into pieces
- 1 red bell pepper, stemmed, seeded, and cut into pieces
- 1 cucumber, coarsely chopped
- 1 onion, coarsely chopped
- ½ cup rice wine vinegar
- ¼ cup apple cider vinegar
- 2 tablespoons sugar
- 3 teaspoons minced garlic
- 1 teaspoon salt

Directions:

1. Using a food processor or blender, combine the jalapeños, Serrano chili, bell pepper, cucumber, and onion, and pulse until coarsely chopped.
2. Mix the the rice wine vinegar, apple cider vinegar, sugar, minced garlic, and salt, and pulse until minced but not puréed.
3. The relish will keep in airtight container in the refrigerator for up to 1 week.

Nutrition: Calories: 30 Carbs: 7g Fat: 0g Protein: 0g

337. Soy Dipping Sauce

Preparation Time: 10 minutes
Cooking Time: 30 minutes
Servings: 4

Ingredients:

- ¼ cup soy sauce
- ¼ cup sugar
- ¼ cup rice vinegar
- ½ cup scallions
- ½ cup cilantro

Directions:

1. Using a blender place all ingredients and blend until smooth
2. Pour smoothie in a glass and serve

Nutrition: Calories: 30 Carbs: 7g Fat: 0g Protein: 0g

- PART 2 -

ELECTRIC SMOKER COOKBOOK

Electric smokers provide very quickly the option to smoke meats through an easy-to-use and accessible interface. This design allows a lot of people to participate in the fun of smoking meats. This is a trend where more people are jumping on, and they are becoming more and more popular.

These smokers are considered to be modern-day smokers. The way it works is it uses electricity to heat the inside of the smoker. That heat is what cooks the meats and creates a smoky flavor.

Also, it uses moist heat instead of dry heat. The result is wet smoke that is considered to be more flavorful than dry smoke. This type of smoker looks like a makeshift wooden box with all kinds of holes and gaps. It has a lid made from clear plastic so people can monitor the meat throughout the cooking process. They can be hung from a ceiling or mounted on a table.

The smoke outlet is a hole in the top of the smoker. Also, there will be a tray that collects the drops of fat and grease. This tray will need to be emptied after the cooking process is complete.

They are also easy to use, which is probably the biggest reason they are becoming more popular. Unlike other smokers, electric smokers don't have to be monitored at all times. They don't need to be held up to burn a fire on the bottom. Everything is done for them once they are plugged in.

The smoking process and the results people get are very enjoyable. Many different types of meats and vegetables can be cooked with this type of smoker. They are very popular with most people because they are relatively inexpensive and easy to use.

In truth, you can smoke meats in just about anything. The pros will say that, for meat smoking, the primarily important thing is temperature control. Rather than smoking meats in a smoker, have a look at some of these options instead. Some of the most popular methods for smoking meats involve ice, water, and wood or charcoal to reduce the meat's temperature below the smoke generator's smoke generator while allowing for ventilation of the heat from the smoke generator.

Electric smokers are a safe way to make sure you have the meat that you want, as well as the smoky flavor you desire. The smokers are sold in a lot of stores like Wal-Mart and Goodwill. They are also sold online. This web site has some reviews about some of these smokers. They can also be purchased from certain internet sources. All of these smokers are very easy to use. They require very little time and patience to learn and operate. This will also allow you to have fun cooking and learning new things.

One thing to remember about these smokers is that they are a lot different from many smokers that you have experienced in the past. For example, you will not have the artificial heat from the fire that comes from a barrel smoker. It does not have to be as hot as a regular smoker. All you have to worry about is having it plugged in and just waiting for the time to pass.

If you are thinking of getting an electric smoker, you should have read this book. There are some very useful facts here.

WHY AN ELECTRIC SMOKER?

Your electric smoker gives you the choice to cook a complete meal all at once or smoke a large quantity of a single meat, fish, or vegetable for batch cooking. The slightly humid environment of an electric smoker means no more dried-out meats, and the low temperature allows connective tissue and fats to melt away, giving you that juicy barbecue you love so much. The smoker's drip tray will catch the drippings, so cleanup isn't a nightmare. The next time you crave a tender pulled pork or thinly sliced brisket, make it yourself!

However, more than allowing you to make delicious smoked foods in a breeze, there are so many benefits of using this type of electric smoker, for example:

- **Ease of operation:** You don't need to be a kitchen expert in order to use the Electric Smoker. It is very simple to operate. Once you have loaded the meat or food inside the smoker, all it takes is to set the temperature and time and wait until it is cooked.

- **An electric energy source is neutral:** Even if this particular electric smoker uses electricity as its energy source, it does not leave a flavor of its own. This means that whatever wood you use will be the only flavoring in the food. The energy source will never contaminate the taste that you desire.

- **Easy cleanup:** This type of electric smoker is effortless to clean. You are not required to use cleaners for the interior of the electric smoker. What makes it trouble-free to clean is that there are no messy charcoal ashes to empty or a propane tank to remove.

- **No need to babysit the cooking:** Since you can digitally control both time and temperature, you don't need to watch your cooking. This means that you can do other things while your food cooks.

- **The smoking unit has a large capacity:** The smoking unit has a hefty capacity, so you can fill it with wood to cook a large volume of foods. This lets you avoid opening the electric smoker to refill with more wood.

The most notable characteristic of an electric smoker is its shape. Electric smokers all share a unique tall and narrow shape, referred to as a box, cabinet, locker, or block smoker, because of the heat source. Located in the base, the electric heating element radiates heat, which naturally rises to the top of the smoker. Then, as the heat cools slightly, it creates convection. Convection moves heat in waves throughout the insulated box. The insulation in an electric smoker is an important quality, as it directly affects the smoker's ability to contain heat. Although high heat is not associated with smoking, consistent heat is paramount. The electrical efficiency, wood consumption, cooking time, and even smoking temperature are all variables controlled by the insulation.

The smoker's electric heating element starts with the press of a button located on the digital control panel, including a cooking timer and precise temperature and smoke regulators. Access to the perforated or wire sliding trays is easy, and the adjustable shelving allows for a variety of food sizes—whether multiple trays of chicken wings or a single large turkey.

THE FUNDAMENTALS OF ELECTRIC SMOKER

Basic Features

While Smokers from different brands are bound to have some tricks of their own! There are some features that almost staple to every Electric Smoker out there.

Having a good knowledge of these base features will give you a clear idea of what you are going into!

- **Considerably spacious:** Most Electric Smokers are usually very spacious to allow you to smoke meat for a large group of people. Generally speaking, the size of the Electric Smoker ranges from 527 square inches to 730 square inches.

- **Light weight:** Regular charcoal smokers tend to be really bulky and even tough to move! Modern Electric Smokers tend to be extremely light in weight, making it easier to move and very mobile. An average Electric Smoker usually weighs somewhere around 40–60 pounds. The inner walls of the smokers are made of stainless steel that makes it lightweight and durable.

- **Construction:** Normally, most Electric Smokers are built with durability kept in mind. The design of an Electric Smoker and the ergonomics are often designed with very high quality imported materials that give it a very long-lasting and safe build. These appliances are 100% safe for both you and your family.

- **Chrome coated racks:** Bigger sized smokers are often divided into 2–4 compartments that are fully plated with high-quality chrome. These racks are very easy to remove and can be used to keep large pieces of meat without making a mess. Even the most basic electric smokers tend to have at least four racks that are chrome coated.

- **Easily cleanable:** As Electric Smokers are getting more and more advanced, they are also becoming more accessible and easy to use. The Stainless Steel walls mean that you will be able to smoke your meat and veggies with ease and easily clean the smoker afterward.
- **Safe to use:** Electric Smokers are generally built with much grace and don't pose any harm. However, a degree of caution is always to be kept. As long as you are following the guidelines and maintain proper safety procedures, there's no risk of any kind of accidental burns or electric shocks from a smoker.

Basics Steps of Using an Electric Smoker

Now, the good news for all of your smoke aficionados out there is that using an Electric Smoker isn't exactly rocket science! This means anyone will be able to use it, following some very basic and simple guidelines. So it is crucial that you go through this section before starting to smoke your meat. After all, you don't want your expensive cut to be ruined just because of some silly mishap, right?

Just follow the basics and you will be fine!

- The first step is to make sure that you always wear safety gloves.
- Take out the chips tray and add your wood chips (before smoking begins).
- However, once the smoking has started, you can easily use the side chip tray for adding your chips.
- The additional chips are required to infuse the meat with more smoky flavor.
- Once the chip bay is ready, load up your marinated meat onto the grill directly.
- The stainless steel rack is made for direct smoking; however, if you wish they you can use a stainless steel container to avoid drippings.
- Once the meat is in place, lock the door of the chamber.
- Turn your smoker "On" using the specified button and adjust the temperature.
- Wait until it is done!

Keep in mind that the above-mentioned steps are merely the basic ones; different recipes might call for different steps to follow.

Either way, they won't be much complicated as well!

APPETIZERS, VEGETABLES, AND SIDES

338. Smashed Potato Casserole

Preparation time: 30minutes.

Cooking time: 45–60minutes.

Servings: 8

Ingredients:

- 1 small red onion, thinly sliced
- 1 small green bell pepper, thinly sliced
- 1 small red bell pepper, thinly sliced
- ¾ cup sour cream
- 1 small yellow bell pepper, thinly sliced
- 3 cups mashed potatoes
- 8–10 bacon
- ¼ cup bacon grease or salted butter, ½ stick
- 1 ½ teaspoons barbecue rub
- 3 cups shredded sharp cheddar cheese, divided
- 4 cups hash brown potatoes, frozen

Directions:

1. Get that bacon cooking over medium heat in a large skillet. Cook till nice and crisp. Aim for 5 minutes on both sides. Then set aside your bacon. Put

the bacon grease into a glass container and set aside.

2. Using the same skillet, warm up the butter or bacon grease over medium heat. When warm enough, sauté bell peppers and red onions. You're aiming for al dente. When done, set it all aside.

3. Grab a casserole dish, preferably one that is 9 by 11 inches. Spray with some nonstick cooking spray, then spread the mashed potatoes out, covering the entire bottom of the dish.

4. Add the sour cream to the next layer over the potatoes. When you're done, season it with some of the barbecue rub.

5. Create a new layer with the sautéed veggies over the potatoes.

6. Sprinkle your sharp cheddar cheese—just 1(½) of the cups. Then add the frozen hash brown potatoes.

7. Scoop out the rest of the bacon grease or butter from the sautéed veggies, all over the hash browns, and then top it all off with some delicious crumbled bacon bits. Set up your wood pellet smoker grill for indirect cooking. Heat to 350°F. Use whatever pellets you like.

8. Add the remaining sharp cheddar cheese (1(½) cups) over the whole thing, and then use some aluminum foil to cover the casserole dish.

9. Set up your wood pellet smoker grill for indirect cooking. Heat to 350°F. Use whatever pellets you like.

10. Let the whole thing bake for 45–60minutes. Ideally, you want the cheese to bubble.

11. Move it out and let it sit for about 10minutes.

Nutrition: Calories: 232 **Fat:** 2g **Carbs:** 48g **Protein:** 9g **Intolerances:** gluten-free, egg-free

339. Buffalo Mini Sausages

Preparation time: 30minutes.

Cooking time: 1 hour 30minutes.

Servings: 10

Ingredients:

- 8 ounces regular cream cheese, room temperature
- ¾ cup cheddar cheese blend and shredded Monterey Jack, not necessary
- 1 teaspoon smoked paprika
- 1 teaspoon garlic powder
- ½ teaspoon red pepper flakes, not necessary
- ¾ cup sour cream
- 20 little smokies sausages

- 10 bacon strips, thinly sliced and halved
- 10 jalapeno peppers, medium
- Cayenne pepper

Directions:

1. Wash the jalapenos, then slice them up along the length. Get a spoon or a paring knife if you prefer, and use that to take out the seeds and the veins.
2. Place the scooped-out jalapenos on a veggie grilling tray and put it all aside.
3. Get a small bowl and mix the shredded cheese, cream cheese, paprika, cayenne pepper, garlic powder, and red pepper flakes. Mix them thoroughly.
4. Get your jalapenos, which you've hollowed out, and then stuff them with the cream cheese and mix.
5. Get your little smokies sausage, and then put it right onto each of the cheese-stuffed jalapenos.
6. Grab some of the thinly sliced and halved bacon strips and wrap them around each of the stuffed jalapenos and their sausage.
7. Grab some toothpicks. Use them to keep the bacon nicely secured to the sausage.
8. Set up your wood pellet smoker grill, so it's ready for indirect cooking. Get it heated to 250°F. Use hickory or blends for your wooden pellets.
9. Put your jalapeno peppers in and smoke them at 250°F for anywhere from 90 minutes to 120 minutes. You want to keep it going until the bacon is nice and crispy.
10. Take out the atomic buffalo turds, and then let them rest for about 5 minutes.
11. Serve!

Nutrition: Calories: 198 **Fat:** 17g **Cholesterol:** 48mg **Carbs**: 3g **Protein:** 8g **Intolerances:** egg-free

340. Brisket Baked Beans

Preparation time: 20 minutes.

Cooking time: 1 hour, 30 minutes.

Servings: 10

Ingredients:

- 1 green bell pepper, medium, diced
- 1 red bell pepper, medium, diced
- 1 yellow onion, large, diced
- 2–6 jalapeno peppers, diced
- 1 can (28-ounces) baked beans
- 2 tablespoons olive oil, extra-virgin
- 3 cups brisket flat, chopped
- 1 can red kidney beans, 1(4-ounces), rinsed, drained
- 1 cup barbecue sauce

- ½ cup brown sugar, packed
- 2 teaspoons mustard, ground
- 3 cloves of garlic, chopped
- 1(½) teaspoon black pepper
- 1 (½) teaspoon kosher salt

Directions:

1. Put a skillet on the fire on medium heat. Warm-up your olive oil. Toss in the diced jalapenos, peppers, and onions. Stir every now and then for 8 minutes.
2. Grab a 4-quart casserole dish. Now, in your dish, mix in the pork and beans, kidney beans, baked beans, chopped brisket, cooked peppers and onions, brown sugar, barbecue sauce, garlic, mustard, salt, and black pepper.
3. Set up your wood pellet smoker grill, so it's ready for indirect cooking.
4. Heat your grill to 325°F, using whatever pellets you want.
5. Cook your brisket beans on the grill for 90 minutes to 120 minutes. Keep it uncovered as you cook. When it's ready, you'll know because the beans will get thicker and will have bubbles as well.
6. Rest the food for 15minutes before you finally move on to step number 5.
7. Serve!

Nutrition: Calories: 200 **Fat:** 2g **Cholesterol:** 10mg **Carbs:** 35g **Protein:** 9g

341. Twice-Baked Spaghetti Squash

Preparation time: 15 minutes.

Cooking time: 45 minutes.

Servings: 2

Ingredients:

- 1 spaghetti squash, medium
- 1 tablespoon olive oil, extra-virgin
- 1 teaspoon salt
- ½ teaspoon pepper
- ½ cup Parmesan cheese, grated, divided
- ½ cup mozzarella cheese, shredded, divided

Directions:

1. Cut the squash lengthwise in half. Make sure you're using a knife that's large enough and sharp enough. Once you're done, take out the pulp and the seeds from each half with a spoon.
2. Season the insides of each half of the squash with some olive oil. When you're done with that, sprinkle the salt and pepper.
3. Heat your grill to 375°F with your preferred wood pellets.
4. Put each half of the squash on the grill. Make sure they're both facing upwards on the grill grates, which should be nice and hot.

5. Bake for 45minutes, keeping it on the grill until the internal temperature of the squash hits 170°F. You'll know you're done when you find it easy to pierce the squash with a fork.

6. Move the squash to your cutting board. Rest for 10minutes, so it can cool a bit.

7. Turn up the temp on your wood pellet smoker grill to 425°F.

8. Using a fork to remove the flesh from the squash in strands by raking it back and forth. Be careful because you want the shells to remain intact. The strands you rake off should look like spaghetti, if you're doing it right.

9. Put the spaghetti squash strands in a large bowl, and then add in half of your mozzarella and half of your Parmesan cheeses. Combine them by stirring.

10. Take the mix, and stuff it into the squash shells. When you're done, sprinkle them with the rest of the Parmesan and mozzarella cheeses.

11. **Optional:** You can top these with some bacon bits, if you like.

12. Allow the stuffed spaghetti squash shells you've now stuffed to bake at 435°F for 15 minutes.

13. Serve and enjoy.

Nutrition: Calories: 214 **Fat:** 3g **Cholesterol:** 17mg **Carbs:** 27g **Protein:** 16g

342. Bacon-Wrapped Asparagus

Preparation time: 15 minutes.

Cooking time: 30 minutes.

Servings: 6

Ingredients:

- 15–20 spears of fresh asparagus (1-pound)
- Olive oil, extra-virgin
- 5 slices bacon, thinly sliced
- 1 teaspoon salt and pepper or your preferred rub

Directions:

1. Break off the ends of the asparagus, then trim it all, so they're down to the same length.

2. Separate the asparagus into bundles—3 spears per bundle. Then spritz them with some olive oil.

3. Use a piece of bacon to wrap up each bundle. When you're done, lightly dust the wrapped bundle with some salt and pepper to taste or your preferred rub.

4. Set up your wood pellet smoker grill so that it's ready for indirect cooking.

5. Put some fiberglass mats on your grates. Make sure they're the fiberglass kind. This will keep your asparagus from getting stuck on your grill gates.

6. Heat your grill to 400°F, with whatever pellets you prefer. You can do this as you prep your asparagus.

7. Grill the wraps for 25 minutes to 30minutes, tops. The goal is to get your asparagus looking nice and tender and the bacon deliciously crispy.

Nutrition: Calories: 71 **Fat:** 3g **Carbs**: 1g **Protein:** 6g

343. Garlic Parmesan Wedges

Preparation time: 15 minutes.

Cooking time: 35 minutes.

Servings: 3

Ingredients:

- 3 russet potatoes, large
- 2 teaspoons garlic powder
- ¾ teaspoon black pepper
- 1(½) teaspoons salt
- ¾ cup Parmesan cheese, grated
- 3 tablespoons fresh cilantro, chopped, optional. You can replace this with flat-leaf parsley
- ½ cup blue cheese per serving, as an optional dip. Can be replaced with ranch dressing
- Olive oil
- Garlic

Directions:

1. Use some cold water to scrub your potatoes as gently as you can with a veggie brush. When done, let them dry.

2. Slice your potatoes along the length in half. Cut each half into a third.

3. Get all the extra moisture off your potato by wiping it all away with a paper towel. If you don't do this, then you're not going to have crispy wedges!

4. In a large bowl, throw in your potato wedges, some olive oil, garlic powder, salt, garlic, and pepper, and then toss them with your hands lightly. You want to make sure the spices and oil get on every wedge.

5. Place your wedges on a nonstick grilling tray, pan, or basked. The single-layer kind. Make sure it's at least 15 x 12 inches.

6. Set up your wood pellet smoker grill, so it's ready for indirect cooking.

7. Heat your grill to 425°F, with whatever wood pellets you like.

8. Set the grilling tray upon your heated grill. Roast the wedges for 15minutes before you flip them. Once you turn them, roast them for another 15minutes, or 20 tops. The outside should be a nice, crispy, golden brown.

9. Sprinkle your wedges generously with the Parmesan cheese. When you're done, garnish it with some parsley or cilantro, if you like. Serve these bad

boys up with some ranch dressing, some blue cheese, or just eat them that way!

Nutrition: Calories: 194 **Fat:** 5g **Cholesterol:** 5mg **Carbs**: 32g **Protein:** 5g

344. Smoked Summer Vegetables

Preparation time: 15 minutes.

Cooking time: 1 hour.

Servings: 4

Ingredients:

- Summer squash
- 2 zucchini
- 1 onion
- 2 cups mushrooms
- 2 cups French-cut green beans

Directions:

1. Wash thoroughly and slice squash, onion, and zucchini, mushrooms, and green beans.
2. Combine all these ingredients and mix well.
3. Heat the electric smoker to 250°F.
4. Make 4 cup-shaped containers from heavy-duty aluminum foil.
5. Put vegetables in these cups.
6. Add herbs and spices to taste.
7. Pinch the top of foil cups together.
8. Make several holes in the foil so that the smoke can circulate around the vegetables. Smoke for 1 hour at 220°F.

Nutrition: Calories: 97 **Protein:** 5.6g **Carbs:** 14g **Fat:** 9g

345. Herby Smoked Cauliflower

Preparation time: 20 minutes.

Cooking time: 2 hours.

Servings: 4

Ingredients:

- 1 head cauliflower
- Olive oil
- Salt
- Pepper
- 2 teaspoons dried oregano
- 2 teaspoons dried basil

Directions:

1. Start by soaking your wood chips for about an hour and heating your smoker to 200°F/93°C.
2. Remove the wood chips from the liquid, then pat dry before using.
3. Then take your cauliflower and chop into medium-sized pieces, removing the core.

4. Place the pieces of cauliflower onto a sheet pan and then drizzle with the olive oil.
5. Sprinkle the seasonings and herbs over the cauliflower, then pop it into the smoker.
6. Smoke for 2 hours, checking and turning often.
7. Serve and enjoy!

Nutrition:

Calories: 31 **Protein:** 1.5g **Carbs:** 6.7g **Fat:** 0.34g

346. Smoked Green Beans with Lemon

Preparation time: 20 minutes.

Cooking time: 1 hour.

Servings: 4

Ingredients:

- 2 pound fresh green beans, trimmed and soaked
- 2 tablespoons apple vinaigrette dressing
- 1 lemon

Directions:

1. Place beans in a colander.
2. Heat the smoker to 140°F and add wood chips (recommended Oak wood chips).
3. Put the beans in the pan in a single layer and lightly coat with the dressing.
4. Put the beans on the upper shelf of the smoker and smoke for 1 hour.
5. Remove from the heat, cover with foil, and let rest for 15 minutes.
6. Pour lemon juice, sprinkle with the lemon zest and serve.

Nutrition:

Calories: 74.31 **Total fat:** 0.66g **Total carbs**: 17.02g **Protein:** 4.19g

347. Smoked Lemony-Garlic Artichokes

Preparation time: 15 minutes.

Cooking time: 35 minutes.

Servings: 4

Ingredients:

- 4 artichokes
- 4 minced garlic cloves
- 3 tablespoons lemon juice
- ½ cup virgin olive oil
- 2 parsley sprigs
- Sea salt

Directions:

1. Put a large pot on your stove with a metal steaming basket inside.
2. Fill with water just to the bottom of the basket and bring to a boil.
3. Slice the artichoke tail and take out the toughest leaves.
4. With cooking shears, cut the pointy ends off of the outermost leaves.
5. Cut the artichokes in half lengthwise. Remove the hairy choke in the center.
6. Put the halves stem side down in the steamer basket. Reduce the heat to a rolling simmer.
7. On the pot, cover and steam for about 20 to 25 minutes, until the inside of the artichoke is tender.
8. **Prepare a dressing:** place in a mortar the garlic, lemon juice, olive oil, parsley, and salt.
9. Take away the basket and let the artichokes come to room temperature.
10. Heat your smoker to 200°F.
11. Place the artichokes in aluminum foil packets and brush garlic mixture all over the artichokes.
12. Smoke the artichokes halves for 1hour.
13. Serve hot.

Nutrition: Calories: 83.22 **Total fat:** 0.29g **Total carbs:** 18.82g **Protein:** 5.54g

348. Smoked Portobello Mushrooms with Herbs de Provence

Preparation time: 10 minutes.

Cooking time: 2 hours.

Servings: 4

Ingredients:

- 12 large Portobello mushrooms
- 1 tablespoon Herbs de Provence
- ¼ cup extra virgin olive oil
- Sea salt
- Black pepper

Directions:

1. Heat the smoker to 200°F and add wood chips (recommended oak wood chips).
2. In a bowl, mix herbs de Provence, olive oil, salt, and pepper to taste.
3. Scrub the mushrooms with a dry cloth or paper towel.
4. Rub the mushrooms all over with herbs mixture.
5. Move the mushrooms, cap side down, directly on the top grill rack. Smoke for approximately 2 hours.
6. Move carefully, so the herbal liquid in the cap remains in place.

Nutrition: Calories: 146.08 **Fat:** 13.63g **Total carbs**: 5.22g **Protein:** 3.03g

349. Smoky Corn on the Cob

Preparation time: 10 minutes.

Cooking time: 2 hours.

Servings: 5

Ingredients:

- 10 ears sweet corn
- ½ cup butter
- Salt
- Black pepper

Directions:

1. Heat your smoker to 225°F and add wood chips (recommended oak or hickory).
2. Put the ears of corn on the top 2 racks of the smoker and smoke for 2 hours.
3. Rotate the corn every 30 minutes.
4. Serve hot with butter, salt, and pepper.

Nutrition:

Calories: 408.72 **Total fat:** 22.27g
Total carbs: 53.5g **Protein:** 9.55g

350. Smoked Potato Salad

Preparation time: 30 minutes.

Cooking time: 2 hours.

Servings: 4

Ingredients:

- 3 eggs, hard-boiled
- 2 tablespoons cider vinegar
- 1 pound russet potatoes
- 1 tablespoon Dijon mustard
- ½ cup red onion
- 1/3 cup light mayonnaise
- Salt
- Black pepper
- Pickles

Directions:

1. Heat the electric smoker to 225°F.
2. Put prepared wood chips in the wood tray—use mesquite chips for the best result.
3. Put peeled potatoes in a saucepan and cover with water. Put on the lid and bring to a boil.
4. Cook for 20 minutes. Pat potatoes dry, and put them on paper towels.
5. Directly smoke potatoes on the racks for 2 hours as you add extra wood chips in a cycle of 45 minutes.

6. Remove potatoes, let them cool.
7. Chop them well for the preparation of the salad.
8. Combine boiled eggs, onion, mayonnaise, pickles, mustard, pepper, salt, and vinegar.
9. Mix all these ingredients well.
10. Add potatoes to the prepared mixture. Put in the fridge for several hours covered.

Nutrition:

Calories: 209 **Total fat:** 9g **Total carbs:** 30g **Protein:** 3g

351. Smoked Volcano Potatoes

Preparation time: 15 minutes.

Cooking time: 1 hour.

Servings: 2

Ingredients:

- 2 russet potatoes
- ¾ cup sour cream.
- 1 cup cheddar cheese
- 2 tablespoons green onion
- 8 bacon strips
- 4 tablespoons butter
- 2 tablespoons olive oil
- Salt

Directions:

1. Heat the electric smoker to 250°F.
2. Wash potatoes, pierce using the fork.
3. Take the oil and salt and rub on the potatoes. Wrap the potatoes in foil and put them in the smoker.
4. Smoke potatoes for 3hours.
5. Cut off the top of each potato and remove the potato flesh, leaving the shell empty.
6. Fry and crumble the bacon. Combine potato flesh with bacon, butter, sour cream, and cheese in a bowl.
7. Put the prepared filling in the potatoes, add some cheese on the top.
8. Wrap the potato with 2 bacon slices—for securing, use toothpicks.
9. Smoke for another 1 hour.
10. Add green onions with a little sour cream on top (sour cream will give a special flavor to the potato).

Nutrition:

Calories: 256 **Total fat:** 39.3g **Total carbs**: 31.7g **Protein:** 32.1g

352. Groovy smoked asparagus

Preparation time: 5 minutes.

Cooking time: 90 minutes.

Servings: 4

Ingredients:

- 1 bunch asparagus
- 2 tablespoons olive oil
- 1 teaspoon chopped garlic
- Kosher salt
- ½ teaspoon black pepper

Directions:

1. Prepare the water pan of your smoker accordingly.
2. Pre-heat your smoker to 275°F/135°C.
3. Fill a medium-sized bowl with water and add 3–4 handfuls of woods and allow them to soak.
4. Add the asparagus to a grill basket in a single layer.
5. Drizzle olive oil on top and sprinkle garlic, pepper, and salt.
6. Toss them well.
7. Put the basket in your smoker.
8. Add a few chips into the loading bay and keep repeating until all of the chips after every 20 minutes.
9. Smoke for 60–90 minutes.
10. Serve and enjoy!

Nutrition:

Calories: 68 **Total fat:** 4.1g **Total carbs:** 7.1g **Protein:** 2.8g

353. Smoked Squash Casserole

Preparation time: 40 minutes.

Cooking time: 1 hour 15 minutes.

Servings: 3

Ingredients:

- 2(½) pounds yellow squash
- 2 tablespoons parsley flakes
- 2 eggs, beaten
- 1 medium yellow onion
- 1 sleeve saltine crackers
- 1 package Velveeta cheese
- ½ cup Alouette sundried tomato basil cheese spread
- ¼ cup Alouette garlic and herb cheese spread
- ¼ cup mayonnaise
- ¾ teaspoon hot sauce
- ¼ teaspoon Cajun seasoning
- ½ cup butter
- ¼ teaspoon salt

- ¼ teaspoon black pepper

Directions:

1. Heat the electric smoker to 250°F.
2. Combine squash and onion in a saucepan and add water to cover. Boil on medium heat until tender.
3. Drain and to this hot mixture, add Velveeta cheese, Alouette cheese, mayonnaise, parsley flakes, hot sauce, Cajun seasoning, salt, and pepper to taste.
4. Stir all together well.
5. Cool a little, add eggs, and stir until mixed.
6. Melt butter in a saucepan.
7. Add crushed crackers to the butter and stir well. Combine ½ cup of butter-cracker mix with the squash mixture. Stir thoroughly.
8. Pour into a disposable aluminum foil pan. Top the squash with the remaining butter and crackers. Cover the pan tightly with aluminum foil.
9. Put on the lower rack of the smoker and cook for 1hour. Put one small handful of prepared wood chips in the wood tray for the best result, use hickory.
10. For an hour, remove the foil from the casserole and cook for another 15 minutes.

Nutrition: Calories: 190 **Total fat:** 8g **Total carbs:** 23g **Protein:** 7g

354. Smoked Eggplant

Preparation time: 20 minutes.

Cooking time: 1 hour.

Servings: 4

Ingredients:

- 2 medium eggplant
- Olive oil

Directions:

1. Heat your smoker to 200°F and soak your wood chips for an hour.
2. Remove the wood chips from the liquid, then pat dry before using.
3. Then carefully peel your eggplant, then slice into rounds of around ¼"/1cm thick.
4. Brush each of these rounds with olive oil, then place directly into the smoker.
5. Smoke for approximately an hour until soft and tender. Serve and enjoy!

Nutrition:

Calories: 85 **Protein:** 1.6g **Carbs:** 9.4g **Fat:** 4.6g

355. Twice Pulled Potatoes

Preparation time: 30 minutes.

Cooking time: 2 hours, 40 minutes.

Servings: 4

Ingredients:

- 1 pound pulled pork
- 2 russet potatoes
- 1/3 cup sour cream
- 4-ounces cream cheese
- 1/3 cup cheddar cheese
- Chives
- BBQ sauce, to taste

Directions:

1. Heat the electric smoker to 225°F. Smoke washed potatoes for 2 hours.
2. Mix potato flesh, cheddar cheese, sour cream, cream cheese, pulled pork, and BBQ sauce in a bowl; stir well.
3. Put prepared mixture back into potatoes skins.
4. Smoke for another 40 minutes.
5. Season with more BBQ sauce, if desired. Sprinkle some cheddar cheese and chives on the top.

Nutrition:

Calories: 285 **Protein**: 3g **Carbs:** 24.1g **Fat:** 8.1g

356. Smoked Apple Crumble

Preparation time: 10 minutes.

Cooking time: 1 hour.

Servings: 15

Ingredients:

For the pie filling:

- 3 teaspoons all-purpose flour
- 1 teaspoon ground cinnamon
- 1 cup sugar
- 3-pound apples

For crumble:

- 2 cups rolled oats
- ½ teaspoon baking powder
- ½ teaspoon baking soda
- 1 cup butter
- 2 cups brown sugar
- Ice cream to serve

Directions:

1. Heat the smoker to 275°F following the manufacturer's instructions.
2. Make the pie filling as follows: Mix together in a bowl all the pie-filling ingredients and toss well.
3. Transfer into 14–15 ramekins (2/3 fill it). Do not grease the ramekins.

4. To make crumble: Mix together in a bowl flour, brown sugar, oats, baking powder, and baking soda. Pour butter over it and mix well.
5. Place about 1/4 cup of this mixture over each of the ramekins (over the apple filling). Place the ramekins on the center rack in the smoker and smoke for an hour.
6. Remove the ramekins from the smoker and invert them onto individual serving bowls. Serve as it is or with a scoop of ice cream.

Nutrition:

Calories: 267 **Protein:** 2.6g **Carbs:** 41.4g **Fat:** 13.3g

357. Smoked Coleslaw

Preparation time: 1 hour, 10 minutes.

Cooking time: 30 minutes.

Servings: 10–12

Ingredients:

- 1 head cabbage, shredded
- 1 carrot, shredded
- 1/4 cup sugar
- 1/2 teaspoon salt
- 1/2 teaspoon freshly ground black pepper
- 1/4 cup white vinegar
- 1 cup heavy whip cream
- 1 teaspoon paprika

Directions:

1. Heat the smoker to 175°F with the maple wood.
2. Spread the cabbage and carrot in a shallow aluminum foil pan. Place the pan in the smoker, then smoke the vegetables for 30minutes. Remove from the smoker and transfer the vegetables to a large bowl.
3. Stir in the sugar, salt, pepper, vinegar, and heavy cream to combine. Refrigerate for 1hour before serving.
4. Sprinkle with paprika.
5. **Variation tip:** This sweet dressing is also excellent for a broccoli-cauliflower salad. Instead of the cabbage and carrot, use 2 cups each of chopped broccoli, cauliflower, and cooked bacon, and add 1 cup each of chopped scallions (white and green parts) and celery.

Nutrition:

Calories: 69 **Total fat:** 3.8g **Saturated fat:** 2.3g **Cholesterol:** 14mg **Sodium:** 115mg **Total carbohydrate:** 8.6g **Dietary fiber:** 1.7g **Total sugars:** 6.4g **Protein:** 1.1g **Vitamin D:** 5mcg **Calcium:** 33mg **Iron:** 0mg **Potassium:** 134mg

358. Smoked Onion Bombs

Preparation time: 15 minutes.

Cooking time: 2 hours.

Servings: 4

Ingredients:

- 4 large Vidalia onions, peeled
- ½ cup (1 stick) butter, divided
- 4 chicken bouillon cubes
- ½ cup grated Parmesan cheese
- 1 teaspoon freshly ground black pepper

Directions:

1. Heat the smoker to 225°F with the maple or mesquite wood.
2. Angle a sharp knife into the onion from the top, cut all the way around, remove the top and create a deep well in the onion. Repeat with the remaining onions. Save the onion tops.
3. Place four pieces of aluminum foil, each about 8 inches square. Place each onion on a sheet of foil. Press 2 tablespoons of butter into the well of each onion and top with a bouillon cube.
4. Mix the Parmesan and pepper. Place 2 tablespoons of the mixture in each onion well.
5. Replace the onion tops tightly (cutting as necessary to fit) and wrap the foil up the sides, but leave the top of the packet open to allow the smoke flavor to permeate the onions.
6. Smoke the onions for about 2 hours, until tender.
7. Ingredient tip: When shopping in the produce section, look for firm bulbs with no bruises or soft areas. Onions can be kept in the refrigerator's crisper drawer, but avoid storing in plastic bags, as the onions are then prone to develop mold.

Nutrition:

Total fat: 24.3g **Saturated fat:** 15.3g **Cholesterol:** 64mg **Sodium:** 814mg **Total carbohydrate:** 15.4g **Dietary fiber:** 3.4g **Total sugars:** 7g **Protein:** 3.6g **Vitamin D:** 16mcg **Calcium:** 77mg **Iron:** 1mg **Potassium:** 249mg

359. Smoked Cabbage

Preparation time: 10 minutes.

Cooking time: 2 hours.

Servings: 4

Ingredients:

- 1 head cabbage, cored completely
- 4 tablespoons butter
- 2 tablespoons rendered bacon fat, or 2 more tablespoons butter, melted
- 1 chicken bouillon cube
- 1 teaspoon freshly ground black pepper

- 1 garlic clove, minced

Directions:

1. Heat the smoker to 240°F with the apple, maple, or oak wood.
2. Fill the hole left by coring the cabbage with the butter, bacon fat, bouillon cube, pepper, and garlic.
3. Wrap the cabbage in aluminum foil two-thirds of the way up the sides to protect the outer leaves, leaving the top open to allow the smoke flavor to permeate the cabbage. Place the cabbage on the grill rack and smoke for about 2 hours.
4. Unwrap and enjoy as a side dish.

Nutrition:

Calories: 202 **Total fat:** 17.6g **Saturated fat:** 11.1g **Cholesterol:** 46mg **Sodium:** 268mg **Total carbohydrate:** 11.2g **Dietary fiber:** 4.6g **Total sugars:** 5.9g **Protein:** 2.7g **Vitamin D:** 8mcg **Calcium:** 81mg **Iron:** 1mg **Potassium:** 322mg

360. Loaded Hasselback Potatoes

Preparation time: 20 minutes.

Cooking time: 1 hour, 30 minutes.

Servings: 4

Ingredients:

- 4 russet potatoes, cut Hasselback style (slice into the potato, all the way across, making your cuts about ¼ inch apart and being careful not to cut all the way through the bottom of the potato; see tip, here)
- 1 cup olive oil, divided
- 2 teaspoons salt
- 2 teaspoons freshly ground black pepper
- 1 small onion, sliced
- 2 jalapeño peppers, seeded and thinly sliced
- 2 cherry peppers, sliced
- 4 ounces block Cheddar cheese, thickly sliced
- 8 bacon slices, cooked and crumbled

Directions:

1. Heat the smoker to 250°F with the hickory wood.
2. Place the potatoes on a grill pan. Drizzle ½ cup of olive oil over the potatoes and sprinkle with the salt and pepper. Move the pan in the smoker and smoke for about 1hour.

3. Take off the potatoes from the smoker and place some onion, jalapeños, cherry peppers, Cheddar slices, and crumbled bacon in between each potato slice and on top.

4. Pour the remaining ½ cup of olive oil overall. Return the potatoes to the smoker for 30 to 40 minutes or so, until the potatoes are tender in a squeeze test.

5. Serve with sour cream, if desired.

Nutrition:

Calories: 2085 **Total fat:** 166.3g **Saturated fat:** 48.2g **Cholesterol:** 309mg **Sodium:** 7644mg **Total carbohydrate:** 41.3g **Dietary fiber:** 6.1g **Total sugars:** 3.6g **Protein:** 105.1g **Vitamin D:** 3mcg **Calcium:** 263mg **Iron:** 6mg **Potassium:** 2390mg

361. Garlic-Rosemary Potato Wedges

Preparation time: 15 minutes.

Cooking time: 1 hour, 30 minutes.

Servings: 6–8

Ingredients:

- 4 to 6 large russet potatoes, cut into wedges
- ¼ cup olive oil
- 2 garlic cloves, minced
- 1 tablespoon dried rosemary
- 2 teaspoons salt
- 1 teaspoon freshly ground black pepper
- 1 teaspoon sugar
- 1 teaspoon onion powder

Directions:

1. Heat the smoker to 250°F with the maple or pecan wood.

2. Toss the potatoes with the olive oil to coat them well.

3. Stir the garlic, rosemary, salt, pepper, sugar, and onion powder. Season this mixture on all sides of the potato wedges. Transfer the seasoned wedges to a grill pan and put it into the smoker.

4. Cook for about 1(½) hours until a fork cuts through the wedges easily.

Nutrition:

Calories: 377 **Total fat:** 13.3g **Saturated fat:** 2g **Cholesterol:** 0mg **Sodium:** 1187mg **Total carbohydrate:** 61.4g **Dietary fiber:** 9.8g **Total sugars:** 5.5g **Protein:** 6.5g **Vitamin D:** 0mcg **Calcium:** 62mg **Iron:** 3mg **Potassium:** 1536mg

362. Smoked Asparagus

Preparation time: 10 minutes.

Cooking time: 1 hour.

Servings: 4–5

Ingredients:

- 2 tablespoons butter, melted
- 2 garlic cloves, minced
- 2 tablespoons freshly squeezed lemon juice
- 1 tablespoon capers
- 1 tablespoon onion powder
- 1 teaspoon salt
- ½ teaspoon freshly ground black pepper
- 1 pound asparagus (about 18 to 20 stalks), woody ends snapped off

Directions:

1. Heat the smoker to 240°F with the maple wood.
2. Stir together the butter, garlic, lemon juice, capers, onion powder, salt, and pepper.
3. Cook the asparagus in a grill pan and drizzle with the seasoned butter. Put the pan in the smoker and smoke for about 1hour until tender.

Nutrition:

Calories: 68 **Total fat:** 4.8g **Saturated fat:** 3g **Cholesterol:** 12mg **Sodium:** 555mg **Total carbohydrate:** 5.4g **Dietary fiber:** 2.2g **Total sugars:** 2.4g **Protein:** 2.4g **Vitamin D:** 3mcg **Calcium:** 33mg **Iron:** 2mg **Potassium:** 213mg

363. Smoked Bacon-Wrapped Onion Rings

Preparation time: 20 minutes.

Cooking time: 1 hour, 30 minutes.

Ingredients:

- 2 large onions, peeled and sliced ½ inch thick (about 4 slices from each onion)
- ¼ cup hot sauce
- 4 tablespoons butter, melted
- 1 pound bacon
- 1 tablespoon cayenne pepper
- 1 tablespoon sugar

Directions:

1. Heat the smoker to 250°F with the hickory, maple, or mesquite wood.
2. Separate the onion rings and remove the smaller internal rings to save for another use. I recommend leaving two rings intact on each to keep them

sturdy. You should get about eight rings out of one large onion, two out of each slice.

3. Mix the hot sauce and melted butter.
4. Dip the onion rings in the butter–hot sauce mixture.
5. Wrap each onion ring tightly with a bacon slice.
6. In another small bowl, stir together the cayenne and sugar. Coat the bacon-wrapped rings well with this mixture. Secure the rings with toothpicks or place them on skewers.
7. Place the onion rings on a grill mat and smoke for about 1½hours until the bacon is done and beyond "chewy" to bite through.

Nutrition:

Calories: 254 **Total fat:** 19.8g **Saturated fat:** 7.6g **Cholesterol:** 52mg **Sodium:** 1028mg **Total carbohydrate:** 4.2g **Dietary fiber:** 0.7g **Total sugars:** 2.2g **Protein:** 14.4g **Vitamin D:** 3mcg **Calcium:** 12mg **Iron:** 1mg **Potassium:** 267mg

364. Smoked Artichokes

Preparation time: 5 minutes.
Cooking time: 2 hours.
Servings: 8

Ingredients:

- ¼ cup olive oil
- 1 garlic clove, minced
- 1 teaspoon salt
- Juice of 1 lemon
- 4 artichokes, stemmed and halved lengthwise

Directions:

1. Preheat the smoker to 225°F with the hickory or maple wood.
2. Whisk together the garlic, olive oil, salt, and lemon juice.
3. Brush the artichoke halves with the seasoned olive oil. Place them directly on the smoker's grate and smoke for about 2hours. The artichoke bottoms should look and feel tender when poked with a fork.

Nutrition:

Calories: 94 **Total fat:** 6.5g **Saturated fat:** 1g **Cholesterol:** 0mg **Sodium:** 368mg **Total carbohydrate:** 8.7g **Dietary fiber:** 4.4g **Total sugars:** 0.9g **Protein:** 2.7g **Vitamin D:** 0mcg **Calcium:** 37mg **Iron:** 1mg **Potassium:** 306mg

365. Hasselback Sweet Potatoes

Preparation time: 15 minutes.

Cooking time: 1 hour, 30 minutes.

Servings: 4–6

Ingredients:

- 4 large sweet potatoes, scrubbed
- ¼ cup canola oil
- 2 tablespoons table salt
- ½ cup (1 stick) butter
- 4 serrano peppers, seeded and sliced
- 1 cup glazed spiced pecans (here), coarsely chopped
- 1 tablespoon sea salt
- ¼ cup honey

Directions:

1. Heat the smoker to 250°F with the pecan wood.
2. Rub the sweet potatoes all over with the oil and table salt.
3. Cut them thick-sliced Hasselback-style: Slice into the sweet potatoes, all the way across, making your cuts about ½ inch apart, and being careful not to cut all the way through the bottom of the sweet potato (see tip).
4. Place the sweet potatoes on a grill pan and put it into the smoker. Smoke for 1hour, then remove from the heat.
5. Place a pat of butter between each slice.
6. Stuff serrano slices and pecans between the slices.
7. Sprinkle well with the sea salt and drizzle with the honey, getting it between the slices.
8. Smoke for 30 to 40minutes more and remove from the smoker when the potatoes pass the squeeze test.

Nutrition:

Calories: 434 **Total fat:** 38.3g **Saturated fat:** 11.7g **Cholesterol:** 41mg **Sodium:** 3446mg **Total carbohydrate:** 24.3g **Dietary fiber:** 2.8g **Total sugars:** 12.9g **Protein:** 2.8g **Vitamin D:** 11mcg **Calcium:** 24mg **Iron:** 1mg **Potassium:** 261mg

366. Smoked Beef Ribs BBQ with Sweet Ginger Tea and Bourbon

Preparation time: 30 minutes.

Cooking time: 4 hours, 30 minutes.

Servings: 10

Ingredients:

- 5-pound (2.3-kilograms) beef ribs

The Rub:

- 2 tablespoons Worcestershire sauce
- 2 tablespoons canola oil
- ¼ cup brown sugar
- 2 tablespoons sweet paprika
- 1 tablespoon black pepper
- 2 tablespoons Kosher salt
- 2 teaspoons garlic powder
- 2 teaspoons onion powder
- 1 teaspoon cayenne powder

The Glaze:

- 1 teaspoon ginger
- 1 cup sweet tea
- ¼ cup beef broth
- ¼ cup bourbon
- 1 teaspoon onion powder

- 1 teaspoon garlic powder
- 3 tablespoons diced parsley
- ½ teaspoon grated lemon zest
- 2 tablespoons lemon juice

Directions:

1. Discard the excess fat from the beef ribs, then baste Worcestershire sauce and canola oil over the beef ribs.
2. Combine the remaining rub ingredients—brown sugar, sweet paprika, black pepper, kosher salt, garlic powder, onion powder, and cayenne pepper, then stir the mixture until well mixed.
3. Sprinkle the spice mixture over the beef ribs, then set aside.
4. Next, plug in the Electric Smoker and press the power button to turn it on.
5. Press the "Temperature" button and set it to 225°F (107°C).
6. After that, add wood chips to the smoker and pour water into the water pan.
7. Once the Electric Smoker is ready, place the seasoned beef ribs on the grill trays that are provided in the Electric Smoker, then close the lid.
8. Smoke the seasoned beef ribs until the internal temperature has reached 130°F (54°C) for medium-rare or 175°F (79°C) for well done. The smoking time will take approximately 4 hours and 30 minutes.
9. Check the wood chips level and add more wood chips if it is necessary.
10. In the meantime, place the entire glaze ingredients—ginger, sweet tea, bourbon, beef broth, onion powder, garlic powder, diced parsley, grated lemon zest, and lemon juice in a saucepan, then bring to a simmer over low heat.
11. Stir the glaze mixture until incorporated and remove from heat.
12. After 6hours of smoking, baste the glaze mixture over the beef ribs, then continue smoking until the smoked beef reaches the desired internal temperature.
13. Once it is done, remove the smoked beef ribs from the Electric Smoker and transfer it to a serving dish. Turn the Electric Smoker off and allow it to cool.
14. Serve the smoked beef ribs and enjoy warm.

Nutrition:

Energy:607kcal **Calcium:** Ca37mg **Magnesium:** Mg36mg **Phosphorus:** P214mg **Iron:** Fe3.45mg **Potassium:** K477mg **Sodium:** Na1543mg **Zinc:** Zn6.19mg

367. Super Spicy Smoked Brisket Garlic

Preparation time: 30 minutes.

Cooking time: 4 hours, 10 minutes.

Servings: 10

Ingredients:

- 4-pound (1.8-kilograms) beef brisket

The Marinade:

- ¼ cup chili powder
- 1 teaspoon cayenne pepper
- 1(¼) tablespoons Kosher salt
- 2 tablespoons brown sugar
- 2 tablespoons garlic powder
- 1 teaspoon cumin
- 1 tablespoon dried oregano

Directions:

1. Combine the marinade ingredients—chili powder, cayenne pepper, kosher salt, brown sugar, garlic powder, cumin, and dried oregano in a bowl.
2. Rub the beef with the spice mixture, then wrap it tightly with plastic wrap.
3. Marinate the beef brisket overnight and store it in the fridge to keep it fresh.
4. On the next day, take the marinated beef brisket out of the fridge, then unwrap and thaw at room temperature.
5. Next, plug in the Electric Smoker and press the power button to turn it on.
6. Press the "Temperature" button and set it to 225°F (107°C).
7. After that, add wood chips to the smoker and pour water into the water pan.
8. Place the seasoned beef brisket in the Electric Smoker and smoke until the internal temperature of the smoked beef brisket has reached 190°F (88°C).
9. The smoking time will take approximately 3 to 4hours. Add more wood chips if it is needed.
10. Once it is done, remove the smoked beef brisket from the Electric Smoker and transfer it to a serving dish.
11. Cut the smoked beef brisket into thick slices, then serve.
12. Enjoy!

Nutrition:

Calcium: 30mg **Magnesium:** 33mg **Phosphorus:** 232mg **Iron:** 3.93mg **Potassium:** 634mg **Sodium:** 3174mg **Zinc:** 5.39mg

368. Juicy Crumbled Smoked Sirloin Beef Steak

Preparation time: 10 minutes.

Cooking time: 5 hours.

Servings: 10

Ingredients:

- 5-pound (2.3-kilograms) beef sirloin

The Marinade:

- ½ cup balsamic vinegar
- ½ cup soy sauce
- ½ cup olive oil
- ¼ cup Worcestershire sauce
- 2 tablespoons honey
- 2 teaspoons Italian seasoning
- 2 teaspoons garlic powder
- 2 teaspoons onion powder
- 2 teaspoons dried mustard
- 1 tablespoon brown sugar
- 2 teaspoons sweet paprika
- 1(½) tablespoons Kosher salt

Directions:

1. Pour balsamic vinegar, soy sauce, olive oil, Worcestershire sauce, and honey into a zipper-lock plastic bag, then season with Italian seasoning, garlic powder, onion powder, dried mustard, sweet paprika, and kosher salt. Stir the spice mixture until well combined.
2. Score the beef sirloin at several places, then put it into the zipper-lock plastic bag.
3. Seal the zipper-lock plastic bag, then shake until the beef sirloin is completely coated with the spice mixture.
4. Marinate the beef sirloin for at least 4 hours and store it in the fridge to keep it fresh.
5. After 4 hours, remove the seasoned beef sirloin from the fridge and thaw at room temperature.
6. Next, plug in the Electric Smoker and press the power button to turn it on.
7. Press the "Temperature" button and set it to 225°F (107°C).
8. After that, add wood chips to the smoker and pour water into the water pan.
9. Place the seasoned beef sirloin in the Electric Smoker and smoke for approximately 5 hours. Add more wood chips if it is necessary.
10. Once the smoked beef sirloin steak is tender or the internal temperature of the smoked beef sirloin steak has reached 190°F (88°C), take it out of the Electric Smoker.
11. Place the smoked beef sirloin steak on a serving dish and cut it into thick slices.
12. Serve and enjoy immediately.

369. Savory Smoked Beef Chuck Roast with Red Wine Sauce

Preparation time: 10 minutes.

Cooking time: 6 hours.

Servings: 10

Ingredients:

- 4-pound (1.8-kilograms) beef chuck roast

The rub:

- 1(¼) tablespoons Kosher salt
- 1 teaspoon pepper

The sauce:

- 2 tablespoons olive oil
- 1 teaspoon onion powder
- 1(½) teaspoons garlic powder
- 2 bay leaves
- ½ cup dried red wine
- 1(½) teaspoons balsamic vinegar
- 2 teaspoons Worcestershire sauce
- 2 teaspoons soy sauce
- ½ cup beef broth
- Brown sugar, a pinch

Directions:

1. Rub the beef chuck roast with salt and pepper, then set aside.
2. Pour beef broth and olive oil into a heavy-duty aluminum pan, then add onion powder, garlic powder, dried red wine, balsamic vinegar, Worcestershire sauce, soy sauce, and brown sugar. Stir the sauce until incorporated.
3. Next, plug in the Electric Smoker and press the power button to turn it on.
4. Press the "Temperature" button and set it to 225°F (107°C).
5. After that, add wood chips to the smoker and pour beer into the water pan.
6. Place the seasoned beef chuck roast in the aluminum pan with sauce and flip until all sides of the beef chuck roast are completely coated with the sauce mixture.
7. Once the Electric Smoker is ready, place the aluminum pan with beef chuck roast in it and set the time to 6 hours. Smoke the beef chuck roast.
8. Regularly check the temperature of the Electric Smoker and add more wood chips if it is necessary.
9. Check the internal temperature of the smoked beef chuck roast and once it reaches 125°F (52°C), remove it from the Electric Smoker.
10. Place the smoked beef chuck roast on a serving dish, then drizzle the sauce on top.
11. Serve and enjoy warm.

370.Butter Garlic Smoked Beef Rib Eye Rosemary

Preparation time: 10 minutes

Cooking time: 3 hours

Servings: 10

Ingredients:

- 5-pound (2.3-kilograms) beef rib eye

The rub:

- 3 tablespoons minced garlic
- ½ teaspoon grated lemon zest
- 1 teaspoon dried thyme
- 1 teaspoon dried rosemary
- 1 teaspoon dried basil
- ¾ teaspoon Kosher salt
- ½ teaspoon black pepper

The topping:

- 1 cup cold butter cubes

Directions:

1. Rub the beef rib eye with minced garlic, grated lemon zest, dried thyme, dried rosemary, dried basil, kosher salt, and black pepper, then place in a heavy-duty aluminum pan.
2. Sprinkle cold butter cubes over the seasoned beef rib eye and put fresh rosemary on top.
3. Plug in the Electric Smoker and press the power button to turn it on.
4. Press the "Temperature" button and set it to 225°F (107°C).
5. After that, add wood chips to the smoker and pour water into the water pan. Add fresh rosemary to the water pan.
6. When the Electric Smoker is ready, place the aluminum pan in it, then smoke the seasoned beef rib eye.
7. Set the time to 3hours and once the internal temperature of the smoked beef rib eye has reached 125°F (52°C), remove it from the Electric Smoker.
8. Place the smoked beef rib eye on a serving dish, then serve.
9. Enjoy warm.

Nutrition:

Calcium: 47mg **Magnesium:** 43mg
Phosphorus: 452mg **Iron:** 5.38mg
Potassium: 769mg **Sodium:** 505mg
Zinc: 17.89mg

371. Oak-Smoked Top Round

Preparation time: 10 minutes.

Cooking time: 5 hours.

Servings: 12

Ingredients:

- 12 hamburger buns
- 1 beef top round
- 3 tablespoon melted butter
- Kosher salt, black pepper

Directions:

1. Add oak wood chips to the Wood Chips box, then plug the smoker and heat it to 275°F.
2. Coat the meat with salt and pepper on the meat. Transfer the meat to the cooking grates and allow to smoke for about 5 hours. Or when the internal temperature of the beef records 145°F.
3. Transfer the meat to an aluminum foil. Let it rest for about 15–20minutes. Brush the sides with melted butter.
4. Slice the meat thinly on the buns. Serve immediately.

Nutrition:

Calories: 310kcal **Carbs:** 32g **Fat:** 50g **Protein:** 46.5g

372. Slam Dunk Brisket

Preparation time: 20 minutes.

Cooking time: 9 hours.

Servings: 8

Ingredients:

- 1/4 cup of Dijon mustard
- 7 pounds of brisket
- 1/4 cup of pickle juice
- 1 teaspoon of onion powder
- 1 teaspoon of garlic powder
- Kosher salt and black pepper

Directions:

1. Put the brisket on a chopping board and trim off the fat. Put the brisket on a baking sheet.
2. Get a bowl to mix the Dijon mustard and pickle juice. Rub the mixture on the brisket, then sprinkle salt and pepper.
3. Also, add onion powder and garlic powder to the brisket.
4. Heat the Electric Smoker to 275°F. Fill up the wood chip box. Transfer the brisket to the smoker to cook for about 8–9 hours until the internal temperature of the meat reads 165°F. Remove the brisket when it is ready.
5. Let it rest for about 1 hour. Slice the brisket in thick inches, then serve.

373. Smoked Beef Tenderloin

Preparation time: 10 minutes.

Cooking time: 2 hours.

Servings: 8

Ingredients:

- Cracked black pepper (ground)
- 2 tablespoon olive oil
- 4 pounds beef tenderloin
- Kosher salt

Directions:

1. Heat the Electric Smoker to 275°F.
2. Put the beef tenderloin on a baking sheet. Sprinkle with salt and pepper, then rub olive oil on all sides.
3. Transfer the beef tenderloin to the cooking grates of the smoker and allow it to smoke for about 2 hours. When the internal temperature of the meat records 180°F, remove them from the smoker and let it rest for 10minutes.
4. Brush olive oil on the beef.
5. Put the beef on the serving place and serve accordingly.

Nutrition:

Calories: 390kcal **Carbs:** 50.5g
Protein: 43.5g **Fat:** 54g

374. Smoked Tri-Tip

Preparation time: 20 minutes.

Cooking time: 2 hours.

Servings: 6

Ingredients:

- 2 teaspoon garlic powder
- 2 pounds beef tri-tip
- 2 teaspoon kosher salt
- 2 teaspoon dried rosemary
- Olive oil
- 1 teaspoon dried oregano
- 2 teaspoon ground pepper

Directions:

1. Heat the smoker to 275°F.
2. Mix the oregano, salt, garlic powder, pepper, and rosemary in a bowl. Brush the seasoning mix on the tri-tip.
3. Transfer the tri-tip to the Electric Smoker cooking grates. Smoke until the internal temperature is at 140°F for about 2hours.
4. Remove the tri-tip from the smoker and place on aluminum foil. Let it rest for about 10minutes.
5. Rub olive oil on all the sides of the tri-tip.
6. Serve it hot.

375. Smoked Bison Sirloin

Preparation time: 10 minutes.

Cooking time: 4 hours.

Servings: 8

Ingredients:

- 2 tablespoon summer savory, thyme, rosemary, fresh herbs, and sage
- 6 pounds bison sirloin
- 3 tablespoon olive oil
- 3 tablespoon spice rub
- Salt and black pepper

Directions:

1. Heat the smoker to a temperature of 275°F.
2. Brush the oil on the meat, then sprinkle with salt and pepper. Mix the other ingredients in a bowl, then rub the seasoning mix on the meat.
3. Transfer the meat to the cooking grates of the Electric Smoker; let it smoke for about 4 hours until the temperature of the beef records 160°F.
4. Remove it from the smoker. Cool for 20 minutes, then slice and serve.

Nutrition:

Calories: 325kcal **Carbs:** 51.2g **Fat:** 47g **Protein:** 48g

376. Cherry Smoked Strip Steak

Preparation time: 10 minutes.

Cooking time: 2 hours, 30minutes.

Servings: 3

Ingredients:

- Kosher salt and black pepper
- 1 teaspoon garlic powder
- 1 teaspoon onion powder
- Olive oil
- 1–1/2 pound of boneless strip steak

Directions:

1. Add cherry wood chips to the Electric Smoker wood chip box, then heat the smoker to 275°F.
2. Sprinkle salt and pepper on the steak. Add garlic powder and onion powder.
3. Place the steak in the smoker. Let it smoke for about 90 minutes until an internal temperature of 180°F is measured.
4. Remove the steak from the smoker and rub it with olive oil. Let it rest for about 15 minutes.
5. Serve immediately.

Nutrition:

Calories: 402kcal **Fat:** 44.5g **Carbs:**5 9g **Protein:** 49g.

377. Smoked Prime Rib

Preparation time: 15 minutes.

Cooking time: 3 hours.

Servings: 6

Ingredients:

- Ground black pepper
- Kosher salt
- Onion powder and garlic powder
- 6 pounds prime rib
- 2 tablespoon olive oil

Directions:

1. Rub the ribs with salt and black pepper, add garlic powder and onion powder, then brush with olive oil.
2. Heat the Electric Smoker to 275°F. Transfer the seasoned rib to the cooking racks.
3. Smoke the rib for 3hours 15 minutes until it records an internal temperature of 135°F
4. Remove the rib from the smoker. Place it on the serving dish and trim off excess fat lumps. Serve.

Nutrition:

Calories: 291kcal **Fat:** 24g **Protein:** 20g **Carbs:** 30g

378. Meaty Chuck Short Ribs

Preparation time: 20 minutes.

Cooking time: 5 hours.

Servings: 4

Ingredients:

- 4 tablespoon olive oil
- 4 pounds beef chuck short rib
- 4 tablespoon Pete's western rub

Directions:

1. Remove excess fat on the beef chuck. Drizzle the beef with oil. Season it with Pete's western rub.
2. Heat the Electric Smoker to 275°F. Transfer the rib to the smoker racks.
3. Let it smoke for about 5 hours until the internal temperature reaches 180°F.
4. Let the rib rest for about 20 minutes before serving.

Nutrition:

Calories: 287kcal **Carbs:** 40g **Protein:** 37g **Fat:** 31g

379. Smoked Pete-zza Meatloaf

Preparation time: 10 minutes.

Cooking time: 8 hours.

Servings: 8

Ingredients:

- 1 cup pizza sauce
- 1 pound ground beef
- 1/2 teaspoon salt
- 1/2 teaspoon garlic powder
- 1 cup bread crumb
- 2 big eggs
- 1/2 teaspoon ground pepper
- 2 tablespoon olive oil
- 3 ounces pepperoni sausage
- 2 cups mozzarella cheese
- 1 cup Portobello mushroom
- 2 cups shredded cheddar
- 2/3 cup green bell pepper
- 1/2 cup red bell pepper
- 2/3 cup red onion, sliced

Directions:

1. In a medium bowl, add the eggs, ground pepper, 1/2 cup of pizza sauce, garlic powder, and salt. Whisk together.
2. Get a skillet, heat the olive oil, fry the red bell pepper, mushroom, green bell pepper, and red onion for about 2minutes. Sprinkle salt and black pepper on the mixture.
3. Get a parchment paper, put the meatloaf on it, then top with pepperoni, place the fried vegetables, and mozzarella. Roll the meatloaf with the parchment paper.
4. Heat the smoker to 275°F. Place the wrapped meatloaf on the smoker rack. Smoke it for about 1hour. Check if the internal temperature is at 180°F.
5. Remove the meatloaf. Let it rest for about 10minutes. Serve it with 1/2 cup of pizza sauce.

Nutrition:

Calories: 345kcal **Carbs:** 31g **Protein:** 48g **Fat:** 36g

380. Smoked Tenderloin Teriyaki

Preparation time: 30 minutes.

Cooking time: 6 hours.

Servings: 10

Ingredients:

- 4(½) pound (2-kilograms) beef tenderloin

The rub:

- 2 cups brown sugar

- ½ cup Worcestershire sauce
- 3 cups Teriyaki sauce
- 1 teaspoon liquid smoke flavoring
- ½ teaspoon meat tenderizer

Directions:

1. Combine brown sugar with Worcestershire sauce, teriyaki sauce, liquid smoke flavoring, and the meat tenderizer in a bowl. Mix well.
2. Rub the tenderloin with the spice, then marinate overnight.
3. In the morning, remove the spiced tenderloin from the refrigerator, then let it sit for about 30minutes.
4. Heat an electric smoker to 225°F (107°C).
5. Wrap the spiced tenderloin with aluminum foil, then place it in the smoker.
6. Smoke the tenderloin for 6hours and check once every hour. Add soaked hickory wood chips as needed.
7. After 6hours and the internal temperature has reached 165°F (74°C), remove the smoked tenderloin from the smoker, then place it on a flat surface. Let it cool.
8. Once it is cool, cut the smoked tenderloin, then arrange it on a serving dish.
9. Serve and enjoy.

Nutrition:

Calcium: 108mg **Magnesium:** 04mg **Phosphorus:** 668mg **Iron:** 9.24mg **Potassium:** 1098mg **Sodium:** 1815mg **Zinc:** 8.72mg

381. Smoked Corned Beef

Preparation time: 30 minutes.

Cooking time: 4 hours.

Servings: 10

Ingredients:

- 5 pound (2.3-kilograms) beef brisket

The brine:

- 2(½) quarts coldwater
- 3 bottles of beer
- ½ cup chopped onion
- 1(¼) cups salt
- ¾ cup brown sugar
- ¼ cup pickling spice
- 2(½) tablespoons minced garlic

The braising:

- ¾ bottle of beer
- ¼ cup chopped onion
- 2(½) tablespoons brown sugar
- 1(½) tablespoons pickling spice

- 2 tablespoons minced garlic
- 1 teaspoon black pepper

Directions:

1. Place chopped onion, salt, brown sugar, pickling spice, and minced garlic in a bowl.
2. Pour beer and cold water over the spices, then mix until incorporated.
3. Soak the beef brisket in the spice liquid, then marinate overnight.
4. In the morning, heat an electric smoker to 250°F (120°C).
5. Place the soaked beef brisket, then place it on the smoker's rack.
6. Smoke the beef brisket for 2(½) hours. Add more soaked wood chips if it is necessary.
7. Meanwhile, combine the entire braising ingredients, then pour into a disposable aluminum foil.
8. After the beef brisket has been smoked for 2(½) years. Take the beef out of the smoker, then place it in the braising mixture.
9. Return the beef brisket to the smoker and smoke for an hour and a half.
10. Once it is done and the internal temperature is 145°F (60°C), remove the smoked beef from the smoker, then place it on a flat surface. Let it cool for a few minutes.
11. When the beef brisket is already cool, cut the smoked corned beef into thin slices, then arrange it on a serving platter.
12. Serve and enjoy.

Nutrition:

Calcium: 52mg **Magnesium:** 12mg **Phosphorus:** 28mg **Iron:** 0.44mg **Potassium:** 114mg **Sodium:** 14350mg **Zinc:** 0.14mg **Copper:** 0.05mg

382. Smoked Beef Ribs

Preparation time: 30 minutes.

Cooking time: 5 hours, 30 minutes.

Servings: 3

Ingredients:

- 10 pound (4.5-kilograms) beef ribs

The sauce:

- 1 tablespoon salt
- ½ cup brown sugar
- 1(½) tablespoons garlic powder
- 1(½) tablespoons onion powder
- 1(¼) tablespoons black pepper

Directions:

1. Heat the electric smoker to 225°F (107°C).

2. Combine the salt, brown sugar, onion powder, garlic powder, and black pepper, then rub the beef ribs with the mixture. Let it sit for about 30minutes.
3. Place the seasoned ribs on a smoker's rack, then smoke for 4hours.
4. After 4 hours, take the beef ribs from the smoker, then carefully wrap them with aluminum foil.
5. Return the beef ribs to the smoker, then smoke again for about 1(1/2) hour or until the internal temperature has reached 160°F (71°C).
6. Remove the smoked beef ribs from the smoker, then let it warm for a few minutes.
7. Unwrap the smoked beef ribs, then transfer them to a serving dish.
8. Serve and enjoy.

Nutrition:

Calcium: 37mg **Magnesium:** 35mg **Phosphorus:** 213mg **Iron:** 3.32mg **Potassium:** 425mg **Sodium:** 794mg **Zinc:** 6.15mg

383. Sweet Smoked Roast Black Pepper

Preparation time: 30 minutes.

Cooking time: 5 hours.

Servings: 10

Ingredients:

- 5 pound (2.3-kilograms) round beef roast

The rub:

- 2(½) teaspoons salt
- 2 tablespoons black pepper
- 1 tablespoon sugar
- 1(½) tablespoons onion powder
- 1(½) tablespoons cayenne pepper

The gravy:

- 1 cup beef broth
- ½ cup chopped onion
- 1 cup apple juice
- 3 tablespoons olive oil

Directions:

1. Heat an electric smoker to 230°F (110°C).
2. Combine the rub ingredients, then rub the beef roast.

3. Wrap the seasoned beef roast with aluminum foil, then place it on the smoker's rack.
4. Smoke the beef roast for 5 hours or until the internal temperature has reached 165°F (74°C).
5. Take the beef roast out from the smoker, then let it warm for a few minutes.
6. Meanwhile, sprinkle chopped onion in a disposable aluminum pan, then pour beef broth, apple juice, and olive oil over the chopped onion.
7. Unwrap the beef roast and place it in the disposable aluminum pan.
8. Return the beef roast back to the smoker, then smoke for 5 hours.
9. Once it is cooked, remove it from the smoker, then let it cool.
10. Cut the smoked beef roast into slices, then serve.
11. Enjoy.

Nutrition:

Calcium: 18mg **Magnesium:** 7mg **Phosphorus:** 13mg **Iron:** 1.12mg **Potassium:** 91mg **Sodium:** 640mg **Zinc:** 0.09mg

384. Smoked Meatloaf Tomato

Preparation time: 10 minutes.

Cooking time: 2 hours.

Servings: 10

Ingredients:

- 2(½) pound (1.3-kilograms) ground beef

The spice:

- ½ cup diced tomato
- 2 organic eggs
- ¼ cup chopped onion
- ¼ cup breadcrumbs
- 3 teaspoons salt
- 2 teaspoons black pepper
- 1 teaspoon paprika

The sauce:

- ¼ cup tomato puree
- 2 tablespoons diced tomato
- 1(½) tablespoons sugar
- ½ teaspoon pepper
- ½ teaspoon nutmeg
- ¼ cup chopped onion

Directions:

1. Heat an electric smoker to 225°F (110°C) and coat a medium loaf pan with cooking spray.
2. Place ground beef in a food processor, then add diced tomato, eggs, chopped onion, and breadcrumbs.
3. Season with salt, black pepper, and paprika, then pulse to combine.
4. Transfer the mixture to the prepared loaf pan, then spread evenly.
5. Place the loaf pan in the smoker, then smoke for 2 hours.
6. Meanwhile, place all of the sauce ingredients in a saucepan, then stir well. Bring to a simmer, then set aside.
7. Monitor the internal temperature and when it reaches 160°F (70°C), remove the meatloaf from the smoker.
8. Let the meatloaf cool, then transfer to a serving dish.
9. Drizzle the sauce over the meatloaf then enjoy.

Nutrition:

Calcium: 43mg **Magnesium:** 30mg **Phosphorus:** 245mg **Iron:** 3.82mg **Potassium:** 437mg **Sodium:** 799mg **Zinc:** 7.36mg

385. Smoked Beef Bites With Brown Sauce

Preparation time: 12 hours.

Cooking time: 1 hour, 10 minutes.

Servings: 10

Ingredients:

- 5 pound (2.3-kilograms) beef steak

The marinade:

- 3 cups red wine vinegar
- 1 cup olive oil

The sauce:

- 3(½) tablespoons butter
- 2 teaspoons sliced shallots
- 1(½) cups dry red wine
- 1(½) cups beef broth
- ½ teaspoon thyme
- 2 tablespoons flour
- 1 teaspoon black pepper
- ¼ teaspoon salt

Directions:

1. Cut the beef steak into medium cubes, then place in a container with a lid.
2. Combine red wine vinegar with olive oil, then pour the mixture over the beef cubes.

3. Marinade the beef cubes for 12hours. Store in a refrigerator to keep them fresh.
4. Heat an electric smoker to 250°F (121°C).
5. After 12 hours, take the beef out from the refrigerator.
6. Transfer the beef cubes to a disposable aluminum pan and discard the marinade.
7. When the smoker has reached the desired temperature, place the pan in the smoker, and smoke the beef cubes for an hour.
8. Meanwhile, cook the sauce.
9. Heat a saucepan over medium heat, then add butter to the saucepan.
10. Once the butter is melted, stir in sliced shallot, then sauté until wilted and aromatic.
11. Pour beef broth and dry red wine over the shallot, then season with salt, black pepper, and thyme. Bring to boil.
12. Once it is boiled, take about ½ cup of the gravy, then add flour into it. Mix well.
13. Pour the flour mixture into the saucepan and stir to combine. Remove from heat.
14. When the smoked beef is done, remove it from the smoker, then transfer it to a serving dish.
15. Serve with the brown sauce, then enjoy.

Nutrition:

Calcium: 16mg **Magnesium:** 12mg **Phosphorus:** 69mg **Iron:** 1.6mg **Potassium:** 174mg **Sodium:** 151mg **Zinc:** 1.46mg

386. Smoked Beef Chuck

Preparation time: 10 minutes.

Cooking time: 5 hours.

Servings: 10

Ingredients:

- 5 pounds beef chuck roll
- 1/3 cup ground black pepper
- ¼ cup kosher salt

Directions:

1. Combine pepper and salt.
2. Season the beef with the mixture generously.
3. Heat the smoker to 275°F.
4. Put beef on the smoker. Smoke for about 4 hours or until the temperature reaches 165°F.
5. Remove the beef from the smoker. Wrap in aluminum foil.

6. Put the beef back on the smoker. Smoke for about 5 hours or until temperature reaches 140°F.
7. Remove the beef from the smoker. Let it cool for 30 minutes.
8. Slice meat thinly.
9. Serve with onion, pickles, and white bread.

Nutrition:

Calories: 422 **Total fat:** 24g **Saturated fat:** 10g **Protein:** 47g **Carbs:** 0g **Fiber:** 0g **Sugar:** 0g

387. Texas Style Smoked Beef Brisket

Preparation time: 30 minutes.

Cooking time: 18 hours.

Servings: 20

Ingredients:

- 14 pounds whole brisket
- 2 tablespoons garlic powder
- 2 tablespoons ground black pepper
- 2 tablespoons kosher salt

Directions:

1. Remove fat or silver skin from brisket.
2. Combine garlic powder, pepper, and salt.
3. Season beef with the mixture generously.
4. Heat the smoker to 225°F.
5. Put the beef on the smoker. Smoke for about 8 hours or until the temperature reaches 165°F.
6. Remove the beef from the smoker. Wrap in aluminum foil.
7. Put the beef back on the smoker. Smoke for about 8hours or until temperature reaches 200°F.
8. Remove the beef from the smoker. Let it cool for 1hour.
9. Slice meat and serve.

Nutrition:

Calories: 250 **Total fat:** 19g **Saturated fat:** 7g **Protein:** 18g **Carbs:** 1g **Fiber:** 0g **Sugar:** 0g

388. Smoked Beef Stew

Preparation time: 30 minutes.

Cooking time: 10 hours.

Servings: 8

Ingredients:

- 2 pounds stewing beef, cubed
- 5 cups beef broth
- 1 can diced tomatoes
- 8 carrots, peeled and diced

- 8 medium potatoes, peeled and diced
- 2 onions, diced
- 2 tablespoons corn starch
- 2 tablespoons water

For the rub:

- 1 tablespoon paprika
- 1 tablespoon sugar
- 2 teaspoon dry oregano
- 1 teaspoon garlic powder
- 1 teaspoon ground black pepper
- 1 teaspoon salt
- ½ teaspoon cayenne pepper
- ½ teaspoon thyme

Directions:

1. Combine the rub ingredients. Mix thoroughly.
2. Season beef with the mixture generously.
3. Heat the smoker to 225°F.
4. Put beef on the smoker. Smoke for about 2hours.
5. Remove beef from the smoker. Put in a slow cooker.
6. Add beef broth, tomatoes, carrots, potatoes, and onions.
7. Cook for 8hours on low heat.
8. Whip corn starch and water together. Add to the slow cooker 15minutes before finishing the stew.
9. Serve with fresh bread.

Nutrition:

Calories: 386 **Total fat:** 18g **Saturated fat:** 6g **Protein:** 52g **Carbs:** 28g **Fiber:** 4g **Sugar:** 2g

389. Smoked Beef Jerky

Preparation time: 10 minutes.

Cooking time: 7 hours.

Servings: 6

Ingredients:

- 2 pounds sirloin, sliced ½ inch thick
- 1 cup soy sauce
- 4 tablespoons ground black pepper
- 1 tablespoon cider vinegar
- 1 dash hot pepper sauce
- 1 dash Worcestershire sauce

Directions:

1. Combine all the ingredients except for the beef. Mix well.
2. Add beef slices. Cover and place in the fridge overnight.
3. Heat the smoker to 170°F.

4. Put beef on the smoker. Smoke for about 7 hours or until jerky edges appear dry.

Nutrition:

Calories: 220 **Total fat:** 4g **Saturated fat:** 2g **Protein:** 28g **Carbs:** 6g **Fiber:** 0g **Sugar:** 3g

390. Smoked Roast Beef

Preparation time: 20 minutes.

Cooking time: 5 hours.

Servings: 8

Ingredients:

- 3 pounds beef rump roast

For the rub:

- 1(½) teaspoon salt
- 1 teaspoon garlic powder
- 1 teaspoon pepper
- 1 teaspoon smoked paprika
- ½ teaspoon onion powder
- Worcestershire sauce

Directions:

1. Combine all the ingredients except for the Worcestershire sauce. Mix well.
2. Rub beef with Worcestershire sauce. Season with the mixture.
3. Heat the smoker to 200°F.
4. Put beef on the smoker. Smoke for about 5hours or until the temperature reaches 150°F.
5. Remove beef from the smoker. Let it cool for 20minutes.
6. Slice and serve.

Nutrition:

Calories: 298 **Total fat:** 10g **Saturated fat:** 4g **Protein:** 46g **Carbs:** 19g **Fiber:** 0g **Sugar:** 10g

391. Smoked Beef Burnt Ends

Preparation time: 15 minutes.

Cooking time: 10 hours.

Servings: 6

Ingredients:

- 3 pounds chuck roast
- ½ cup barbecue sauce
- ¼ cup brown sugar
- 2 tablespoons brown sugar

For the rub:

- 1 tablespoon garlic powder
- 1 tablespoon ground black pepper
- 1 tablespoon kosher salt

Directions:

1. Heat the smoker to 275°F.
2. Combine the rub ingredients.
3. Season the beef generously with the rub mixture.
4. Put beef on the smoker. Smoke for about 5 hours or until the temperature reaches 165°F.
5. Remove the beef from the smoker. Wrap in aluminum foil.
6. Put the beef back on the smoker. Smoke for about 1 hour or until the temperature reaches 195°F.
7. Remove the beef from the smoker. Let it cool for 20 minutes.
8. Slice and transfer to a foil pan.
9. Sprinkle with ¼ cup sugar and drizzle with barbecue sauce.
10. Put the pan on the smoker, close lid, and cook for 2 hours.
11. Sprinkle with 2 tablespoons sugar and drizzle with barbecue sauce.
12. Stir and grill for a few minutes.

Nutrition:

Calories: 208 **Total fat:** 5g **Saturated fat:** 2g **Protein:** 30g **Carbs:** 11g **Fiber:** 0g **Sugar:** 11g

392. Grass-Fed Beef Sirloin Kebabs

Preparation time: 25 minutes.

Cooking time: 3 minutes.

Servings: 4

Ingredients:

- 1 pound grass-fed top sirloin steak, trimmed
- 8 skewers
- Cooking spray
- 2 tablespoons olive oil
- 1 teaspoon ground coriander
- 1 teaspoon black pepper
- ¾ teaspoon kosher salt

For the sauce:

- ½ cup plain 2% reduced-fat Greek yogurt
- 2 tablespoons fresh dill, chopped
- 1 tablespoon fresh lemon juice
- 1 tablespoon lemon rind, grated
- ¼ teaspoon kosher salt

Directions:

1. Heat the smoker to 550°F.
2. Combine the sauce ingredients. Mix thoroughly.

3. Cut steak into 16 strips. Mix with oil, coriander, pepper, and salt.
4. Thread 2 steak strips into each skewer.
5. Put skewers on the smoker coated with cooking spray. Smoke for about 90 seconds on each side.
6. Serve with yogurt sauce.

Nutrition:

Calories: 244 **Total fat:** 14g **Saturated fat:** 6g **Protein:** 27g **Carbs:** 4g **Fiber:** 0g **Sugar:** 3g

393. Smoky Caramelized Onion Burgers

Preparation time: 25 minutes.

Cooking time: 20 minutes.

Servings: 4

Ingredients:

For the patties:

- 1 pound 90% lean ground sirloin
- 1(½) cups yellow onion, thinly sliced
- 1(½) tablespoons minced garlic
- ½ tablespoon ground cumin
- ½ tablespoon olive oil
- 1 teaspoon smoked paprika
- 1 teaspoon kosher salt
- 1 teaspoon ground black pepper
- Cooking spray

For the burgers:

- 4 whole-wheat hamburger buns
- 4 tomato slices
- 2 red onion slices
- 1 cup baby arugula leaves
- 2 tablespoons canola mayonnaise
- 2 tablespoons roasted red bell pepper, finely chopped
- 1 tablespoon minced fresh chives

Directions:

1. Heat the smoker to 450°F.
2. In a skillet, put oil over medium heat. Sauté yellow onion until golden brown.
3. Add garlic, cook until fragrant.
4. Remove skillet from heat. Let cool. Put onion mixture in a large bowl.
5. Add the rest of the patty ingredients. Mix thoroughly.
6. Shape beef mixture into 4 patties.
7. Coat smoker with cooking spray. Put patties on the smoker. Smoke for 4minutes on each side.
8. Remove patties from the smoker.

9. In a bowl, mix bell pepper, mayonnaise, and chives. Mix thoroughly.
10. Put patties on bottom halves of buns. Layer with tomato, red onion, arugula, mayonnaise mixture, and top halves of buns.

Nutrition:

Calories: 340 **Total fat:** 26g **Saturated fat:** 10g **Protein:** 23g **Carbs:** 29g **Fiber:** 1g **Sugar:** 5g

394. Smoked Pulled Beef

Preparation time: 10 minutes.

Cooking time: 12 hours.

Servings: 10

Ingredients:

- 6 pounds chuck roast
- 3 cups beef stock
- 1 yellow onion, sliced

For the rub:

- 2 tablespoons black pepper
- 2 tablespoons garlic powder
- 2 tablespoons kosher salt

Directions:

1. Heat the smoker to 225°F.
2. Combine the rub ingredients. Mix thoroughly.
3. Season the beef and roast generously with the rub mixture.
4. Put beef on the smoker. Smoke for about 3 hours.
5. Spray with 1 cup of beef stock every hour while smoking.
6. Spread sliced onions in an aluminum pan. Pour 2 cups of beef stock. Place the beef roast on top of the onions.
7. Put the pan on the smoker. Increase the temperature to 250°F.
8. Smoke for about 3 hours or until the temperature reaches 165°F.
9. Cover the pan tightly with aluminum foil. Continue smoking for 5 hours or until the temperature reaches 202°F.

Nutrition:

Calories: 422 **Total fat:** 24g **Saturated fat:** 10g **Protein:** 47g **Carbs:** 0g **Fiber:** 0g **Sugar:** 0g

395. Crowd Pleasing Meatloaf

Preparation time: 20 minutes.

Cooking time: 4 hours.

Servings: 6

Ingredients:

For meatloaf:

- 2 pounds ground beef
- ½ cup panko bread crumbs
- 2 eggs, lightly beaten
- ¼ cup milk
- ½ medium red onion, grated
- 2 garlic cloves, minced
- 2 tablespoon whiskey
- 1 tablespoon Worcestershire sauce
- 1 tablespoon steak rub
- 6-ounce pepper jack cheese, cut into strips

For sauce:

- ½ cup ketchup
- 1/3 cup brown sugar
- ¼ cup whiskey
- 1 tablespoon steak rub
- 2 teaspoon red pepper flakes, crushed

Directions:

1. Heat the smoker to 225°F.
2. Soak oak wood chips in water for at least 1hour.
3. **For the meatloaf:** in a large bowl, add all the ingredients and mix until well combined.
4. Place half of the meat mixture in the bottom of a grill basket.
5. Place the cheese over the meatloaf, leaving about 1-inch on all sides.
6. Move the remaining meat mixture on top and press the edges together to seal completely.
7. **For the sauce:** in a small bowl, add all the ingredients and mix until well combined.
8. Spread the sauce on top of the meatloaf evenly.
9. Place the meatloaf into the smoker and cook, covered for about 4hours.
10. Transfer the meatloaf onto a wire rack for about 5minutes before slicing.
11. Cut into the desired sized slices and serve.

Nutrition:

Calories: 552 **Carbohydrates:** 22.6g **Protein:** 56.9g **Fat:** 20.9g **Sugar:** 14.5g **Sodium:** 727mg **Fiber:** 0.9g

396. 2 Meats Combo Meatloaf

Preparation time: 20 minutes.

Cooking time: 4 hours.

Servings: 6

Ingredients:

- 1(½) pounds ground pork
- 1(½) cups BBQ sauce, divided
- 2 pounds ground beef chuck
- 2 roasted bell peppers, chopped
- 1/3 cup onion, chopped finely
- 4 garlic cloves, minced
- 2 eggs, beaten
- ¾ cup fresh breadcrumbs
- 1 tablespoon dried oregano, crushed
- Salt and freshly ground black pepper

Directions:

1. Heat the smoker to 225°F, using charcoal.
2. In a large bowl, add ½ cup of BBQ sauce and the remaining all ingredients and mix until well combined.
3. Arrange a 24-inch piece of foil in a small baking sheet, doubling it over by folding in half.
4. Mold the sides of foil upwards to make a loaf pan.
5. Place the meat mixture in loaf pan and press to form a meatloaf.
6. Place the loaf pan over smoker rack and cook for about 3-4hours.
7. In the last hour of coking, coat the meatloaf with the remaining BBQ sauce.
8. Transfer the meatloaf onto a wire rack for about 5 minutes before slicing.
9. Cut into desired sized slices and serve.

Nutrition:

Calories: 736 **Carbohydrates:** 36.8g **Protein:** 59.4g **Fat:** 37.2g **Sugar:** 19.5g **Sodium:** 1012mg **Fiber:** 1.7g

LAMB RECIPES

397. Orange Smoked Lamb Leg With Honey Beer Marinade

Preparation time: 30 minutes.

Cooking time: 2 hours.

Servings: 10

Ingredients:

- 4-pound (1.8-kilograms) boneless lamb leg

The marinade:

- 1(½) cups beer
- ½ cup honey
- 2 tablespoons minced garlic
- 1 tablespoon onion powder
- ¼ cup kosher salt
- 1 tablespoon black pepper
- ¼ cup Worcestershire sauce
- 2 tablespoons mustard
- 5 sprigs of fresh rosemary

The rub:

- 1 tablespoon onion powder
- 1 tablespoon garlic powder
- 3 tablespoons lemon juice
- 2 tablespoons apple cider vinegar
- 2 tablespoons olive oil
- 1 tablespoon kosher salt
- 1 teaspoon cumin
- 1 teaspoon black pepper
- 1/2 teaspoon cayenne pepper
- 1 teaspoon cinnamon
- 1 teaspoon nutmeg
- ¼ teaspoon ground clove

The glaze:

- 1 cup orange marmalade
- ½ cup honey

Directions:

1. Pour beer and honey into a container, then season with minced garlic, onion powder, kosher salt, black pepper, Worcestershire sauce, mustard, and fresh rosemary. Stir well.
2. Score the boneless lamb leg at several places, then add to the beer mixture.
3. Marinate the lamb leg overnight and store it in the fridge to keep it fresh.
4. On the next day, remove the lamb leg from the fridge, then take it out of the marinade.
5. Combine the rub ingredients—onion powder, garlic powder, lemon juice, apple cider vinegar, olive oil, kosher salt, cumin, black pepper, cayenne pepper, cinnamon, nutmeg, and ground clove then mix well.
6. Rub the marinated lamb leg with the spice mixture, then set aside.
7. Next, plug in and turn the Electric Smoker on, then set the temperature to 250°F (121°C).
8. Wait until the Electric Smoker has reached the desired temperature, then add wood chips to the chip tray.
9. Pour orange juice into the water pan and add ginger to the water pan. Wait until the smoke is ready.
10. Place the seasoned lamb leg on the grill tray inside the Electric Smoker and smoke for 4hours.
11. In the meantime, mix orange marmalade with honey and stir until incorporated.
12. After 2 hours of smoking, baste the honey orange mixture over the lamb leg once every 30 minutes and continue smoking until the internal temperature of the smoked lamb leg has reached 145°F (63°C).
13. Remove the smoked lamb leg from the Electric Smoker and let it rest for approximately 10 minutes.

14. Cut the smoked lamb leg into slices and serve.

15. Enjoy.

Nutrition:

Calcium: 169mg **Magnesium:** 68mg
Phosphorus: 435mg **Iron:** 5.36mg
Potassium: 866mg

398. Garlic Mint Smoked Lamb Chops Balsamic

Preparation time: 10 minutes.

Cooking time: 4 hours.

Servings: 10

Ingredients:

- 6-pound (2.7-kilograms) lamb chops

The rub:

- 1 teaspoon dried mint leaves
- 1/4 cup olive oil
- 1/4 cup minced garlic
- 1 teaspoon black pepper
- 1 teaspoon dried rosemary
- 1/2 teaspoon oregano
- 1/2 teaspoon dried thyme
- 1(1/2) teaspoon Kosher salt

The glaze:

- 1 cup balsamic vinegar
- 6 tablespoons brown sugar
- 1/2 teaspoon black pepper

Directions:

1. Combine the rub ingredients—dried mint leaves, olive oil, minced garlic, black pepper, dried rosemary, oregano, dried thyme, and kosher salt, then stir well.
2. Rub the lamb with the spice mixture, then set aside.
3. Turn the Electric Smoker on, then set the temperature to 225°F (107°C).
4. Wait until the Electric Smoker has reached the desired temperature, then add wood chips to the chip tray. Pour water into the water pan.
5. Put the seasoned lamb chops on the grill tray provided by the Electric Smoker and smoke for 2hours.
6. Combine balsamic vinegar with brown sugar and black pepper, then stir until incorporated.
7. After 2hours, transfer the lamb chops to a disposable aluminum pan, then drizzle the balsamic vinegar mixture over the lamb chops.
8. Continue smoking the lamb chops for another 2 hours or until the internal temperature of

the smoked lamb chops has reached 135°F (57°C).

9. Remove the smoked lamb chops from the Electric Smoker, then transfer to a serving dish.

10. Serve and enjoy.

Nutrition:

Calcium: 58mg **Magnesium:** 66mg **Phosphorus:** 520mg **Iron:** 4.35mg **Potassium:** 923mg **Sodium:** 216mg **Zinc:** 7.34mg

399. Sweet Cherry Smoked Lamb Ribs Barbecue

Preparation time: 10 minutes.
Cooking time: 3 hours.
Servings: 10

Ingredients:

- 4-pound (1.8-kilograms) lamb ribs

The rub:

- ½ cup brown sugar
- 1 tablespoon ginger
- ½ teaspoon dried thyme
- ½ teaspoon dried tarragon
- ½ teaspoon dried marjoram
- 1(½) tablespoons cinnamon
- ½ teaspoon black pepper
- ¾ tablespoon Kosher salt
- 2 tablespoons lemon juice

The glaze:

- 1 cup cherry cola
- 1 cup fresh cherry
- ½ cup ketchup
- ½ teaspoon onion powder
- ½ teaspoon garlic powder
- ¼ cup Worcestershire sauce
- 3 tablespoons lemon juice
- 2 tablespoons white vinegar
- 2 tablespoons soy sauce
- 3 tablespoons brown sugar
- 1 tablespoon mustard
- 1(½) tablespoons olive oil
- ½ teaspoon black pepper
- ½ teaspoon liquid smoke

Directions:

1. Cut and trim the excess fat from the lamb ribs, then rub with the mixture of brown sugar, ginger, dried thyme, dried marjoram, dried tarragon, cinnamon, black pepper, kosher salt, and lemon juice. Set aside.

2. Plug in and turn the Electric Smoker on, then set the temperature to 225°F (107°C).

3. Wait until the Electric Smoker has reached the desired temperature, then

add wood chips to the chip tray. Pour water into the water pan.

4. Once the Electric Smoker is ready, place the seasoned lamb ribs on the grill tray inside the Electric Smoker and smoke for an hour.

5. In the meantime, place the entire glaze ingredients—fresh cherries, cherry cola, ketchup, garlic powder, onion powder, Worcestershire sauce, lemon juice, white vinegar, soy sauce, brown sugar, mustard, olive oil, black pepper, and liquid smoke in a blender then blend until smooth.

6. After an hour of smoking, baste the glaze mixture over the lamb ribs, then wrap it with aluminum foil.

7. Return the wrapped lamb ribs to the Electric Smoker and smoke for about 2hours or until the internal temperature has reached 63°C

8. Once it is done, remove the smoked lamb ribs from the Electric Smoker and unwrap it.

9. Transfer the smoked lamb ribs to a serving dish and enjoy!

Nutrition:

Calcium: 18mg **Magnesium:** 8mg **Phosphorus:** 22mg **Iron:** 0.89mg **Potassium:** 152mg **Sodium:** 256mg **Zinc:** 0.14mg

400. Minty Apple Smoked Pulled Lamb

Preparation time: 30 minutes.

Cooking time: 5 hours, 30 minutes.

Servings: 10

Ingredients:

- 4-pound (1.8-kilograms) boneless lamb shoulder

The rub:

- ½ cup brown sugar
- 1 tablespoon Kosher salt
- 1 teaspoon black pepper
- 1 teaspoon dried thyme
- 1(½) teaspoons smoked paprika

The spray:

- 1(½) cups apple juice
- 3 tablespoons apple cider vinegar
- 3 sprigs fresh mint leaves

The sauce:

- 1 cup apple juice
- 1 tablespoon apple cider vinegar
- 3 tablespoons olive oil
- 1 teaspoon onion powder
- 1 teaspoon garlic powder

- 3 tablespoons brown sugar
- ¼ cup mint sauce

Directions:

1. Pour the spray ingredients—apple juice and apple cider vinegar into a spray bottle, then add fresh mint leaves. Shake until combined, then set aside.
2. Score the lamb shoulder at several places, then rub with brown sugar, kosher salt, black pepper, dried thyme, and smoked paprika. Set aside.
3. Plug in and turn the Electric Smoker on, then set the temperature to 225°F (107°C).
4. Wait until the Electric Smoker has reached the desired temperature, then add wood chips to the chip tray. Pour apple juice into the water pan.
5. Place the seasoned lamb shoulder on the grill tray provided by the Electric Smoker, then smoke for approximately 5hours. Spray the apple juice mixture over the lamb shoulder once every 30 minutes.
6. In the meantime, place the entire sauce ingredients—apple juice, apple cider vinegar, olive oil, garlic powder, brown sugar, and mint sauce in a bowl, then stir until incorporated.
7. Once the internal temperature of the smoked lamb has reached 165°F (74°C), baste half of the sauce over it and wrap it with aluminum foil.
8. Return the lamb shoulder to the Electric Smoker and smoke for about 30 minutes.
9. Remove the smoked lamb shoulder from the Electric Smoker, then unwrap it. Using a fork or a sharp knife, shred the smoked lamb shoulder into pieces.
10. Transfer the smoked lamb to a serving dish, then drizzle the remaining sauce on top. Mix well.
11. Serve and enjoy.

Nutrition:

Calcium: 19mg **Magnesium:** 43mg **Phosphorus:** 327mg **Iron:** 2.28mg **Potassium:** 688mg **Sodium:** 167mg **Zinc:** 6.94mg

401. Spicy Brown Smoked Lamb Ribs

Preparation time: 30 minutes.

Cooking time: 4 hours.

Servings: 10

Ingredients:

- 3,5-pound (1.6-kilograms) lamb ribs

The rub:

- ¼ cup brown sugar
- 1 tablespoon Kosher salt
- 1 teaspoon pepper

The sauce:

- 2 cups apricot jam
- 2 tablespoons dried chilies
- 2 tablespoons diced onion
- 1 tablespoon minced garlic
- 1/4 teaspoon ground cloves
- 1/2 teaspoon black peppercorns
- 1/2 teaspoon ground coriander
- 3/4 teaspoon cumin
- 1 teaspoon oregano
- 1/2 teaspoon Kosher salt
- 3 tablespoons canola oil
- 1 teaspoon ground cinnamon

Directions:

1. Pour water over the dried chilies, then soak for approximately 10 minutes or until softened. Discard the water.
2. Place the softened chilies in a blender, add apricot jam, diced onion, ground cloves, minced garlic, black peppercorns, ground coriander, cumin, oregano, kosher salt, canola oil, and ground cinnamon. Blend until smooth, then set aside.
3. Plug in and turn the Electric Smoker on, then set the temperature to 225°F (107°C).
4. Wait until the Electric Smoker has reached the desired temperature, then add wood chips to the chip tray. Pour apple juice into the water pan.
5. Rub the lamb ribs with brown sugar, kosher salt, and pepper, then place in the Electric Smoker. Smoked the lamb ribs for an hour.
6. After an hour of smoking, take the lamb ribs out of the Electric Smoker and baste apricot sauce over it.
7. Wrap the glazed lamb ribs with aluminum foil, then continue smoking for 3hours or until the internal temperature has reached 63°C
8. Once it is done, remove the smoked lamb ribs from the Electric Smoker and let it rest for approximately 30 minutes.
9. Unwrap the smoked lamb ribs, then transfer them to a serving dish.
10. Serve and enjoy.

Nutrition:

Calcium: 48mg **Magnesium:** 52mg **Phosphorus:** 338mg **Iron:** 3.81mg **Potassium:** 573mg **Sodium:** 319mg **Zinc:** 7.03mg

402. Chinese Style Lamb Shanks

Preparation time: 15 minutes.

Cooking time: 10 hours.

Servings: 2

Ingredients:

- 2 (1(¼)pound) lamb shanks
- 1-2 cups water
- ½ cup brown sugar
- ½ cup rice wine
- ½ cup soy sauce
- 3 tablespoon dark sesame oil
- 4 (1½x½-inch) orange zest strips
- 2 (3-inch long) cinnamon sticks
- 1½ teaspoon Chinese-five spice powder

Directions:

1. Soak applewood chips in water for at least 1hour.
2. Heat the smoker to 225–250°F, using charcoal and soaked applewood chips.
3. With a sharp knife, pierce each lamb shank at many places.
4. In a bowl, add the remaining ingredients and mix until sugar is dissolved.
5. In a large roasting pan, place lamb shanks and top with the sugar mixture evenly.
6. In a foil pan, transfer the lamb shanks with the sugar mixture.
7. Place the foil pan into the smoker and cook for about 8–10 hours, flipping after every 30 minutes. (If required, add enough water to keep the liquid ½-inch over.
8. Serve hot.

Nutrition:

Calories: 1500 **Carbohydrates:** 68g **Protein:** 163.3g **Fat:** 62g **Sugar:** 52.3g **Sodium:** 4000mg **Fiber:** 0.5g

403. Cola Flavored Lamb Ribs

Preparation time: 15 minutes.

Cooking time: 3 hours.

Servings: 8

Ingredients:

- 4 (1–1½pound) lamb rib racks, trimmed
- 1 tablespoon unsweetened cocoa powder
- 1 tablespoon brown sugar
- 1 tablespoon smoked paprika
- 1 tablespoon salt

- 1 tablespoon cracked black pepper
- 1 cup cherry cola

Directions:

1. Heat the smoker to 225°F, using charcoal.
2. With a sharp knife, make ½ x ¼-inch cuts in each rib rack.
3. In a bowl, add the remaining ingredients except for cherry cola and mix well.
4. Generously rub the rib racks with sugar mixture.
5. In a spray bottle, place the cherry cola.
6. Arrange the rib racks into the smoker and cook for about 2(½)–3hours, coating with cherry cola after every 1 hour.

Nutrition:

Calories: 547 **Carbohydrates:** 2.5g **Protein:** 64.3g **Fat:** 29.5g **Sugar:** 1.2g **Sodium:** 1060mg **Fiber:** 0.8g

404. Glorious Lamb Shoulder

Preparation time: 15 minutes.

Cooking time: 7 hours.

Servings: 10

Ingredients:

- 1(7-pound) boneless lamb shoulder, excess fat trimmed
- 1(½) tablespoon ancho chili powder
- ½ tablespoon dried oregano
- ½ tablespoon dry mustard powder
- ½ tablespoon ground allspice
- ½ tablespoon ground coriander
- ½ tablespoon garlic powder
- ½ tablespoon celery salt
- ½ tablespoon smoked sweet paprika
- 2 tablespoon canola oil
- 1–2 cups BBQ sauce
- Salt and freshly ground black pepper

Directions:

1. Add all the ingredients except for the lamb, canola oil, and BBQ sauce and mix well.
2. Rub the lamb shoulder with the spice mixture generously.
3. Roll the meat and with the kitchen string, tie at 1-inch intervals.
4. Cover the lamb shoulder and refrigerate overnight.
5. Remove the lamb shoulder from the refrigerator and coat with oil evenly.
6. Set aside at room temperature for about 30 minutes before cooking.
7. Heat the smoker to 225°F, using charcoal and cherry wood chips.

8. Place the lamb shoulder into the smoker, fat side up and cook for about 6–7hours.
9. In the last 30minutes, coat the shoulder with some of the BBQ sauce.
10. Transfer the lamb shoulder onto a cutting board and discard the kitchen twine.
11. Set aside for about 20minutes before serving.
12. With a sharp knife, cut the leg of lamb into the desired sized slices and serve.

Nutrition:

Calories: 699 **Carbohydrates:** 19.7g **Protein:** 89.6g **Fat:** 26.6g **Sugar:** 13.3g **Sodium:** 830mg **Fiber:** 1g

405. Divine Lamb Chops

Preparation time: 15 minutes.

Cooking time: 33 hours.

Servings: 4

Ingredients:

- 4 (10-ounce) lamb shoulder chops
- 4 cups buttermilk
- 1 cup cold water
- ¼ cup kosher salt
- 2 tablespoon olive oil, as required
- 2 tablespoon Texas style rub

Directions:

1. In a large bowl, add buttermilk, water, and salt and stir until salt is dissolved.
2. Add chops and coat with the mixture evenly.
3. Refrigerate for at least 4 hours.
4. Remove the chops from the bowl and rinse under cold water.
5. Coat the chops with olive oil and then sprinkle with rub evenly.
6. Heat the smoker to 240°F, using charcoal.
7. Arrange the chops on the smoker and cook for about 25–30 minutes or until the desired doneness.
8. Meanwhile, heat the broiler of the oven.
9. Broil the chops for about 2–3 minutes or until browned.

Nutrition:

Calorie: 585 **Carbohydrates:** 11.7g **Protein:** 63.3g **Fat:** 31.7g **Sugar:** 11.7g **Sodium:** 7500mg **Fiber:** 0g

406. Foolproof Leg of Lamb

Preparation time: 15 minutes.

Cooking time: 2 hours, 30 minutes.

Servings: 8

Ingredients:

For the leg of lamb:

- 1 (4–5pound) leg of lamb, butterflied
- 2–3 tablespoon olive oil

For filling:

- 1(8-ounce) package cream cheese, softened
- ¼ cup cooked bacon, crumbled
- 1 jalapeño pepper, seeded and chopped

For spice mixture:

- 1 tablespoon dried rosemary, crushed
- 2 teaspoon garlic powder
- 1 teaspoon onion powder
- 1 teaspoon paprika
- 1 teaspoon cayenne pepper
- Salt, to taste

Directions:

1. **For filling:** In a bowl, add all the ingredients and mix until well combined.
2. **For spice mixture:** In another small bowl, mix together all the ingredients.
3. Place the leg of lamb onto a smooth surface.
4. Sprinkle the inside of the leg with some spice mixture.
5. Place filling mixture over the inside surface evenly and roll tightly.
6. With a butcher's twine, tie the roll to secure the filling.
7. Coat the outer side of the roll with olive oil evenly and then sprinkle with spice mixture.
8. Heat the smoker to 225–240°F, using charcoal and cherry wood chips.
9. Arrange the leg of lamb into the smoker and cook for about 2–2(½) hours.
10. Place the leg of lamb onto a cutting board and loosely cover with a piece of foil for about 20–25 minutes before serving.
11. With a sharp knife, cut the leg of lamb into the desired sized slices and serve.

Nutrition:

Calories: 758 **Carbohydrates:** 2.1g **Protein:** 86g **Fat:** 43.1g **Sugar:** 0.5g **Sodium:** 640mg **Fiber:** 0.5g

407. Rosemary-Smoked Lamb Chops

Preparation time: 15 minutes.

Cooking time: 2 hours and 5 minutes.

Servings: 4

Ingredients:

- **Wood Pellet Flavor:** Mesquite
- 4(½) pounds bone-in lamb chops
- 2 tablespoons olive oil
- Salt
- Freshly ground black pepper
- 1 bunch fresh rosemary

Directions:

1. Supply your smoker with wood pellets and follow the manufacturer's specific start-up procedure. Heat the grill to 180°F.
2. Rub the lamb generously with olive oil and season on both sides with salt and pepper.
3. Spread the rosemary directly on the grill grate, creating a surface area large enough for all the chops to rest on. Place the chops on the rosemary and smoke until they reach an internal temperature of 135°F.
4. Increase the grill's temperature to 450°F, remove the rosemary, and continue to cook the chops until their internal temperature reaches 145°F.
5. Take off the lamb from the grill and let them rest for 5 minutes before serving.

Nutrition:

Calcium: 57mg **Magnesium:** 123mg **Phosphorus:** 1011mg **Iron:** 8.43mg **Potassium:** 1789mg **Sodium:** 327mg **Zinc:** 16.14mg

408. Classic Lamb Chops

Preparation time: 10 minutes.

Cooking time: 30 minutes.

Servings: 4

Ingredients

- **Wood Pellet Flavor:** Alder
- 4 (8-ounce) bone-in lamb chops
- 2 tablespoons olive oil
- 1 batch Rosemary-Garlic Lamb Seasoning

Directions:

1. Supply your smoker with wood pellets and follow the manufacturer's specific start-up procedure. Heat the grill to 350°F. Close the lid
2. Rub the lamb generously with olive oil and coat them on both sides with the seasoning.
3. Place the chops on the grill grate and grill until their internal temperature

reaches 145°F. Remove the lamb from the grill and serve immediately.

Nutrition:

Calcium: 7mg **Magnesium:** 14mg **Phosphorus:** 112mg **Iron:** 0.98mg **Potassium:** 200mg **Sodium:** 36mg **Zinc:** 1.79mg

4. Take off the lamb from the grill and let it rest for 20 to 30 minutes before removing the netting, slicing, and serving.

Nutrition:

Calcium: 92mg **Magnesium:** 209mg **Phosphorus:** 1769mg **Iron:** 15.91mg **Potassium:** 2979mg **Sodium:** 726mg **Zinc:** 31.76mg

409. Roasted Leg of Lamb

Preparation time: 15 minutes.

Cooking time: 1–2 hours.

Servings: 4

Ingredients:

- **Wood Pellet Flavor:** Hickory
- 1(6 to 8-pound) boneless leg of lamb
- 2 batches Rosemary-Garlic Lamb Seasoning

Directions:

1. Supply your smoker with wood pellets and follow the manufacturer's specific start-up procedure. Heat the grill to 350°F. Close the lid
2. Using your hands, rub the lamb leg with the seasoning, rubbing it under and around any netting.
3. Put the lamb directly on the grill grate and smoke until its internal temperature reaches 145°F.

410. Smoked Rack of Lamb

Preparation time: 25 minutes.

Cooking time: 4–6 hours.

Servings: 4

Ingredients:

- **Wood Pellet Flavor:** Hickory
- 1 (2-pound) rack of lamb
- 1 batch Rosemary-Garlic Lamb Seasoning

Directions:

1. Supply your smoker with wood pellets and follow the manufacturer's specific start-up procedure. Heat the grill to 225°F. Close the lid
2. Using a boning knife, score the bottom fat portion of the rib meat.
3. Using your hands, rub the rack of lamb all over with the seasoning, making sure it penetrates the scored fat.

4. Place the rack directly on the grill grate, fat-side up, and smoke until its internal temperature reaches 145°F.

5. Take off the rack from the grill and let it rest for 20 to 30 minutes before slicing it into individual ribs to serve.

Nutrition:

Calcium: 92mg **Magnesium:** 209mg **Phosphorus:** 1769mg **Iron:** 15.91mg **Potassium:** 2979mg **Sodium:** 726mg **Zinc:** 31.76mg

411. Roast Rack of Lamb

Preparation time: 10 minutes.

Cooking time: 1 hour.

Servings: 6–8

Ingredients:

- **Wood Pellet Flavor:** Alder
- 1(2-pound) rack of lamb
- 1 batch Rosemary-Garlic Lamb Seasoning

Directions:

1. Supply your smoker with wood pellets and follow the manufacturer's specific start-up procedure. Heat the grill to 450°F.

2. Using a boning knife, score the bottom fat portion of the rib meat.

3. Using your hands, rub the rack of lamb with the lamb seasoning, making sure it penetrates the scored fat.

4. Place the rack directly on the grill grate and smoke until its internal temperature reaches 145°F.

5. Take off the rack from the grill and let it rest for 20 to 30 minutes before slicing into individual ribs to serve.

Nutrition:

Calcium: 92mg **Magnesium:** 209mg **Phosphorus:** 1769mg **Iron:** 15.91mg **Potassium:** 2979mg **Sodium:** 726mg **Zinc:** 31.76mg

CHICKEN RECIPES

412. Smoked Chicken Legs

Preparation time: 30 minutes.

Cooking time: 2 hours.

Servings: 3

Ingredients:

- 6 chicken legs
- 1 cup olive oil
- 1 tablespoon cayenne pepper
- 1 tablespoon paprika
- 2 teaspoon salt
- 1 tablespoon onion powder
- 1 tablespoon dried thyme
- 1 tablespoon garlic powder
- 1 tablespoon pepper

Directions:

1. Start your smoker up half an hour before you start cooking.
2. Stir all the dry ingredients. Be sure to rub the olive oil over all the chicken legs.
3. Now rub the seasoning over the chicken legs until they're fully coated.
4. Place the chicken legs on the smoking rack.
5. Turn them over, so they cook evenly. You may have to add more smoker

chips as you do this. Smoke temperature: 220°F.

Nutrition:

Calories: 1053 **Total fat:** 100g **Saturated fat:** 9.8g **Cholesterol:** 0mg **Sodium:** 1555mg **Total carbohydrate:** 12.1g **Dietary fiber:** 2.6g **Total sugars:** 2g **Protein:** 35.6g **Vitamin D:** 0mcg **Calcium:** 45mg **Iron:** 3mg **Potassium:** 177mg

413. Smoked Chicken Wings with Herbs

Preparation time: 10 minutes.

Cooking time: 1 hour.

Servings: 5

Ingredients:

- 5 pounds chicken wings
- ½ cup extra-virgin olive oil
- 1/3 cup chopped fresh basil leaves
- 2 tablespoons chopped fresh rosemary leaves
- 2 tablespoons thyme leaves or fresh oregano
- 2 large minced garlic cloves
- The juice of 1 small lime or lemon
- 1 and ½ teaspoons of sea salt
- 1 teaspoon freshly ground black pepper

Directions:

1. Start by rinsing the chicken wings under cool water and pat dry with clean paper towels
2. Cut the tips of the wings and cut the wings into half at the joints
3. Place the portions of the wings in a large bowl or a baking dish.
4. Mix the herbs with the garlic, the lemon, the salt, and the pepper
5. Add half the marinade to the wings and mix it very well with both your hands; then set the chicken wings and mix with both your hands; then set the other half of the marinade aside for the serving
6. Close the dish with a plastic wrap, then places it in the refrigerator for about 1 to 4 hours.
7. Move the chicken away from the refrigerator about 30 minutes before smoking it.
8. Take a rack or 2 racks of your smoker and place it on top of the counter; then add the mild wood chips; like alder chips to the tray of your electric smoker
9. Fill the water bowl halfway, then open the top of the vent and heat the smoker to a temperature of about 250°F.
10. Place the racks of the wings in your electric smoker and smoke for about 1 hour.

11. Check the temperature with a meat thermometer for an internal temperature that is about 165°F.
12. Serve and enjoy your delicious dish!

Nutrition:

Calories: 1051 **Total fat:** 54.2g **Saturated fat:** 12.3g **Cholesterol:** 404mg **Sodium:** 953mg **Total carbohydrate:** 3.8g **Dietary fiber:** 1.8g **Total sugars:** 0.4g **Protein:** 131.8g **Calcium:** 123mg **Iron:** 7mg **Potassium:** 1176mg

414. Smoky Wrap Chicken Breasts

Preparation time: 2 hours.

Cooking time: 3 hours.

Servings: 6

Ingredients:

- 6 chicken breasts, skinless and boneless
- 18 bacon slices
- 3 tablespoon chicken rub

For brine:

- 1/4 cup brown sugar
- 1/4 cup kosher salt
- 4 cups water

Directions:

1. Combine all the brine ingredients into the glass dish.
2. Place chicken into the dish and coat well.
3. Soak chicken for about 2 hours.
4. Rinse chicken well and coat with chicken rub.
5. Wrap each chicken breast with three bacon slices.
6. Heat the smoker to 230°F/110°C using soaked wood chips.
7. Place wrapped chicken breasts into the smoker and smoke for about 3 hours or until internal temperature reaches 165°F/73°C.
8. Serve and enjoy.

Nutrition:

Calories: 441 **Total fat:** 25.3g **Saturated fat:** 8.3g **Cholesterol:** 128mg **Sodium:** 6340mg **Total carbohydrate:** 6.7g **Dietary fiber:** 0g **Protein:** 46.1g **Vitamin D:** 0mcg **Calcium:** 19mg **Iron:** 2mg **Potassium:** 333mg

415. American Style Chicken Thighs

Preparation time: 15 minutes.

Cooking time: 2 hours.

Servings: 6

Ingredients:

- 6 (6-ounce) skinless, boneless chicken thighs
- 6 cups water
- 1 (12-ounce) can beer
- ¼ cup brown sugar
- ¼ cup kosher salt

For rub:

- 2 tablespoon brown sugar
- 2 tablespoon cornstarch
- ½ teaspoon cayenne pepper
- Salt and freshly ground black pepper

Directions:

1. In a large bowl, add water, beer, brown sugar and salt and mix until sugar is dissolved.
2. Add the chicken thighs and mix well.
3. Cover and refrigerate overnight.
4. Remove the chicken thighs from the brine and with paper towels, pat them dry.
5. Heat the smoker to 180°F, using charcoal.
6. **For the rub:** in a bowl, mix together all the ingredients.
7. Rub the chicken with the mixture generously.
8. Place the chicken thighs into the smoker and cook for about 1 hour.
9. Now, set the temperature of the smoker to 350°F and cook for about ¾–1 hour

Nutrition:

Energy: 178kcal **Carbohydrate:** 12.54g **Calcium:** 129mg **Magnesium:** 19mg **Phosphorus:** 133mg **Iron:** 0.48mg **Potassium:** 148mg

416. Smoked Chicken Cutlets In Strawberries-Balsamic Marinade

Preparation time: 2 hours.

Cooking time: 2 hours, 15 minutes.

Servings: 6

Ingredients:

- 3 tablespoon balsamic vinegar
- 20 medium strawberries
- 1/4 cup extra-virgin olive oil
- 2 tablespoon chopped fresh basil

- Kosher salt and freshly ground black pepper
- 2 pounds boneless, skinless chicken breast cutlets

Directions:

1. Whisk balsamic vinegar, strawberries, olive oil, and fresh basil in your blender.
2. Sprinkle marinade on and rub into the tops, bottoms, and sides of the chicken cutlets.
3. Refrigerate for 2 hours.
4. Heat Electric Smoker. Allow the smoker temperature to reach 225°F.
5. When it is ready, add some water to the removable pan that is usually on the bottom shelf.
6. Fill the side "drawer" with dry wood chips.
7. Smoke chicken for about two hours or until the internal temperature reaches 165°F.
8. Serve hot.

Nutrition:

Calories: 127 **Total fat:** 9g **Saturated fat:** 1.2g **Cholesterol:** 22mg **Sodium:** 26mg **Total carbohydrate:** 3.2g **Dietary fiber:** 0.8g **Total sugars:** 2g **Protein:** 9g **Vitamin D:** 0mcg **Calcium:** 15mg **Iron:** 0mg **Potassium:** 69mg

417. Beer Can Chicken

Preparation time: 5 minutes.

Cooking time: 3–4 hours.

Servings: 4

Ingredients:

- 1 can (12 ounces) beer
- 2 tablespoons apple cider vinegar
- 2 garlic cloves (minced)
- 1 whole chicken (4 to 5 pounds)
- 1 to 2 teaspoons chili powder
- 1 teaspoon salt
- 1 teaspoon onion powder
- 1/2 teaspoon freshly ground black pepper

Directions:

1. Pour 2 cups of water into the smoker's water pan. Place oak or pecan wood chips in the smoker's wood tray and heat the smoker to 225°F.
2. Drink half of the can of beer. Pop two more holes in the top of the can with a can opener. Add apple cider vinegar and garlic to beer and set aside until beer comes to room temperature.
3. Remove gizzards and neck from chicken cavity if necessary. Rinse chicken inside and out with cold water and pat dry with paper towels. Mix chili powder, salt, onion powder, and

pepper and rub over inside and outside of the chicken.

4. Set the beer can on a sturdy surface and slide the chicken cavity over the can so the chicken is standing up. Transfer the chicken with the can onto the smoker grate and smoke until the internal temperature of the meat reaches 165°F, 3 to 4 hours. Add wood chips to the wood tray as necessary.
5. Remove the chicken from the smoker and remove the can. Cover chicken loosely with aluminum foil and let rest for about 10 minutes. Carve chicken as desired, serve and enjoy!

Nutrition:

Calories116 **Total Fat** 2.8g **Saturated Fat** 0.8g **Cholesterol** 32mg **Sodium** 624mg **Total Carbohydrate** 4.7g **Dietary Fiber** 0.4g **Total Sugars** 0.3g **Protein** 11.2g **Vitamin D** 0mcg **Calcium** 18mg **Iron** 1mg **Potassium** 146mg

418. Herbed and Smoked Chicken

Preparation time: 3 hours

Cooking time: 1 hour, 30 minutes

Servings: 3

Ingredients:

- 15 cups filtered water
- 4 cups nonalcoholic beer
- Salt, to taste
- 1 cup brown sugar
- 1 tablespoon rosemary
- 1 teaspoon sage
- 2 pounds whole chicken, trimmed and giblets removed
- 4 tablespoons butter
- 3 tablespoons olive oil, for basting
- 1 cup Italian seasoning
- 2 tablespoons garlic powder
- Zest of 3 small lemons

Directions:

1. Add water to a large cooking pot, then add salt and sugar.
2. Let it boil until dissolved.
3. Add the herbs and let it cook for a few minutes until aromatic.
4. Pour the beer and then immerse the chicken in it.
5. Let it refrigerate for a few hours.
6. Remove the chicken from the brine, then dry with a paper towel.
7. Uncover and let it sit for one more hour in-room temperate.
8. Next, butter the chicken.
9. Massage the chicken for fine coating.

10. Next, rub the chicken with Italian seasoning, garlic powder, and lemon zest.

11. Load electric smoker with the wood chips, and heat to 250°F until smoke starts to build.

12. Then, slow roast it for 1.5 hours to 2 hours, and keep basting with olive oil every 30 minutes.

13. Once the internal temperature reaches 165°F and juices run clear, serve and enjoy.

Nutrition:

Calories: 1284 **Total fat:** 74.2g **Saturated fat:** 21.5g **Cholesterol:** 362mg **Sodium:** 461mg 20% **Total carbohydrate:** 65g **Dietary fiber:** 1g **Total sugars:** 53.6g **Protein:** 88.6g **Calcium:** 119mg **Iron:** 5mg **Potassium:** 861mg

419. Smoked Chicken Thighs

Preparation time: 2 hours, 10 minutes.
Cooking time: 90 minutes.
Servings: 2

Ingredients:

- 2 pounds chicken thighs
- 6 tablespoons soy sauce
- 3 teaspoons sesame oil
- 4 garlic cloves
- 4 scallions
- 1 tablespoon thyme
- 1 teaspoon allspice
- 1/3 teaspoon cinnamon
- 1/3 teaspoon crushed red pepper

Directions:

1. Combine soy sauce and oil in a bowl and rub it gently over the chicken thighs.
2. In a food processor, blend together garlic, scallion, thyme, cinnamon, allspice, and red pepper.
3. Blend it until smooth.
4. Rub it all over the thighs and seal the chick in the zip-lock plastic bag.
5. Let it marinate for about 2 hours.
6. Heat the smoker to 250°F, by adding the cherry wood to the smoker, and let wait for the smoke to release.
7. Smoke the chicken for 90 minutes.
8. Adjust the thermometer to read the internal temperate.
9. Once the internal temperature reaches 165°F.
10. Serve it and enjoy.

Nutrition:

Calories: 974 **Total fat:** 40.8g **Saturated fat:** 10.3g **Cholesterol:** 404mg **Sodium:** 3103mg **Total carbohydrate:** 9.9g **Dietary fiber:** 2.3g **Total sugars:** 1.6g **Protein:** 135.4g **Calcium:** 146mg **Iron:** 9mg **Potassium:** 1342mg

420. BBQ Chicken Wings Recipe

Preparation time: 2 hours.
Cooking time: 2 hours.
Servings: 4

Ingredients:

- 4 pounds turkey wings
- 1 cup of BBQ sauce

Directions:

1. Cut the chicken wings and discard the tips.
2. Marinate the wings in the BBQ sauce for about 2 hours.
3. Now, heat the smoker for a few minutes at 250°F.
4. Add the cherry wood chip to the smoker and let the smoker release smoke.
5. Place the chicken into the smoker.
6. Cook for 2 hours until the internal temperature reaches 165°F.
7. Use the digital meat thermometer to measure the temperature.
8. Serve and enjoy.

Nutrition:

Calories: 276 **Total fat:** 10.2g **Saturated fat:** 0g **Cholesterol:** 0mg **Sodium:** 699mg **Total carbohydrate:** 22.7g **Dietary fiber:** 0.4g **Total sugars:** 16.3g **Protein:** 23g **Calcium:** 8mg **Iron:** 0mg **Potassium:** 130mg

421. Lemon Garlic Chicken Breast Recipe

Preparation time: 2 hours.

Cooking time: 90 minutes.

Servings: 2

Ingredients:

- 2-pound chicken breasts, boneless and skinless
- 4 cloves minced garlic
- 2-inches ginger, minced
- 4 lemons, juice only
- 4 tablespoons olive oil
- Salt, to taste
- Black pepper, to taste
- 1 teaspoon turmeric

Directions:

1. Take a bowl and combine salt, pepper, lemon juice, olive oil, turmeric, ginger, and garlic in a bowl.
2. Mix well and rub the chicken with the prepared mix.
3. Let the chicken marinate for 2 hours in the refrigerator.
4. Now, heat the smoker to 250°F.
5. Add the cherry wood chip to the smoker and let the smoke release.
6. Place the chicken into the smoker.

7. Cook for 90 minutes until the internal temperature reaches 165°F.

8. Use the digital meat thermometer to measure the temperature.

9. Serve and enjoy.

Nutrition:

Calories: 1155 **Total fat:** 62.2g **Saturated fat:** 13.4g **Cholesterol:** 404mg **Sodium:** 472mg **Total carbohydrate:** 14.8g **Dietary fiber:** 3.9g **Total sugars:** 3.1g **Protein:** 133.1g **Calcium:** 113mg **Iron:** 7mg **Potassium:** 1339mg

422. Authentic Citrus Smoked Chicken

Preparation time: 15 minutes.

Cooking time: 18 hours, 5 minutes.

Servings: 12

Ingredients:

- 1 whole chicken
- 4 cups lemon-lime flavored carbonated beverage
- 1 tablespoon garlic powder
- 2 cups soaked wood chips

Directions:

1. Transfer the whole chicken to a large-sized zip bag.
2. Sprinkle garlic powder and pour lemon-lime soda mix into the bag.
3. Seal the bag, then allow it to marinate overnight.
4. Pre-heat your electric smoker to 225°F.
5. Take off the chicken from the bag and transfer it to your smoker rack.
6. Discard the marinade.
7. Smoker for 10 hours, making sure to keep adding more wood chips after every hour.
8. Serve and enjoy!

Nutrition:

Calories: 644 **Fats:** 34g **Carbs:** 19g **Fiber:** 0.1g

423. Amazing Mesquite Maple and Bacon Chicken

Preparation time: 20 minutes.

Cooking time: 2 hours.

Servings: 7

Ingredients:

- 4 boneless and skinless chicken breast
- Salt as needed
- Freshly ground black pepper
- 12 slices of uncooked bacon

- 1 cup maple syrup
- ½ a cup melted butter
- 1 teaspoon liquid smoke

Directions:

1. Pre-heat your smoker to 250°F.
2. Season the chicken with pepper and salt.
3. Wrap the breast with 3 bacon slices and cover the entire surface.
4. Secure the bacon with toothpicks.
5. Take a medium-sized bowl and stir in maple syrup, butter, liquid smoker and mix well.
6. Reserve 1/3rd of this mixture for later use.
7. Submerge the chicken breast into the butter mix and coat them well.
8. Place a pan in your smoker and transfer the chicken to your smoker.
9. Smoker for 1 to 1 and a ½ hours.
10. Brush the chicken with reserved butter and smoke for 30minutes more until the internal temperature reaches 165°F.
11. Enjoy!

Nutrition:

Calories: 458 **Fats:** 20g **Carbs:** 65g **Fiber:** 1g

424. Smoked Paprika Chicken

Preparation time: 20 minutes.

Cooking time: 2–4 hours.

Servings: 4

Ingredients:

- 4–6 chicken breast
- 4 tablespoon olive oil
- 2 tablespoon smoked paprika
- ½ a tablespoon kosher salt
- ¼ teaspoon ground black pepper
- 2 teaspoon garlic powder
- 2 teaspoon garlic salt
- 2 teaspoon black pepper
- 1 teaspoon cayenne pepper
- 1 teaspoon rosemary

Directions:

1. Pre-heat your smoker to 220°F using your favorite wood chips.
2. Prepare the chicken breast according to your desired shapes and transfer to a greased baking dish.
3. Take a medium bowl and add spices, stir well.
4. Press the spice mix over the chicken and transfer the chicken to the smoker.
5. Smoke for 1–1 and a ½ hours.

6. Turn-over and cook for 30 minutes more.
7. Once the internal temperature reaches 165°F.
8. Remove from the smoker and cover with foil.
9. Rest for 15 minutes.
10. Enjoy!

Nutrition:

Calories: 237 **Fats:** 6.1g **Carbs:** 14g **Fiber:** 3g

425. Fully Smoked Herbal Chicken

Preparation time: 10 minutes.

Cooking time: 60 minutes.

Servings: 8

Ingredients:

- 4–6 chicken breast
- 2 tablespoon of olive oil
- Salt as needed
- Freshly ground black pepper
- 1 pack of dry Hidden Valley Ranch dressing (or your preferred one)
- ½ a cup of melted butter

Directions:

1. Pre-heat your smoker to 225°F using hickory wood.
2. Season the chicken with olive oil and season with salt and pepper.
3. Place the in your smoker and smoke for 1 hour.
4. Take a small bowl and add ranch dressing mix and melted butter.
5. After the first 30minutes of cooking, brush the chicken with the ranch mix.
6. Repeat again at the end of the cooking time.
7. Once the internal temperature of the chicken reaches 145°F, they are ready!

Nutrition:

Calories: 209 **Fats:** 13g **Carbs:** 0g **Fiber:** 3g

426. Orange Crispy Chicken

Preparation time: 8 hours, 30 minutes.

Cooking time: 2 hours

Servings: 4

Ingredients:

For the poultry spice rub:

- 4 teaspoon paprika
- 1 tablespoon chili powder

- 2 teaspoon ground cumin
- 2 teaspoon dried thyme
- 2 teaspoon salt
- 2 teaspoon garlic powder
- 1 teaspoon freshly ground black pepper

For the marinade

- 4 chicken quarters
- 2 cups frozen orange-juice concentrate
- ½ a cup soy sauce
- 1 tablespoon garlic powder

Directions:

1. Take a small bowl and add paprika, chili powder, cumin, salt, thyme, garlic powder, pepper and mix well.
2. Transfer the chicken quarters to a large dish.
3. Take a medium bowl and whisk in orange-juice concentrate, soy sauce, garlic powder, half of the spice-rub mix.
4. Place the marinade over the chicken, then cover.
5. Refrigerate for 8 hours.
6. Pre-heat your smoker to 275°F.
7. Discard the marinade and rub the surface of the chicken with the remaining spice rub.
8. Transfer the chicken to smoker and smoker for 1 and a ½ to 2 hours.
9. Remove the chicken from the smoker and check using a digital temperature that the internal temperature is 160°F.
10. Allow it to rest for 10 minutes.
11. Enjoy!

Nutrition:

Calories: 165 **Fats:** 8g **Carbs:** 14g **Fiber:** 2g

427. Standing Smoked Chicken

Preparation time: 15 minutes.

Cooking time: 2 hours.

Servings: 4

Ingredients:

- 12 garlic cloves, minced
- 3 whole onions, quartered
- ½ of a quartered lemon
- 1 tablespoon salt
- 1 teaspoon black pepper
- 1 and a ½ tablespoon ground sage
- 1 and a ½ tablespoon dried thyme
- 1 and a ½ tablespoon dried rosemary
- 1 teaspoon paprika
- 1 whole chicken 4–6 pounds

- 3 tablespoon vegetable oil

Directions:

1. Remove one or two of the top racks from the smoker to make room for your standing chicken.
2. Smash 8 pieces of garlic cloves and add them into the water pan alongside the onion and lemon pieces.
3. Pre-heat your smoker to a temperature of 250°F.
4. Finely mince up the rest of the 4 garlic cloves and combine them in a small-sized bowl with sage, rosemary, thyme, pepper, salt, paprika, and set it aside for later use.
5. Remove the innards from the chicken, then rinse up the bird finely.
6. Pat it dry and rub it up with oil and then with the seasoning mixture created previously.
7. Set your chicken in a vertical position on top of your smoker and add in just a handful of soaked chips in the chip loading area.
8. Keep adding the chips every 30 minutes.
9. The chicken should be done after about 2 hours when the internal temperature registers 165°F.
10. Let it cool for 15 minutes and serve.

Nutrition: Calories: 128 **Fats:** 3g **Carbs:** 12g **Fiber:** 1g

428. Equally Worthy Cinnamon Cured Smoked Chicken

Preparation time: 1 hour, 15 minutes.

Cooking time: 1 hour, 30 minutes.

Servings: 4

Ingredients:

- 1-quart water
- ¼ cup salt
- ¼ cup firmly packed brown sugar
- 4 chicken breast
- 1 sliced onion
- 1 sliced lemon
- 2 halved cinnamon stick
- 1 tablespoon ground cinnamon
- 1 tablespoon red pepper flakes
- 1 tablespoon seasoned salt

Directions:

1. Take a large bowl and stir in water, brown sugar, and salt. Keep stirring until dissolved well.
2. Add chicken, lemon, onion, and cinnamon stick to the bowl and cover with plastic wrap.
3. Allow it to chill for 1 hour.
4. Pre-heat your smoker to 250°F with your desired wood.

5. Remove the chicken and discard the marinade.
6. Sprinkle chicken with cinnamon, pepper flakes, and seasoning salt.
7. Transfer to your smoker rack and smoke for 1 and a ½hours until the internal temperature reaches 165°F.
8. Take it out and serve!

Nutrition:

Calories: 240 **Fats:** 5g **Carbs:** 2g **Fiber:** 1g

429. Dadgum Good Gumbo

Preparation time: 1 hour.

Cooking time: 4 hours.

Servings: 5

Ingredients:

- 4 pounds roaster chicken, cut into pieces (wings, legs, etc.)
- 2 tablespoon extra-virgin olive oil, divided
- 2 teaspoon garlic, minced
- 2 packages (5 ounces each) Louisiana fish fry gumbo mix
- 1 teaspoon liquid shrimp boil
- 1 pound turkey sausage
- 1 cup uncooked rice
- 1 cup celery, sliced
- 1¾ cups okra, sliced
- ½ cup sweet onion, chopped
- ½ tablespoon meat tenderizer, divided
- ½ cup green onions, thinly sliced
- French bread or crackers

Directions:

1. Heat the electric smoker to 250F.
2. Punch through each piece of chicken with a fork. Sprinkle each with 1 tablespoon olive oil and ¼ teaspoon meat tenderizer. Smoke on the middle rack for 2 to 3(½) hours, till the internal temperature reaches 165°F.
3. In a large stockpot, prepare the gumbo mix, following the package directions. Add celery, okra, garlic, green onions, sweet onion, and liquid shrimp boil, mixing well. Boil then simmer for 15 minutes, covered. Leave it to cool.
4. Cut 1/4-inch thick slices of turkey sausage.
5. When chicken is cooked, let it rest and pull meat from the bone. Combine it with sausage slices and gumbo mixture and simmer on low for 20 minutes
6. Then add rice, mix all well and simmer for additional 15 minutes. Leave the pot covered, turn the heat off and let stand until rice is ready.
7. Serve with bread or crackers.

430. Smoked Chicken Tenders

Preparation time: 8 hours, 20 minutes.

Cooking time: 1 hour.

Servings: 4

Ingredients:

- 4-pound chicken tenders, rinsed and patted dry
- 2 teaspoon minced garlic
- 1(½) tablespoon sesame seeds
- ¾ teaspoon ginger root, peeled, freshly grated.
- ½ cup vegetable oil
- ½ cup soy sauce
- ¼ teaspoon Cajun seasoning
- ¼ cup water
- Jane's Krazy Mixed-Up Salt

Directions:

1. Put soy sauce, vegetable oil, water, sesame seeds, garlic, ginger, Cajun seasoning, and ½ teaspoon Jane's Krazy Salt in a bowl and mix all these ingredients well to make the marinade.
2. Place the chicken tenders in a plastic bag and add the prepared marinade. Seal the bag.
3. Place the meat in the refrigerator for about 8 hours. Make sure that the chicken is well marinated. Turn the plastic bag several times.
4. Heat the electric smoker to 250°F.
5. Remove the tenders from the bag, then clean the meat of any clinging marinade. Smoke for 50 minutes to 1 hour on the middle rack. The internal temperature should reach 160°F.

Nutrition:

Calcium: 77mg **Phosphorus:** 1041mg **Iron:** 5.09mg **Magnesium:** 145mg **Zinc:** 3.99mg **Potassium:** 1125mg **Sodium:** 2519mg

431. Smoked Chicken Salad Sandwiches

Preparation time: 1 hour.

Cooking time: 6 hours.

Servings: 8

Ingredients:

- 8 soft hoagie buns
- 3 large chicken breasts, boneless, skinless
- 3/4 cup pecans, very coarsely chopped
- 3 bay leaves
- 2 teaspoon meat tenderizer
- 1/4 teaspoon kosher salt
- 1/4 teaspoon Cajun seasoning
- 1/4 cup butter, melted
- 1 cup finely chopped celery

- 1 1/3 cups mayonnaise
- 1 (5-ounces) bag dried cranberries
- 1/2 teaspoon black pepper, freshly ground
- 1/2 teaspoon salt
- Red leaf lettuce

Directions:

1. Heat the electric smoker to 225°F. For the best result, use hickory chips.
2. Pierce the chicken with a fork and rub with the meat tenderizer.
3. Smoke the chicken breasts for 45 minutes per pound or until the internal temperature reaches 165°F.
4. Add melted butter to pecans in an aluminum foil pan. Season with kosher salt and mix the ingredients well. Coat the pecans thoroughly.
5. Put pecans on the top rack of the smoker for the last 30 minutes of smoking the chicken breasts.
6. Remove the pecans and chicken. Let the breasts rest for a while and then chop the chicken well.
7. Combine the chicken, pecans, celery, dried cranberries, and mayonnaise. Mix well and add Cajun seasoning, salt, and pepper—as much as desired.
8. Spread the mayonnaise mixture on the buns and add a lettuce leaf. Divide the chicken salad among the buns and put another lettuce leaf on the top.

432. Southwestern Smoker Chicken Wrap

Preparation time: 8 hours 40 minutes.

Cooking time: 6 hours.

Servings: 5

Ingredients:

Rub:

- 3 teaspoon chili powder
- 2 teaspoon paprika
- 1 teaspoon onion powder
- 1 teaspoon kosher salt
- 1 teaspoon fresh ground pepper
- ¾ teaspoon cumin
- ½ teaspoon cayenne pepper
- ½ teaspoon coriander
- 1½ teaspoon garlic powder

Wrap:

- 1 pound chicken breasts
- 5 flour tortillas
- 2 tomatoes, diced
- 1 onion, diced
- 2 teaspoon olive oil
- 1 red bell pepper, halved
- 1 green bell pepper, halved

- Lettuce

Directions:

1. Mix all the rub ingredients in a medium bowl. Coat the chicken breasts with the rub and refrigerate for 8 hours.
2. Heat the electric smoker to 275F.
3. Put chicken breasts, bell peppers, and onion on the middle rack of the smoker.
4. Smoke the chicken breasts until the internal temperature reaches 160°F. Remove the chicken and let it cool.
5. Slice the chicken breasts. Mix chicken breasts together with lettuce, tomatoes, peppers, and onions. Divide this mixture between the tortillas.

Nutrition:

Calcium: 80mg **Magnesium:** 37mg **Phosphorus:** 218mg **Iron:** 2.47mg **Potassium:** 396mg **Sodium:** 661mg **Zinc:** 1mg **Copper:** 0.139mg

433. Blue Cheese Wings

Preparation time: 10 minutes.

Cooking time: 1 hour, 45 minutes.

Servings: 8

Ingredients:

- 5-pound raw chicken wings
- Pepper and celery salt, to taste

Sauce Ingredients:

- 2 cup blue cheese chunky dressing
- ½ cup molasses
- 1 cup hot sauce
- 2 tablespoon of each:
- Minced garlic
- Extra-virgin olive oil

Directions:

1. Add some flavor to the wings with the celery salt and pepper. Arrange the wings on the smoker rack. Set the temperature to 250F and cook for one hour. Meanwhile, prepare the sauce in a saucepan using the low heat setting. Whisk all of the ingredients and simmer for 15 minutes.
2. Take the chicken from the cooker and add it to a disposable aluminum pan. Empty the blue cheese sauce over the wings and place it back in the cooker for 30 minutes. Serve hot and enjoy!

434. Smoked Spicy Pulled Chicken

Preparation time: 25 minutes.

Cooking time: 3 hour, 30 minutes.

Servings: 10

Ingredients:

- 5-pound boneless chicken
- ¾ cup butter
- ½ cup chili powder
- ½ cup brown sugar
- ½ tablespoon salt
- 2 tablespoons black pepper
- 2(½) tablespoons mustard
- 2(½) tablespoons cumin
- 1(½) tablespoons garlic powder
- ¾ tablespoons onion powder
- ¾ tablespoon paprika
- ¾ teaspoon cayenne pepper

Directions:

1. Heat an electric smoker to 230°F (110°C). Add applewood chips to the smoker.
2. Place butter in a mixing bowl; using a hand mixer whisk it until softened.
3. Add chili powder, brown sugar, salt, black pepper, mustard, cumin, garlic powder, onion powder, paprika, and cayenne pepper in the bowl, then continue whisking until combined.
4. Brush all sides of the chicken with the mixture, then wrap with aluminum foil.
5. Place your chicken in the smoker and smoke for 3 hours.
6. Check the internal temperature and when it has reached 165°F (74°C), take the smoked chicken out from the smoker.
7. Unwrap the chicken, then return it back to the smoker.
8. Smoke the chicken for about 30 minutes until brown.
9. Once ready, remove from the smoker, then place on a flat surface.
10. Shred the chicken, then transfer to a serving dish.
11. Enjoy.

Nutrition:

Calories: 243 **Total fat:** 3g **Net carbs:** 17g **Protein:** 27g

435. Simple Buttery Smoky Chicken

Preparation time: 20 minutes.

Cooking time: 3 hours.

Servings: 10

Ingredients:

- 5-pound whole chicken
- 4 tablespoons salt
- ½ cup butter

Directions:

1. Rub the chicken with salt, then let it sit for an hour.
2. Meanwhile, heat the electric smoker to 250°F (120°C).
3. After an hour, rinse the chicken, then pat dry using paper towels.
4. Melt the butter, then brush the chicken with the melted butter.
5. Wrap the chicken with aluminum foil, then place in the electric smoker.
6. Cook the chicken for 3 hours, then take it out from the smoker.
7. Unwrap the chicken, then cook again without cover for approximately 3 hours or until the internal temperature has reached 165°F (74°C).
8. Once it is done, remove it from the electric smoker, then let it sit until warm.
9. Slice the smoked chicken and serve.
10. Enjoy.

Nutrition:

Calories: 469 **Total fat:** 31g **Net carbs:** 9.8g **Protein:** 37g

436. Sweet Smoked Chicken Wings with Cinnamon

Preparation time: 12 hours, 20 minutes.

Cooking time: 2 hours.

Servings: 4

Ingredients:

- 4-pound chicken wings
- 1(¼) cup brown sugar
- ½ cup salt
- 1(½) teaspoons. black pepper
- 2(¼) tablespoons garlic powder
- 1(¾) tablespoon cinnamon
- ¾ tablespoons cumin
- 1(¼) tablespoons cayenne pepper
- ½ cup almond butter

Directions:

1. Place brown sugar, salt, black pepper, garlic powder, cinnamon, cumin, and cayenne pepper in a zipper-lock plastic bag. Stir well.

2. Take half of the spice mixture, then place in a container with a lid. Store in the refrigerator.
3. Add the chicken wings to the plastic bag, then shake to season the chicken wings.
4. Place the chicken wings in the refrigerator overnight or more—a maximum of 12 hours.
5. Heat an electric smoker to 200°F (93°C).
6. Remove the chicken wings from the plastic bag, then transfer them to a disposable aluminum pan.
7. Once the electric smoker has reached 200°F (93°C), place the pan in the smoker, then smoke for 2 hours.
8. After 2 hours, take the pan out, then brush the chicken wings with almond butter.
9. Sprinkle the remaining spice mixture, then return the pan back to the smoker.
10. Smoke the chicken for an hour or until the internal temperature is already 165°F (73°C).
11. Take the chicken wings out from the smoker and transfer them to a serving dish.
12. Serve and enjoy warm.

Nutrition:

Calories: 356 **Total fat:** 22.7g **Net carbs:** 18g **Protein:** 71.2g

437. Smoked Chicken Barbecue

Preparation time: 20 minutes.
Cooking time: 1 hour.
Servings: 5

Ingredients:

- 2(½) pound chicken fillet
- 4 cup barbecue sauce
- 2 teaspoons pepper
- 1 teaspoon salt

Directions:

1. Heat the electric smoker to 200°F (93°C). Pour beer into the water pan, then place in the smoker.
2. Cut the chicken fillet into thick slices, then place in a disposable aluminum pan.
3. Sprinkle pepper and salt over the chicken, then pour barbecue sauce into the pan.
4. Place the pan in the smoker, then smoke the chicken for an hour.
5. When the internal temperature has reached 165°F (74°C), remove the pan, then transfer the smoked chicken to a serving dish.
6. Place the sauce over the smoked chicken, then serve.
7. Enjoy.

Nutrition: Total fat: 13g **Calories:** 270 **Net carbs:** 0g **Protein:** 78g

438. Crispy Chicken Legs and Thighs

Preparation time: 1 hour, and 30 minutes.

Cooking time: 2 hours.

Servings: 6

Ingredients:

- 3-pound chicken legs, thigh, and legs separated
- 6 tablespoons poultry seasoning
- 4 tablespoons olive oil

Directions:

1. In a re-sealable plastic bag, put the chicken thighs and legs. Add the poultry seasoning, zip, and thoroughly shake to make sure all the pieces are sufficiently coated and allow them to sit for 1 hour and 30 minutes.
2. Heat your electric smoker to 250F.
3. Smoke the chicken for 2 hours.
4. Heat the boiler on high and place the chicken. Make sure that the skin side is facing up.
5. Remove from the boiler once the skin starts to crackle.
6. Serve with rice.

Nutrition:

Net carbs: 1.5g **Calories:** 261 **Total fat:** 21g **Protein:** 16g

439. Whole Chicken with Herbs

Preparation time: 24 minutes.

Cooking time: 3 hours.

Servings: 10

Ingredients:

- 3 lemon, quartered
- 3 garlic cloves
- Rosemary, parsley, and sprigs of thyme
- 3 tablespoons butter
- 1 whole chicken
- 2 tablespoons ground sage
- 2 tablespoons fresh thyme
- 2 tablespoons kosher salt
- 2 tablespoons fresh rosemary

Directions:

1. Heat your smoker to 250°F.
2. Fill the chicken using the stuffing ingredients.
3. Combine the rubbing ingredients, then rub on the chicken evenly.
4. Using indirect heat, smoke the chicken for 3 hours and remove.
5. Rest for 10 minutes and serve.

Nutrition: Calories: 331 **Total fat:** 13g **Net carbs:** 27g **Protein:** 27g

440. Fruits and Nut Stuffed Chicken Breast

Preparation time: 9 hours.

Cooking time: 3 hours.

Servings: 4

Ingredients:

- 4 skinless chicken breasts
- 12 ounces apple-cherry concentrate
- ¾ cup red wine vinegar
- 1 tablespoon ground cinnamon
- ½ cup diced apple
- ½ cup diced cherries
- ½ cup chopped cranberries
- ½ cup walnuts

Directions:

1. Create a pocket into each chicken breast.
2. Put the breasts in a re-sealable plastic bag.
3. Mix the marinade ingredients and place them in a microwave for 15 seconds.
4. Pour the marinade mixture into the plastic bags, seal and refrigerate for 9 hours.
5. After refrigeration, combine the stuffing ingredients in a bowl. Stuff the breasts and tie with a butcher twine.
6. Heat your electric smoker until temperatures reach 250°F.
7. Place the breasts on the rack and maintain the same temperature.
8. Smoke for 3 hours.
9. Allow to rest for a few minutes and serve.

Nutrition:

Calories: 579 **Total fat:** 41.6g **Net carbs:** 7.6g **Protein:** 44.3g

441. Electric Smoker Chicken Breast Sandwich

Preparation time: 2 hours, 40 minutes.

Cooking time: 3 hours.

Servings: 3

Ingredients:

- 3 pieces chicken breasts
- 16 cup water
- ½ tablespoon sea salt
- 4 tablespoons olive oil
- ½ cup soy sauce
- ½ cup sugar
- Crushed garlic cloves
- Pepper cones
- A handful of spinach

Directions:

1. Remove fat from the chicken breast and clean in cold water.
2. To make the chicken juicer and tender, brine it by adding sea salt and water into a bowl, then stir until the salt dissolves.
3. Place the chicken breasts in a plastic bag full of brine.
4. Extract out the excess air, seal, and refrigerate for 2 hours.
5. Drain the breasts and dry with a paper towel after 2 hours and leave to air dry for 40 minutes.
6. Heat the electric smoker to 250° F.
7. Put the breasts on the grill and cook for 60 minutes.
8. Check the meat temperature after 60 minutes and wait for 45 more minutes before checking again.
9. After the breasts are at the temperature of 165° F, pull them out of the smoker.
10. Place the breasts on aluminum foil and let them cool for 12 minutes at room temperature.
11. Add lettuce, tomatoes, and sandwich the breasts in between the slices of bread.
12. Enjoy.

Nutrition: Calories: 115 **Total fat:** 1.1g **Net carbs:** 0.1g **Protein:** 25g

442. Sweet Chili Chicken Wings

Preparation time: 2 hours.

Cooking time: 3 hours, 30 minutes.

Servings: 4

Ingredients:

- 5-pound chicken wings
- 3 tablespoons black pepper
- 2 tablespoons onion powder
- 2 tablespoons chili powder
- 2 tablespoons garlic powder
- 2 tablespoons salt
- 1½ cup honey
- ½ cup spicy barbecue sauce
- 4 tablespoons apple juice

Directions:

1. Heat the smoker to 225°F.
2. Prepare 2 aluminum baking sheets.
3. Mix all the ingredients for the rub.
4. Divide wings and rub among 2 re-sealable plastic bags. Seal the bags, then shake well until the pieces are evenly coated.
5. Refrigerate for 2 hours.
6. In the meantime, place your wood chips in the smoker.

7. Remove marinated wings from the bags and transfer them to the baking sheets, one portion of each sheet.
8. Place the pans in the smoker, and cook for 3 hours, turning once after the first 2 hours.
9. Meanwhile, combine sauce ingredients in a saucepan over medium heat until heated through.
10. Coat wings evenly with sauce, and return to smoker for 30 more minutes

Nutrition:

Protein: 6.4g **Calories:** 42 **Total fat:** 1.7g **Net carbs:** 0g

TURKEY RECIPES

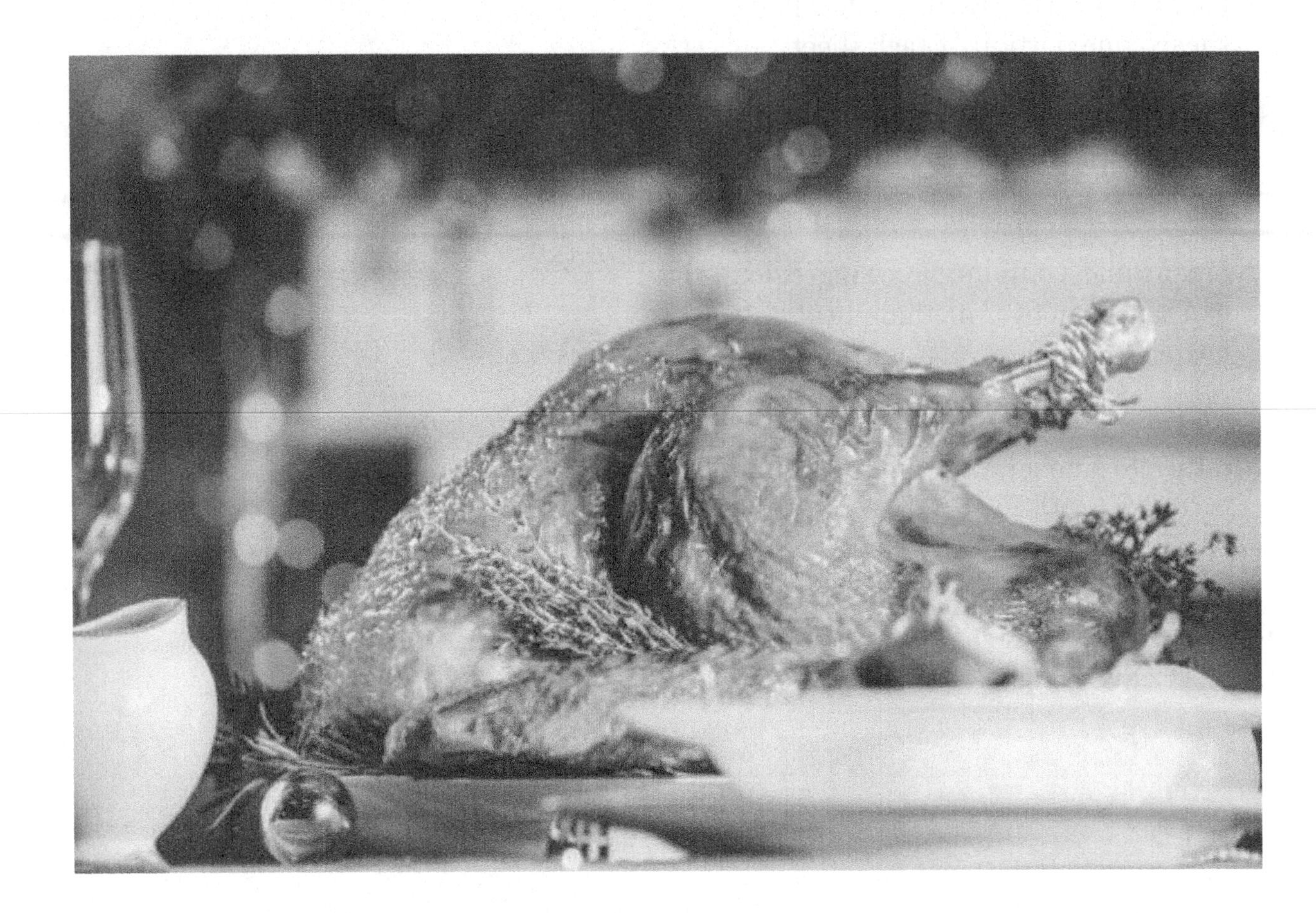

443. Smoked Turkey

Preparation time: 1 hour, 30 minutes.

Cooking time: 4 hours, 20 minutes.

Servings: 5

Ingredients:

- 5 pounds trimmed turkey
- 5 tablespoons Ras el Hanout (Moroccan Spice Blend) seasoning
- Kosher salt
- Zest of 3 lemons
- 1 cup olive oil
- Mint leaves

Directions:

1. Cut the backbone of the turkey and remove the spine and discard the fat.
2. Flip the turkey breast-side up and press into the breastbone to flatten it
3. Then rub the turkey with oil and then massage the seasoning along with salt and lemon zest.
4. Seal the turkey with plastic wrap and marinate it for 30 minutes.

5. Heat the smoker for 20 minutes.
6. Soak the wood chip in water one hour before smoking.
7. Remove plastic wrap and cook the turkey for 4 hours 20 minutes at 250°F.
8. Garnish it with mint leaves.

Nutrition:

Calories: 1062 **Total fat:** 72.7g **Saturated fat:** 15.9g **Cholesterol:** 284mg **Sodium:** 1660mg **Total carbohydrate:** 3.8g **Dietary fiber:** 2.2g **Total sugars:** 1.5g **Protein:** 85.2g **Calcium:** 45mg **Iron:** 1mg **Potassium:** 10mg

444. Turkey with Chimichurri

Preparation time: 1 hour, 10 minutes.

Cooking time: 4 hours.

Servings: 5

Ingredients:

- 5 pounds bone-in, skin-on turkey pieces
- Salt and pepper
- 1teaspoon paprika
- ½ teaspoon cayenne
- 2 tablespoons olive oil
- 1 pepper
- 1 onion
- 2 carrots, chopped
- 2 scallions
- 2 tomatoes, chopped

Homemade chimichurri sauce:

- ½ cup olive oil
- 1 teaspoon parsley
- 1 teaspoon red pepper flakes
- 2 garlic cloves
- 2 red onions

Directions:

1. Season the washed and clean turkey with salt, pepper, paprika, and cayenne pepper.
2. Rub it gently all over.
3. Arrange the wood chip inside the smoker and heat the smoker to 230°F.
4. Transfer the turkey to the sheet pan and arrange peppers, onions, carrots, scallion, and tomatoes beside it.
5. Drizzle the olive oil on top.
6. Place the pan sheet inside the smoker.
7. Close the electric smoker door and then cook for 4 hours at 250°F.
8. Check the turkey to an internal temperature of 165°F.
9. Now, it is time to make the chimichurri.

10. Blend all the homemade chimichurri ingredients in a blender and puree until combined.

11. Serve the cooked turkey and veggie with the sauce.

Nutrition:

Calories: 807 **Total fat:** 35.9g **Saturated fat:** 4.5g **Cholesterol:** 283mg **Sodium:** 920mg **Total carbohydrate:** 11.7g **Dietary fiber:** 2.9g **Total sugars:** 5.8g **Protein:** 94.8g **Calcium:** 32mg **Iron:** 6mg 35 **Potassium:** 311mg

445. Whole Smoked Turkey Recipe

Preparation time: 16 hours.

Cooking time: 10 hours.

Servings: 14

Ingredients:

- ½ cup salt
- 1/3 cup molasses
- 1/3 cup granulated sugar
- ½ cup Worcestershire sauce
- 6 cloves smashed garlic
- 4 dried bay leaves
- Black pepper to taste
- 14 pounds whole turkey
- 2 cups bourbon
- 1 cup canola oil for coating

Directions:

1. Pour a gallon of water, salt, sugar, molasses, garlic, Worcestershire sauce, bourbon, pepper, and bay leaves in a large pot.
2. Boil for a few minutes and then cool it down completely.
3. Submerge the turkey completely in the brine using a large bucket.
4. Brine it in the liquid for 15 hours.
5. The next day, take the turkey out of the brine and pat dry with a paper towel.
6. Rub the turkey with oil and additional pepper.
7. Load the smoker with soaked wood chips, and place the turkey inside the smoker for cooking.
8. Set temperature to 250°F.
9. Once the internal temperate is 165°F, the turkey is ready.
10. Note: It took about 10 hours of cooking.

Nutrition:

Calories: 770 **Total fat:** 29.2g **Saturated fat:** 1.7g **Cholesterol:** 249mg **Sodium:** 9583mg **Total carbohydrate:** 47.8g **Dietary fiber:** 1.9g **Total sugars:** 29g **Protein:** 61.3g **Calcium:** 95mg **Iron:** 10mg **Potassium:** 1705mg

446. Classic Smoked Turkey Recipe

Preparation time: 30 minutes.

Cooking time: 12 hours.

Servings: 16

Ingredients:

- 16 pounds turkey
- 2 tablespoons dried thyme
- 1 tablespoon dried sage
- 2 teaspoons dried oregano
- 2 teaspoons paprika
- 1 tablespoon sea salt
- Black pepper, to taste
- 1 teaspoon dried rosemary
- Zest of 1 orange
- 1/3 cup extra-virgin olive oil
- 1/3 cup apple cider
- 1/3 cups water

Directions:

1. Heat the electric smoker to 250°F.
2. Take a small bowl and mix all the dry spices and ingredients.
3. Rub it gently over the rekey meat.
4. In the end, drizzle olive oil on top.
5. Now pour water along with apple cider in the large water pan in the bottom of the electric smoker.
6. Place a drip pan on the next rack or shelf of the smoker.
7. Fill the sides with the applewood chips.
8. Move the turkey on the top rack of the smoker.
9. Close the rack and then cook for approximately 12 hours.
10. Add more wood if smoke stops coming.
11. Use the digital probe thermometer to get an internal temperature of 165°F.
12. Remove the turkey and serve.

Nutrition:

Total fat: 27g **Saturated fat:** 8.1g **Cholesterol:** 343mg **Sodium:** 669mg **Total carbohydrate:** 2.6g **Dietary fiber:** 0.7g **Total sugars:** 1.7g **Calories:** 818 **Protein:** 133.1g **Calcium:** 31mg **Iron:** 46mg **Potassium:** 1393mg

447. Turkey in the Electric Smoker

Preparation time: 1 hour, 10 minutes.

Cooking time: 10 hours.

Servings: 10

Ingredients:

- 1(10 pounds) whole turkey
- 4 cloves garlic, crushed
- 2 tablespoons salt, seasoned
- ½ cup butter
- 1(12 fluid ounce) cola-flavored carbonated beverage
- 1 apple, quartered
- 1 onion, quartered
- 1 tablespoon garlic powder
- 1 tablespoon salt
- 1 tablespoon black pepper

Directions:

1. Heat the electric smoker to 225°F and then rinse the turkey well underwater, pat dry, and then rub it with seasoned salt.
2. Place it inside a roasting pan.
3. Combine cola, butter, apples, garlic powder, salt, and pepper in a bowl.
4. Fill the cavity of the turkey with cola, apples, garlic powder, salt, and pepper.
5. Rub the butter and crushed garlic outside of the turkey as well.
6. Cover the turkey with foil.
7. Smoke the turkey for 10 hours at 250°F.
8. Once it's done, serve.

Nutrition:

Calories: 907 **Total fat:** 63.2g **Saturated fat:** 22.3g **Cholesterol:** 4256mg **Sodium:** 2364mg **Total carbohydrate:** 17.9g **Dietary fiber:** 1.1g **Total sugars:** 9.7g **Protein:** 62.7g **Calcium:** 462mg **Iron:** 19mg **Potassium:** 710mg

448. Smoked Turkey Legs

Preparation time: 15 minutes.

Cooking time: 4 hours.

Servings: 6

Ingredients:

- 6 turkey legs
- 3 teaspoon Worcestershire sauce
- 1 tablespoon vegetable oil
- Dry rub, recipe follows
- Mop, recipe follows
- Sweet & Spicy BBQ sauce

Dry rub:

- 1/4 cup chipotle seasoning, (recommended: North of the Border Chipotle Seasoning)
- 1 to 2 teaspoon mild dried ground red chili or paprika
- 1 tablespoon packed brown sugar

Mop mixture:

- 1 cup white vinegar
- 1 tablespoon BBQ Sauce, (recommended: North of the Border Chipotle Barbecue Sauce)

Directions:

1. Loosen the skin of the turkey legs beforehand. You can do this by running your fingers under it.
2. Mix together the oil and Worcestershire sauce. Rub the mixture over the turkey legs. Be sure to put some of it under the skin. Next, sprinkle the rub over the legs.
3. Put the legs in a plastic bag. Store in the fridge.
4. Take out the turkey legs from the fridge. Allow them to sit out for ½ hour, so they come up to room temperature.
5. Warm up the mop mixture.
6. Place the turkey legs in the smoker. Be sure to mop the legs every 45 minutes.
7. Once the turkey legs are done, serve them hot. They're best eaten with your fingers with some barbeque sauce. Smoke Temperature: 220°F.

Nutrition:

Calories: 203 **Total fat:** 9.6g **Saturated fat:** 2.7g **Cholesterol:** 63mg **Sodium:** 233mg **Total carbohydrate:** 5.4g **Dietary fiber:** 0.2g **Total sugars:** 4.7g **Protein:** 20.6g **Vitamin D:** 0mcg **Calcium:** 29mg **Iron:** 2mg **Potassium:** 244mg

449. Brined Whole Turkey

Preparation time: 24 hours.

Cooking time: 12 hours.

Servings: 15

Ingredients:

- 2 cups kosher salt
- 6 cups water
- ¼ cup black peppercorns
- 3 tablespoons chopped garlic cloves
- 2 tablespoons chopped basil leaves
- 2 tablespoons onion powder
- ½ cup soy sauce
- ½ cup Worcestershire sauce
- ¼ cup extra virgin olive oil

- 10 pounds whole turkey

Directions:

1. In a large stockpot, mix together salt, water, peppercorn, garlic, and basil. Heat until the salt dissolves. Let it cool.
2. Submerge the turkey into the brine solution and put it in the fridge for 24 hours.
3. Pour out the brine and rinse the water completely. Pat dry.
4. In a bowl, combine the onion powder, soy sauce, Worcestershire sauce, and olive oil.
5. Brush onto the entire surface of the turkey.
6. Heat the smoker to 225°F.
7. Place water in the water pan, then add maple wood chips into the side tray. Place turkey in the smoker. Cook for 12 hours.

Nutrition:

Calories: 525 **Total fat:** 25g **Saturated fat:** 7.2g **Cholesterol:** 189mg **Sodium:** 16543mg **Total carbohydrate:** 2g **Dietary fiber:** 0.2g **Total sugars:** 0.5g **Protein:** 57.4g **Vitamin D:** 0mcg **Calcium:** 21mg **Iron:** 0mg **Potassium:** 39mg

450. Honey Smoked Turkey

Preparation time: 6 minutes.

Cooking time: 6 hours.

Servings: 7

Ingredients:

- 1-gallon hot water
- 1 pound kosher salt
- 2 quarts vegetable broth
- 8-ounce jars of honey
- 1 cup orange juice
- 7-pound bag of ice cubes
- 15-pound whole turkey with giblets and neck removed
- ¼ cup vegetable oil
- 1 teaspoon poultry seasoning
- 1 granny smith apples cored and cut up into large chunks
- 1 celery stalk cut up into small chunks
- 1 small sized onion cut up into chunks
- 1-quartered orange

Directions:

1. Take a 54-quart cooler and add kosher salt and hot water.
2. Mix them well until everything dissolves.

3. Add vegetable broth, orange juice, and honey.
4. Pour ice cubes into the mix and add the turkey into your brine, keeping the breast side up.
5. Lock up the lid of your cooler and let it marinate overnight for 12 hours.
6. Make sure that the brine temperature stays under 40°F.
7. Remove the turkey from the brine, then discard the brine.
8. Dry the turkey using a kitchen towel.
9. Take a bowl, then mix vegetable oil and poultry seasoning.
10. Rub the turkey with the mixture.
11. Place apple, onion, celery, and orange pieces inside the cavity of the turkey.
12. Pre-heat your smoker to a temperature of 400°F and add 1 cup of hickory wood chips.
13. Set your turkey onto your smoker and insert a probe into the thickest part of your turkey breast.
14. Set the probe to 160°F.
15. Smoke the turkey for 2 hours until the skin is golden brown.
16. Cover the breast, wings and legs using aluminum foil and keep smoking it for 2–3 hours until the probe thermometer reads 160°F.
17. Make sure to keep adding some hickory chips to your heat box occasionally.
18. Remove the vegetables and fruit from the cavity of your turkey and cover it up with aluminum foil.
19. Let it rest for 1 hour and carve it up!

Nutrition:

Calories: 353 **Fats:** 16g **Carbs:** 29g **Fiber:** 2g

451. Turkey Salad

Preparation time: 24 hours, 20 minutes.

Cooking time: 5 hours.

Servings: 8

Ingredients:

- 1(4 pound)turkey breast
- 6 tablespoon extra-virgin olive oil
- 3 tablespoon fresh thyme, chopped
- 3 tablespoon white wine vinegar
- 2 cups apple juice
- 2 cups water
- 2 tablespoon lemon juice
- 2(½) cups dried cherries, chopped
- 1 cup kosher salt
- 1 cup brown sugar

- 1 cup green onions, chopped
- ½ cup maple syrup
- ¾ cup celery, chopped
- ½ cup mayonnaise
- 8-ounces package mixed baby greens
- ½ cup toasted hazelnuts, coarsely chopped
- Salt and black pepper, to taste

Directions:

1. To prepare the brine, mix the salt, brown sugar, syrup, apple juice, and water in a stockpot. Put the turkey breast in the prepared mixture and refrigerate for 12-24 hours.
2. Heat the electric smoker to 225°F. Put one small handful of prepared wood chips in the wood tray.
3. Clean the meat of the remaining brine mixture. Rinse well.
4. Put the turkey breast on the middle rack of the smoker. Cook for 25 to 30 minutes until the internal temperature reaches 165°F.
5. Let it cool for 15 minutes.
6. To prepare the salad, combine sliced or chopped turkey breast, green onions, celery, mayonnaise, 2 tablespoon thyme, and lemon juice and stir well.
7. For the vinaigrette, blend oil, vinegar, and the remaining thyme in a salad bowl. Add salt and pepper to taste.
8. Toss the greens in this mixture. Serve the turkey salad on top of the greens and top with cherries and nuts.

Nutrition:

Calcium: 142mg **Magnesium:** 114mg **Phosphorus:** 586mg **Iron:** 5.83mg **Potassium:** 1162mg **Sodium:** 14575mg **Zinc:** 5.28mg

452. Smoked Turkey & Sausage Gumbo

Preparation time: 25 hours.

Cooking time: 8 hours.

Servings: 12

Ingredients:

- 1 (10–12 pound) smoked turkey, bought
- 8 cups water
- 3 cloves garlic, minced
- 2 tablespoon green onions, chopped
- 2 tablespoon fresh parsley leaves, chopped
- 2 cups coarsely chopped yellow onions
- 2 celery stalks
- 1 teaspoon fresh thyme, chopped

- 1 cup celery, coarsely chopped
- 1 cup green bell peppers, chopped
- 1(½) pound Andouille or smoked sausage, sliced
- 1(½) cups canola oil
- 1(½) cups bleached all-purpose flour
- Cayenne

Poultry rub:

- Hot cooked white rice (for serving)
- Salt, pepper

Directions:

1. Brine the turkey for no longer than 24 hours, using your favorite brine, bought or homemade. Spice Hunter Turkey Brine is advisable. Clean the poultry from the remaining brine mixture. Rinse well inside and out.
2. Heat the electric smoker to 250°F. Put one small handful of prepared wood chips in the wood tray; for the best result, use a combination of hickory and apple.
3. Smoke the turkey until the temperature inside the turkey meat reaches 180°F.
4. Remove the turkey and cut the meat from the bones. Reserve the meat aside.
5. Mix water, celery stalks, onion, garlic cloves in a large stockpot.
6. Put the turkey carcass in this mixture and boil for 1 hour. Strain the broth.
7. Combine oil and flour in another pot. Cook the mixture on low for about 25 minutes, often stirring until a dark roux is formed. Add the onions, bell peppers, celery, and garlic. Mix the ingredients well and cook for another 10 minutes. Add the thyme and cayenne.
8. Add the smoked turkey broth and combine well, stirring constantly. Cook on medium heat for 1(½) hours. Add water, if needed.
9. Chop the smoked turkey meat. Add the turkey and the sausage to the gumbo. Cook for another 30 minutes. Season with salt and pepper to taste.
10. Set aside, then skim off the fat from the surface.
11. Serve over hot white rice in gumbo or soup bowls.

Nutrition:

Calcium: 78mg **Magnesium:** 38mg **Phosphorus:** 201mg **Iron:** 3.65mg **Potassium:** 290mg **Sodium:** 532mg **Zinc:** 1.7mg

453. Sweet Smokey Turkey with Herbs

Preparation time: 30 minutes.

Cooking time: 1 hour, 20 minutes.

Servings: 10

Ingredients:

- 1 whole turkey
- 2(½) tablespoons sage
- 1(½) tablespoons black pepper
- 1(½) teaspoons. salt
- 2(½) tablespoons basil
- 2(½) tablespoons olive oil
- ¾ cup raw honey

Directions:

1. Heat the electric smoker to 225°F (107°C). Don't forget to soak the wood chips in water before using them.
2. Carefully wash and clean the turkey, then place it in the roasting pan.
3. Combine sage with black pepper, salt, basil, and olive oil, then mix until incorporated.
4. Rub the turkey with the spice mixture, then wrap it with aluminum foil.
5. Place the wrapped turkey on the electric smoker rack, then close the smoker's lid.
6. Cook the turkey for 50 minutes, then take it out from the smoker.
7. Unwrap the turkey, then brush all sides of the turkey with raw honey.
8. Re-wrap the glazed turkey with aluminum foil, then return back to the electric smoker.
9. Add about 2 handfuls of soaked mesquites wood chips to the fire and cook the turkey for 2 hours more or until the internal temperature has reached 180°F (80°C).
10. Once the turkey is done, remove it from the electric smoker, then unwrap it.
11. Brush the turkey with the remaining honey, then cook again without cover for about 20 minutes. This process is done to darken the turkey.
12. Once it is done, take the turkey out from the electric smoker, then place it on a serving dish.
13. Serve and enjoy.

Nutrition:

Calories: 312 **Total fat:** 9g **Net carbs:** 1g **Protein:** 55g

454. Apple Cola Smoked Turkey

Preparation time: 15 minutes.

Cooking time: 9 hours.

Servings: 10

Ingredients:

- 5-pound whole turkey
- 3 tablespoons minced garlic
- 2(½) tablespoons salt
- ¾ cup butter
- 3 cans carbonated beverage with cola flavor
- 2 fresh apples
- ½ cup chopped onion
- 3 teaspoons garlic powder
- 1(½) teaspoons black pepper

Directions:

1. Heat the electric smoker to 225°F (110°C).
2. Meanwhile, rub the turkey with rub ingredients, then place in a disposable pan. Set aside.
3. Cut the apple into cubes, then place in a bowl.
4. Add chopped onion, garlic powder, butter, and black pepper to the bowl, then pour the carbonated beverage. Mix well.
5. Fill the turkey with the apple mixture, then cover with aluminum foil.
6. Place the pan in the smoker, then smoke the turkey for 9 hours.
7. Brush the turkey with juice from the filling every 2 hours.
8. After 9 hours or when the internal temperature has reached 180°F (80°C), remove the smoked turkey from the smoker.
9. Cool for a few minutes, then serve.
10. Cut into slices, then enjoy.

Nutrition:

Calories: 625 **Total fat:** 31.7g **Net carbs:** 9.8g **Protein:** 71.2g

455. Savory Herb Rubbed and Aromatic Stuffed Smoked Turkey Recipe

Preparation time: 15 minutes.

Cooking time: 7 hours.

Servings: 16

Ingredients:

- 3 tablespoons extra virgin olive oil (EVOO)
- 3 tablespoons unsalted butter at room temperature
- 1/2 cup apple cider (might need more)
- 1 lemon or orange cut in quarters

- 2 cloves fresh garlic minced
- 2 tablespoons dried thyme
- 12 to 14-pound turkey
- 1 tablespoon powdered sage
- 2 teaspoons dried oregano
- 2 teaspoons paprika
- 1–1/2 teaspoons cracked black pepper
- 1 teaspoon dried rosemary
- 1 apple cut in quarters
- 1 medium onion cut in half
- 1/2 cup water
- 2 teaspoons sea salt

Directions:

1. Line a drip pan and water bowl with aluminum foil for easier cleanup. Heat the smoker to 225F.
2. In a small bowl, cream together the EVOO and softened butter. Mix in the garlic, herbs, and spices.
3. Rub the interior cavity of the turkey with 1/3 of this mixture. Stuff the cavity with the fruits and onion. Rub the outside of the bird with the remaining fat and herb blend.
4. Place the water and apple cider to fill the water pan halfway. Place the drip pan on the next rack just above the water pan to collect drippings from the turkey. Fill the side tray with the wood chips.
5. Tuck the tips of the wings tightly beneath the turkey. Place the seasoned turkey directly on the middle rack of the smoker. Insert the digital thermometer into the thigh of the bird, if your smoker has one. Set a timer for approximately 6.5 hours. A turkey generally smokes for 30 to 40 minutes per pound. You want to achieve an inside temperature of 165F.
6. Check the vent every hour for a smoke. Add more wood chips if the smoke has died down. Also, check the water pan and add additional cider and water as needed.
7. Start checking the internal temperature of the bird after 3 or 4 hours and every 45 minutes thereafter with either the digital thermometer or a good meat thermometer.
8. Remove the cooked turkey to a cutting board and allow it to rest for a minimum of 20 minutes before carving. You can tent it with aluminum foil to keep more moisture in.

Nutrition:

Calories: 618 **Total fat:** 19.8g **Saturated fat:** 6g **Cholesterol:** 258mg **Sodium:** 473mg **Total carbohydrate:** 4.3g **Dietary fiber:** 0.9g **Total sugars:** 2.6g **Protein:** 99.9g **Vitamin D:** 0mcg **Calcium:** 24mg **Iron:** 35mg **Potassium:** 1065mg

PORK RECIPES

456. Smoked Pork Ribs

Preparation time: 30 minutes.

Cooking time: 6 hours.

Servings: 4

Ingredients:

Rub:

- 1/4 cup ancho chili powder
- 2 teaspoon Spanish paprika
- 2 teaspoon freshly ground black pepper
- 2 teaspoon dry mustard
- 2 teaspoon kosher salt
- 2 teaspoon ground coriander
- 1 tablespoon dried oregano
- 1 tablespoon ground cumin
- 2 teaspoon chile de arbol
- 2 racks St. Louis-style pork ribs, 12 ribs each, membrane removed
- 1/4 cup canola oil

Mop mixture:

- 2 cups cider vinegar
- 2 teaspoon light brown sugar
- 1/2 tablespoon cayenne powder
- Few dashes hot pepper sauce (recommended: Tabasco)
- 1 tablespoon kosher salt
- 1/4 tablespoon freshly ground black pepper
- 1-quart apple cider

Directions:

1. Get a small bowl, and combine all the spices inside it. Brush each side of all the racks with a little bit of oil and your spice mixture. Wrap it all in plastic and refrigerate overnight for 12 hours.
2. Get a large pot heated up over low heat. Add all the mop ingredients into it. You're going to want to bring it all to a simmer and just cook it until the sugar has completely dissolved. Allow it cool at room temperature.
3. Take the rips out of the fridge one hour before it's time to start smoking them. Also, at this time, put your apple cider in a heatproof pan and place it within the smoker.
4. Place your ribs on the rack in the smoker. Once every hour for the first five hours, you're going to brush your ribs with the mop. In the last hour, brush your ribs with the barbecue sauce every 10 minutes.
5. Afterwards, take your ribs off the smoking rack and serve.

Nutrition:

Calcium: 77mg **Magnesium:** 51mg **Phosphorus:** 251mg **Iron:** 3.89mg **Potassium:** 655mg

457. Apple-Injected Smoked Pork

Preparation time: 10 minutes.

Cooking time: 6 hours.

Servings: 12

Ingredients:

- 2 cups apple cider
- 2 teaspoon dry rub seasoning
- 2 teaspoon apple cider vinegar
- 2 teaspoon honey
- 1/2 tablespoon cayenne pepper
- 1/4 cup orange juice
- 1/2 cup lemon juice

Dash Worcestershire sauce:

- 2 teaspoon kosher salt
- 1 (6 to 8-pound) pork butt

Directions:

1. Whisk all the ingredients for your marinade.

2. Put your pork in a large dish. Using a syringe, inject your marinade 3/4 of the way inside of the pork. You're going to want to do this several times in a different place each time you eat.
3. Close the pork in plastic wrap and store in the fridge between 4–12 hours.
4. Place pork in the smoker.
5. Drain any liquid that has remained on the meat, and be sure to pat it dry. Season the pork with your dry rub seasoning, so it has a better taste. You're going to want to cover both sides.
6. Cool the pork for a few minutes before serving.

Nutrition:

Calcium: 48mg **Magnesium:** 61mg **Phosphorus:** 680mg **Iron:** 3.26mg **Potassium:** 1336mg **Sodium:** 894mg **Zinc:** 9.41mg

458. Smoked Pork Chops

Preparation time: 15 minutes.

Cooking time: 1 hour, 10 minutes.

Servings: 4

Ingredients:

- 4 tablespoon salt
- 2 teaspoon freshly ground black pepper
- 2 teaspoon dark brown sugar
- 2 teaspoon ground thyme
- 2 teaspoon onion powder
- 1 tablespoon cayenne pepper
- 4 center-cut, bone-in pork chops

Buttermilk BBQ sauce:

- 1 cup apple cider
- 1 tablespoon brown sugar
- 1/2 preferred BBQ sauce
- 1 tablespoon buttermilk

Directions:

1. Mix together salt, onion powder, black pepper, thyme, cayenne pepper, and brown sugar.
2. After this, you're going to want to rub your pork chops with this mixture. Wrap the chops in plastic wrap and store in the fridge for four hours.
3. Place your chops in the smoker.
4. Get a Saucepan and heat it up to medium-low heat.
5. Add in the brown sugar and apple cider. Stir everything together.
6. Let the mixture reduce for only 25 minutes.
7. Turn the heat down to low, and pour your preferred barbecue sauce. Stir everything well.

8. Once the sauce has been thoroughly cooked, turn off the heat source.
9. Add in the buttermilk and stir it all together. Serve this over the pork chops, and you're done!

Nutrition:

Calcium: 36mg **Magnesium:** 39mg
Phosphorus: 266mg **Iron**: 0.91mg
Potassium: 577mg

459. Smoked Boston Butt Roast

Preparation time: 20 minutes.

Cooking time: 4 hours.

Servings: 6

Ingredients:

- 1(5-pounds) pork butt roast
- 4 teaspoon house seasoning, recipe follows
- 2 teaspoon seasoned salt
- 1 medium onion, sliced
- 1 cup water
- 3 bay leaves
- Sweet or Smoky BBQ sauce

House Seasoning:

- 1 cup salt
- 1/4 cup black pepper
- 1/4 cup garlic powder

Directions:

1. On one side of your roast, sprinkle two teaspoon of the House Seasoning. Be sure to rub it in well with your fingers. Flip your roast over and rub in the remaining two teaspoon of the House Seasoning. Repeat this process with your seasoned salt.
2. Place your roast on a large pan for roasting. Add in the bay leaves, onion, and water. Place the roast in your smoker for 4 hours.
3. The internal temperature for the meat should be 170°F. Once it is, allow it to cool for a few minutes.

1. Serve this roast with sweet or smoky BBQ sauce.

Nutrition:

Calcium: 31mg **Magnesium:** 42mg
Phosphorus: 342mg **Iron:** 1.58mg
Potassium: 596mg

460. Smoked Pork Shoulder

Preparation time: 30 minutes.

Cooking time: 5 hours, 30 minutes.

Servings: 6

Ingredients:

- 1(5–6pound) pork shoulder/Boston butt pork roast

- 2 teaspoon salt
- Sweet BBQ Sauce

Directions:

1. Season with salt your pork shoulder. Close it up and chill it in the fridge for half an hour.
2. Place the pork inside the smoker. Close the lid.
3. Cook the meat for Five and a half hours. The temperature of your pork inside should be 165°F. Rotate the pork over for the last two hours of its smoking.
4. Remove the pork, give it a few minutes to cool. Serve with sweet BBQ sauce.

Nutrition:

Calcium: 101mg **Magnesium:** 89mg **Phosphorus:** 809mg **Iron:** 6.81mg **Potassium:** 1186mg **Sodium:** 890mg

461. Smoked Pork Sausage

Preparation time: 60 minutes.

Cooking time: 3 hours.

Servings: 30 sausage.

Ingredients:

- 20 pounds of home-dressed lean pork meat
- 10 pounds of clear fat pork
- 1/2 pound fine salt (best quality)
- 2 teaspoon sugar
- 1 tablespoon ginger
- 2 teaspoon pepper
- 1 tablespoon sage
- 2 teaspoon cure (either Instacure #1 or Prague Powder #1)

Directions:

1. Cut the meat into cubes. Grind them all together in an extra-large bowl with the spices.
2. Using a sausage grinder, pass the mixed meat and spices through a medium plate on the grinder.
3. Stuff the sausages in natural pork casings so you'll be able to smoke them.
4. Place them in the smoker.

Nutrition:

Calcium, Ca55 mg **Magnesium**, Mg103 mg **Phosphorus**, P1260 mg **Iron**, Fe5.27 mg **Potassium**, K1529 mg

462. Sweet Smoked Pork Ribs

Preparation time: 20 minutes.

Cooking time: 4 hours.

Servings: 15

Ingredients:

- 1/4 cup salt
- 1/4 cup white sugar
- 2 teaspoon packed brown sugar
- 2 teaspoon ground black pepper
- 2 teaspoon ground white pepper
- 2 teaspoon onion powder
- 1 tablespoon garlic powder
- 1 tablespoon chili powder
- 1 tablespoon paprika
- 1 tablespoon ground cumin
- 10 pounds baby back pork ribs
- 1 cup apple juice
- 1/4 cup packed brown sugar
- 1/4 cup sweet BBQ sauce

Directions:

1. Mix salt, cumin, white sugar, paprika, 2 teaspoon of brown sugar, chili powder, black pepper, garlic powder, white pepper, and onion powder. Rub this mixture onto the back of the rips on all of the sides. Wrap your ribs in plastic wrap. Place the pork fridge for half an hour.
2. Start your smoker. Put your ribs on the rack.
3. Combine the barbeque sauce, apple juice, and 1/4 brown sugar. Every 30 to 45 minutes, brush the ribs with the barbeque sauce for the first hour. Brush the sauce onto the rips during its last half hour of cooking.
4. When the ribs have finished, wrap them up in aluminum foil. Allow them to sit for an additional 15 minutes. You may serve them afterward.

Nutrition:

Calcium: 82mg **Magnesium:** 79mg **Phosphorus:** 636mg **Iron:** 3.47mg **Potassium:** 1141mg **Sodium:** 2139mg **Zinc:** 9.26mg

463. Smoked Pork Tenderloin

Preparation time: 15 minutes.

Cooking time: 3 hours.

Servings: 4

Ingredients:

- ½ cup sweet BBQ sauce
- 2 pork tenderloins (1.5–2 pounds)
- ½ cup marinade

Directions:

1. Put your pork loin in your chosen marinade. Marinade for 3 hours or overnight.
2. Drain the marinade. Place your pork loin on the rack.
3. Half an hour before taking the loin out of the smoker, pout it with barbeque sauce
4. When you finish smoking remove the pork loin and wrap it in foil for ten minutes
5. Slice the pork loin thinly.

Nutrition:

Calcium: 29mg **Magnesium:** 80mg **Phosphorus:** 638mg **Iron:** 2.97mg **Potassium:** 1105mg **Sodium:** 930mg **Zinc:** 5.68mg

464. Smoked Asian Style Pork Tenderloin

Preparation time: 2 hours.

Cooking time: 2 hours, 15 minutes.

Servings: 10

Ingredients:

- 1 cup brown sugar
- 1/4 cup tamari sauce
- 1 cup apple cider vinegar
- 1 teaspoon fresh ginger grated
- 1 teaspoon salt and black pepper to taste
- 5 pounds pork tenderloin

Directions:

1. Whisk brown sugar, tamari sauce, apple cider vinegar, grated ginger and salt and pepper to taste in a mixing bowl.
2. Place the tenderloin in a large container and pour apple cider mixture; toss well. Place in the fridge for 4–5 hours (preferably overnight).
3. Heat your electric smoker to 225°F.
4. When it is ready, add some water to the removable pan that is usually on the bottom shelf.
5. Fill the side "drawer" with dry wood chips.

6. Remove tenderloin from fridge and pat dry on the kitchen paper.

7. Place pork in a smoker, smoke for 2(½)–3 hours, or until the internal temp reaches 150°F.

8. Take off from the smoker and wrap with aluminum foil. Place back into the smoker for 30 minutes or until internal temp reaches 145°F.

9. When ready, let cool for 15 minutes before slicing and serving.

Nutrition:

Calories: 294,47 **Calories from fat:** 44 **Total fat:** 4,94g **Saturated fat:** 1,59g **Cholesterol:** 147,42mg **Sodium:** 335,22mg **Potassium:** 960,75mg **Total carbohydrates:** 11,32g **Fiber:** 0,08g **Sugar:** 9,68g **Protein:** 47,87g

465. Smoked Hot Pepper Pork Tenderloin

Preparation time: 10 minutes.

Cooking time: 3 hours.

Servings: 6

Ingredients:

- 3/4 cup chicken stock
- 1/2 cup tomato-basil sauce
- 2 teaspoon hot red chili pepper
- 1 tablespoon oregano
- Salt and pepper
- 2 pounds pork tenderloin

Directions:

1. Whisk together the chicken stock, tomato-basil sauce, hot red chili pepper, oregano, salt, and pepper.

2. Brush generously all over the tenderloin.

3. Heat your electric smoker to 225°F.

4. When it is ready, add some water to the removable pan that is usually on the bottom shelf.

5. Fill the side "drawer" with dry wood chips.

6. Place meat in the smoker and smoke until the internal temperature of 145°F, for about 2(½)–3 hours.

7. Before slicing, let it rest for 10 minutes.

8. Serve.

Nutrition:

Calories: 360,71 **Total fat:** 14,32g **Saturated fat:** 5,1g **Cholesterol:** 159,8mg **Sodium:** 331,83mg **Potassium:** 905,83mg **Total carbohydrates:** 3,21g **Fiber:** 1,46g **Protein:** 52,09g **Sugar:** 1,01g

466. Smoked Pork Chops in Garlic Soy Marinade

Preparation time: 1 hour.

Cooking time: 3 hours, 15 minutes.

Servings: 8

Ingredients:

- 1 cup soy sauce
- 1/4 cup lemon juice freshly squeezed
- 1 tablespoon brown sugar
- 5 pounds bone-in pork loin
- 3 cloves garlic minced

Directions:

1. Whisk minced garlic, soy sauce, fresh lemon juice, and brown sugar in a large resealable plastic bag: toss to combine well.
2. Place in the pork chops. Seal bag and refrigerate overnight.
3. Remove the pork chops from the marinade; reserve the marinade for basting.
4. Heat your electric smoker to 225°F.
5. When it is ready, add some water to the removable pan that is usually on the bottom shelf. Fill the side "drawer" with dry wood chips.
6. Arrange pork chops on racks and smoke for 3 hours.
7. After 3 hours, remove ribs, baste generously with reserved marinade and wrap in aluminum foil.
8. Return the smoker and cook for an additional 1 to 1(1/2) hours, or until the internal temperature reaches 160°F.
9. Remove meat from the smoker and let it rest for 10–15 minutes. Serve hot.

Nutrition:

Calories: 554,09 **Total fat:** 31,35g **Saturated fat:** 6,71g **Potassium:** 1027,42mg **Total carbohydrates:** 5,3g **Fiber:** 0,3g **Sugar:** 2,41g **Protein:** 59,24g

467. Smoked Pork Loin with Beer-Anise Marinade

Preparation time: 30 minutes.

Cooking time: 3 hours.

Servings: 6

Ingredients:

Marinade:

- 1/4 cup honey
- 1(1/2) cups dark beer
- 2 teaspoon Anise seeds
- 1 tablespoon fresh thyme finely chopped
- Salt and pepper to taste

Pork:

- 3 pounds pork loin

Directions:

1. In a casserole, mix all the marinade ingredients.
2. Place the pork with the marinade mixture in a resealable plastic bag. Refrigerate for several hours or overnight.
3. Place the water in the pan at the bottom of your smoker. Fill the drawer or tray with wood chips.
4. Heat your smoker (use a 2-zone or indirect setup) to about 225°F.
5. Remove the pork from the marinade (reserve marinade for later) and place it on the kitchen towel.
6. Put the meat in the smoker, then smoke till the internal temperature is 145F, about 2(1/2) to 3 hours.
7. Remove the meat, then let rest for 10 minutes before slicing.
8. Serve hot or cold.

Nutrition:

Calories: 360,86 **Total fat:** 7,92g **Saturated fat:** 2,76g **Cholesterol:** 149,69mg **Sodium:** 211,47mg **Potassium:** 917,04mg **Total carbohydrates:** 14,56g **Fiber:** 0,4g **Sugar:** 11,61g **Protein:** 51,33g

468. Smoked Pork Loin with Sweet Habanero Rub

Preparation time: 30 minutes.

Cooking time: 3 hours.

Servings: 8

Ingredients:

- 4 pounds pork loin
- Tamari sauce
- Mandarin habanero seasoning or any other hot sauce
- 1 cup honey
- 1 cup mustard
- 1 tablespoon salt and white pepper to taste

Directions:

1. Combine the Habanero seasoning, honey, mustard and tamari sauce in a mixing bowl.
2. Rub lots of spice mix all over the meat.
3. Heat your electric smoker to 225°F.
4. When it is ready, add some water to the removable pan that is usually on the bottom shelf.
5. Fill the side "drawer" with dry wood chips (hickory or maple).
6. Put meat in the smoker and smoke till the internal temperature is 145°F, about 2(1/2) to 3 hours.

7. When the meat reaches a temperature around 145°F, remove the meat, then cover it for about 5 to 10 minutes.

8. Slice and serve hot.

Nutrition:

Calories: 429,91 **Total fat:** 8,4g **Saturated fat:** 2,78g **Cholesterol:** 149,69mg **Potassium:** 921,88mg **Total carbohydrates:** 36,36g **Fiber:** 0,83g **Sugar:** 34,93g **Protein:** 51,73g

469. Smoked Pork Ribs with Avocado Oil

Preparation time: 4 hours.

Cooking time: 2 hours.

Servings: 7

Ingredients:

- 1 cup avocado oil
- 1 teaspoon garlic salt, or to taste
- 2 teaspoon garlic and onion powder
- 1/2 cup fresh parsley finely chopped
- 4 pounds spare ribs

Directions:

1. Whisk avocado oil, garlic salt, garlic powder, onion powder, fresh chopped parsley in a mixing bowl.

2. Put pork ribs in a shallow container and pour avocado mixture over; toss to combine well. Refrigerate for at least 4 hours, or overnight.

3. Heat your electric smoker to 225°F.

4. When it is ready, add some water to the removable pan that is usually on the bottom shelf. Fill the side "drawer" with dry wood chips.

5. Remove pork ribs from marinade (reserve marinade) and arrange the pork chops on the rack.

6. Smoke for 1(1/2) hours at 225°F.

7. Remove the ribs, baste generously with reserved marinade, and wrap in heavy-duty aluminum foil.

8. Return the meat to the smoker, then cook for an additional 1 hour, or until internal temp reaches 160°F.

9. Transfer pork chops on serving plate and let rest for 15–20 minutes before serving.

Nutrition:

Calories: 760,68 **Total fat:** 76,26g **Saturated fat:** 21,32g **Cholesterol:** 207,36mg **Sodium:** 505,73mg **Potassium:** 661,57mg **Total carbohydrates:** 1,06g **Fiber:** 0,37g **Sugar:** 0,06g **Protein:** 40,37g

470. Smoked Pork Ribs with Fresh Herbs

Preparation time: 30 minutes.

Cooking time: 5 hours.

Servings: 6

Ingredients:

- 1/2 cup olive oil
- 1 teaspoon fresh parsley finely chopped
- 1 teaspoon fresh sage finely chopped
- 1 teaspoon fresh rosemary finely chopped
- Salt and ground black pepper to taste
- 3 pound bone-in pork rib roast

Directions:

1. Combine the olive oil, garlic, parsley, sage, rosemary, salt, and pepper in a bowl; stir well.
2. Generously rub the herbs mix all over the meat.
3. Heat your electric smoker to 225°F.
4. When it is ready, add some water to the removable pan that is usually on the bottom shelf.
5. Fill the side "drawer" with dry wood chips (Hickory and mesquite).
6. Smoke the meat directly on the racks for 3 hours at 225°F.
7. Remove the ribs from the racks and tightly wrap them in aluminum foil
8. Move them back in the smoker for 2 hours.
9. Transfer to a serving platter; let it rest 10–15 minutes before serving.

Nutrition:

Calories 532,35 **Total fat** 40,14g **Saturated fat** 7,24g **Potassium** 678,54mg **Total carbohydrates** 0,11g **Fiber** 0,07g **Sugar** 0g **Protein** 40,65g

471. Pork Belly Burnt Ends

Preparation time: 10 minutes.

Cooking time: 5 hours.

Servings: 4

Ingredients:

- 1 Pork belly slab
- BBQ pork rub as needed
- BBQ sauce as needed

Directions:

1. Cut the pork belly into 1-inch long strips, starting from the skinless side, and then place the strips in a large bowl.
2. Add rub, toss until coated, and then let it marinate for a minimum of 30 minutes in the refrigerator.

3. Prepare the grill, and for this, fill it hopper with apple blend wood pallets, go to Electric Smoker app, set the grill temperature to 250°F, and let it heat.

4. When the grill has heated, open its lid, place pork belly strips on the pellet grill by using a tong, shut with lid and let it grill for 3 to 4 hours or set the food temperature to 198°F in the app and let it grill until the food reaches the set food temperature.

5. Once the app shows that the internal temperature of the pork has reached 198°F, open the grill and transfer pork strips to a cutting board.

6. Let pork cool for 15 minutes, then cut them into 1-inch cubes and place them in a disposable aluminum foil tray.

7. Add BBQ sauce, toss until well coated, cover the tray with foil, return it on the grill and continue grilling for 1 hour.

8. Serve straight away.

Nutrition:

Calories: 300 **Fat:** 13g **Carbs:** 21g **Protein:** 26g **Fiber:** 1g

472. Beer Brats

Preparation time: 10 minutes.

Cooking time: 3 hours.

Servings: 6

Ingredients:

- 6 uncooked German brats
- 1 medium sweet onion, peeled, chopped into thick slices
- 3 tablespoons brown sugar
- 8 ounces butter, salted, cut into 8 pieces
- 22 ounces amber ale

Directions:

1. Switch on the grill, go to the Wifi setting on your cell phone, and then connect with the grill using your serial number as the password.

2. Prepare the grill, and for this, fill it hopper with apple blend wood pallets, go to Electric Smoker app, set the grill temperature to 250°F, and let it heat.

3. Meanwhile, take a large pan, scatter onion in its bottom, sprinkle with sugar, top with butter slices and sausages, and then pour ale over the sausages.

4. When the grill has heated, open its lid, place the pan on the pellet grill using a tong, shut with lid, and let it grill for 3 hours until thoroughly cooked.

5. Serve straight away.

Nutrition:

Calories: 230.4 **Fat:** 21g **Carbs:** 2g **Protein:** 12g **Fiber:** 0.4g

473. Cherry Chipotle Ribs

Preparation time: 10 minutes.

Cooking time: 6 hours.

Servings: 4

Ingredients:

- 1 large rack of baby back ribs, at room temperature, membrane removed
- ½ cup brown sugar
- 2 tablespoons mustard paste
- Pork rub as needed
- 2 tablespoons butter, unsalted
- 3 tablespoons honey
- Cherry chipotle sauce as needed

Directions:

1. Prepare the grill, and for this, fill it hopper with Texas blend wood pallets, go to Electric Smoker app, set the grill temperature to 225°F, and let it heat.
2. Meanwhile, prepare the ribs, and for this, rub with mustard paste and then sprinkle with pork rub until coated on all sides.
3. When the grill has heated, open its lid, place ribs on the pellet grill, shut with lid, then grill for 2 hours.
4. Then take a large piece of foil, place ribs on it, spread butter and honey on top, sprinkle with sugar, wrap the ribs and then continue grilling for 2 hours or until tender.
5. Then remove ribs from the grill, unwrap them, and spread evenly with chipotle sauce.
6. Switch the temperature of the grill to 275°F, return the ribs on the grill and continue grilling for 1 hour or 1 hour and 30 minutes until done.
7. Let the meat rest for 15 minutes and then serve.

Nutrition:

Calories: 292.5 **Fat:** 17g **Carbs:** 20g **Protein:** 13g **Fiber:** 2g

474. Sweet Espresso Ribs

Preparation time: 10 minutes.

Cooking time: 6 hours.

Servings: 6

Ingredients:

- 1 large rack of baby back ribs, at room temperature, membrane removed
- ½ cup brown sugar
- ½ cup ground espresso beans
- 3 tablespoons ground black pepper
- 2 tablespoons olive oil
- 2 tablespoons butter, unsalted, cubed
- BBQ sauce as needed

Directions:

1. Prepare the grill, and for this, fill it hopper with Texas blend wood pallets, go to Electric Smoker app, set the grill temperature to 225°F, and let it heat.
2. Meanwhile, brush ribs with oil, then stir together sugar, espresso, and black pepper and sprinkle this mixture on the ribs until coated on all sides.
3. When the grill has heated, open its lid, place ribs on the pellet grill by using a tong, shut with lid, and let it grill for 2 hours.
4. Then take a large piece of foil, place ribs on it, spread butter cubes on top, sprinkle with remaining espresso mixture, wrap the ribs and continue grilling for 2 hours or until tender.
5. Then remove the ribs from the grill, unwrap them, and spread evenly with sauce.
6. Return the ribs to the grill and then continue grilling for 1 hour or 1 hour and 30 minutes until done.
7. Rest for 15 minutes, and serve.

Nutrition:

Calories: 416 **Fat:** 21g **Carbs:** 35g **Protein:** 22g **Fiber:** 1g

475. BBQ Pulled Pork

Preparation time: 20 minutes.

Cooking time: 15 hours.

Servings: 7

Ingredients:

- 7 to 10 pounds pork butt, fat trimmed, at room temperature
- ½ cup apple cider vinegar
- Sea salt as needed
- Ground black pepper as needed
- ½ cup apple juice

For the Marinate:

- 2 tablespoons red pepper flakes
- 2 tablespoons onion powder
- 1 cup brown sugar
- 3 tablespoons paprika
- 5 tablespoons mustard paste
- 8 tablespoons olive oil

Directions:

1. Prepare the marinade, and for this, take a medium bowl, place all the ingredients in it and then whisk until well combined.
2. Rub the marinade on all sides of the pork butt and then let it marinate for 12 to 24 hours in the refrigerator.

3. Then switch on the grill, go to the WiFi setting on your cell phone, and connect with the grill by using your serial number as the password.

4. Go to the app of Electric Smoker, press the 'connect' button, and when connected, go to its setting and select the WiFi mode option and after few minutes, select the connect option again.

5. Prepare the grill, and for this, fill it hopper with gold blend wood pallets, go to Electric Smoker app, set the grill temperature to 380°F, and let it heat.

6. Meanwhile, remove pork putt from the refrigerator, bring it to room temperature and season it well with salt and black pepper.

7. When the grill has heated, open its lid, place pork butt fat-side-up on the pellet grill, shut with lid, and let it grill for 1 hour.

8. Meanwhile, pour apple juice into a small spray bottle, add apple cider vinegar, and then shake well.

9. Switch the temperature of the grill to 225°F, spray with the apple juice mixture and then continue grilling for 5 hours or until the internal temperature reaches 160°F.

10. Return pork butt from the grill, place it on a large piece of foil, spray with apple juice mixture, wrap it with foil and then continue grilling the pork until the internal temperature reaches 195°F.

11. Once the app shows that the internal temperature of the chicken has reached 195°F, open the grill and then transfer wrapped pork to a cutting board.

12. Let the pork rest for 30 minutes, uncover it, and then shred it by using two forks.

13. Serve straight away.

Nutrition:

Calories: 484 **Fat:** 21.6g **Carbs:** 46.2g **Protein:** 26.2g **Fiber:** 10.4g

476. Baby Back Ribs

Preparation time: 10 minutes.

Cooking time: 9 hours.

Servings: 6

Ingredients:

- 2 racks of baby back ribs, at room temperature
- Cherry chipotle sauce as needed

For the Marinade:

- 2 tablespoons minced garlic
- 1 teaspoon onion powder
- 2 teaspoons ground black pepper
- 1 tablespoon brown sugar
- 1 cup soy sauce

- 2 tablespoons red wine vinegar
- ¼ cup olive oil
- ½ teaspoon Tabasco sauce
- ¼ cup white wine

Directions:

1. Take a small bowl, place all the ingredients for the marinade and whisk until combined.
2. Pour the marinade in a large plastic bag, add ribs, seal the bag, turn it upside down to coat ribs with the marinade and then let it marinate for a minimum of 4 hours.
3. When ready to grill, switch on the grill, go to the WiFi setting on your cell phone, and then connect with the grill by using your serial number as the password.
4. Go to the app of Electric Smoker, press the 'connect' button, and when connected, go to its setting and select the WiFi mode option and after few minutes, select the connect option again.
5. Prepare the grill, and for this, fill it hopper with gold blend wood pallets, go to Electric Smoker app, set the grill temperature to 165°F, and let it heat.
6. When the grill has heated, open its lid, place ribs on the pellet grill by using a tong, shut with lid, and let it grill for 4 to 6 hours, turning halfway through.
7. Switch the temperature to 225°F, continue grilling for 2 hours, brush with chipotle sauce and continue grilling for 1 hour until glazed.
8. When done, let ribs rest for 30 minutes, then cut it into slices and serve.

Nutrition:

Calories: 668 **Fat:** 45g **Carbs:** 13g **Protein:** 48g **Fiber:** 0.3g

477. Perfect Smoked Pork Butt

Preparation time: 10 minutes.

Cooking time: 6 hours.

Servings: 8

Ingredients:

- 6 pound bone-in pork butt
- 1 tablespoon chili powder
- 1/2 tablespoon cayenne pepper
- 1 tablespoon paprika
- 1 teaspoon garlic powder
- 1 tablespoon onion powder
- 2 tablespoon dark sugar
- Coarse salt and freshly ground pepper

Directions:

1. In a bowl, blend all ingredients and rub the butt.

2. Place in the aluminum tray—uncovered—and smoke for 6 hours (1 hour per pound) at 250°F.
3. Remove from smoker, then strain the juice into a small bowl.
4. Place the butt in double foil.
5. Pour juice over pork and wrap 4 more times.
6. Place back on the smoker at 250°F until internal temp reaches at 195°F for pulled pork.
7. Serve hot. Enjoy!

Nutrition:

Calories: 469 **Fat:** 35g **Protein:** 54g **Carbs:** 5.6g

478. Rodeo Drive Baby Ribs

Preparation time: 4–12 hours.

Cooking time: 3 hours.

Servings: 8

Ingredients:

- 4 pound baby ribs (without membrane)
- Salt and ground pepper
- 1 can (15 ounces) tomato sauce
- 2 tablespoon maple syrup
- 2 tablespoon apple cider vinegar
- 2 teaspoon smoked paprika
- 1/2 teaspoon garlic powder
- Onion powder

Directions:

1. Cut the Baby Ribs into two-three pieces, place them in a shallow container and rub them with salt and pepper.
2. Combine the tomato sauce, maple syrup, apple cider vinegar, paprika, garlic, and onion powder.
3. Pour sauce over spareribs evenly.
4. Refrigerate for about 4 hours or overnight.
5. Heat the smoker to 225°F, then add wood chips.
6. Remove the pork chops from marinade, then dry.
7. Smoke for 3—4 hours or until the internal temperature reaches 165°F.
8. Serve hot. Enjoy!

Nutrition:

Calories: 567.6 **Fat:** 53.3g **Protein:** 35.9g **Carbs:** 6.8g

479. Shredded Pork Loin with Chipotle

Preparation time: 4–5 hours.

Cooking time: 2 hours, 20 minutes.

Servings: 6

Ingredients:

- 4 pound pork loin
- 1 spring onion, finely chopped
- 1 tablespoon minced garlic
- 1/4 cup fresh lime juice
- 1/4 cup apple cider vinegar
- 2 tablespoon chipotles in adobo sauce
- 3 bay leaves
- 1 tablespoon dried oregano
- 1 tablespoon ground cumin
- 1 tablespoon kosher salt

Directions:

1. Whisk the garlic, chipotles, onion, lime juice, vinegar, bay leaves, oregano, cumin, and salt in a bowl.
2. Cut the pork loin into four pieces and place in a large container; rub both sides of tenderloin with salt.
3. Pour the chipotle mixture over the pork and toss to combine well.
4. Cover and refrigerate for 4–5 hours or overnight.
5. Heat the smoker to 225°F.
6. Remove the pork from the marinade, then dry with the kitchen towel.
7. Smoke unwrapped for 1 1/2 hours, or until internal temp reaches 150°F.
8. Remove from the smoker and wrap with heavy-duty aluminum foil.
9. Put back into the smoker for an additional 30 minutes, or until internal temp reaches 165°F.
10. Transfer pork to the plate and let cook for 15 minutes.
11. When cool, use two forks or your fingers to shred the meat into pieces.
12. Serve and enjoy!

Nutrition:

Calories: 396.1 **Fat:** 10.6g **Protein:** 68.2g **Carbs:** 2.6g

480. Smoked Avocado Pork Ribs Appetizer

Preparation time: 4–12 hours.

Cooking time: 4 hours, 30 minutes.

Servings: 7

Ingredients:

- 3 pound spare ribs
- 1 cup avocado oil
- 1 teaspoon garlic salt, to taste

- 1 teaspoon garlic powder
- 1/2 teaspoon onion powder
- Fresh parsley, finely chopped
- Salt and pepper, to taste

Directions:

1. In a bowl, combine the avocado oil, garlic salt, garlic powder, onion powder, chopped parsley, salt, and pepper.
2. Place pork ribs in a shallow container and pour the avocado mixture evenly.
3. Cover and refrigerate for at least 4 hours, or overnight.
4. Remove pork ribs from marinade (reserve marinade) and smoke for 3 hours at 225°F in the heated smoker.
5. Remove the ribs from the smoker.
6. Baste generously with reserved marinade, and wrap in heavy-duty aluminum foil.
7. Return to smoker then cook for an additional 1 to 1(1/2) hours, or until internal temp reaches 160°F.
8. Transfer pork chops on serving plate and let rest for 15–20 minutes.
9. Serve hot. Enjoy!

Nutrition:

Calories: 678.8 **Fat:** 61.1g **Protein:** 30.2g **Carbs:** 0.6g

481. Smoked Chops in Sweet Soy Marinade

Preparation time: 10 minutes.

Cooking time: 4 hours, 30 minutes.

Servings: 8

Ingredients:

- 2 garlic cloves, minced
- 3/4 cup soy sauce
- 1/4 cup lemon juice, freshly squeezed
- 1 tablespoon chili sauce
- 1 tablespoon brown sugar
- 5 pound bone-in pork loin

Directions:

1. In a large, resealable plastic bag, combine minced garlic, soy sauce, fresh lemon juice, chili sauce, and brown sugar; stir.
2. Place inside the pork chops. Seal bag and refrigerate overnight.
3. Remove the pork chops from the marinade; reserve the marinade for basting.
4. Heat the smoker to 225°F and add wood chips.
5. Arrange the pork chops on the grill and smoke for 3 hours.
6. After 3 hours, remove ribs, baste generously with reserved marinade and wrap in heavy-duty aluminum foil.

7. Return the smoker, then cook for an additional 1 to 1(1/2) hours, or until internal temperature reaches 160° F.

8. Serve hot. Enjoy!

Nutrition:

Calories: 549.5 **Fat:** 31.4 g **Protein:** 58.8 g **Carbs:** 4.5 g

482. Smoked Pork Chops in Lavender Balsamic Marinade

Preparation time: 10 minutes.

Cooking time: 1 hour.

Servings: 4

Ingredients:

- 2 tablespoon olive oil
- 3 teaspoon cooking dry lavender
- 2 teaspoon light brown sugar
- 2 teaspoon fresh thyme, chopped
- Salt and grated black pepper
- 2 garlic cloves, finely chopped
- 4 pork chops or fillets
- 1/4 cup honey
- 2 teaspoon balsamic vinegar

Directions:

1. In a bowl, combine oil, lavender, sugar, thyme, salt, pepper, and garlic; mix well
2. Place the pork chops in a container, then pour with the oil-lavender mixture.
3. Marinate at room temperature for one hour.
4. Meanwhile, in a pan, pour the honey and balsamic vinegar and cook over moderate to low heat until boiling. Set aside.
5. Take off the meat from marinade and pat dry with the kitchen towel.
6. Smoke pork chops for about 1 hour at 225°F.
7. Remove pork chops from grill and brush with honey balsamic-mixture.
8. Roll the chops in aluminum foil and smoke the chops for about 30 minutes or until the internal temperature reaches 145°F.
9. Let rest for 5 minutes and serve. Enjoy!

Nutrition:

Calories: 257.4 **Fat:** 31.4g **Protein:** 58.8g **Carbs:** 21.1g

483. Smoked Pork Ribs with Hoisin Sauce

Preparation time: 30 minutes.

Cooking time: 4 hours.

Servings: 10

Ingredients:

- 5 pound pork ribs
- 1/4 cup Hoisin sauce
- 3 tablespoon dry sherry
- 2 tablespoon soy sauce
- 2 tablespoon honey
- 1/4 cup water
- 2 garlic cloves, minced

Directions:

1. Cut the ribs into serving-size portions.
2. Place the large plastic bag in a large bowl and combine Hoisin sauce, dry sherry, soy sauce, honey, water, and minced garlic; stir.
3. Add the ribs, close bag tightly, and shake several times.
4. Refrigerate 6 hours or overnight.
5. Drain ribs and reserve the marinade.
6. Smoke for 3 hours at 225°F in the heated smoker.
7. After 3 hours, remove ribs, baste generously with reserved marinade and wrap in heavy-duty aluminum foil.
8. Return the meat to the smoker, then cook for an additional 1 to 1(1/2) hours until internal temperature reaches 160°F.
9. Serve hot. Enjoy!

Nutrition:

Calories: 553.9 **Fat:** 53.3g **Protein:** 35.5g **Carbs:** 7.1g

484. Smoked Spiced Pork Tenderloin

Preparation time: 15 minutes.

Cooking time: 2 hours, 15 minutes.

Servings: 6

Ingredients:

- 2 pound pork tenderloin
- 1 teaspoon garlic salt
- 1 teaspoon onion powder
- 1 teaspoon garlic powder
- 1 teaspoon dry ginger
- Sea salt and fresh ground pepper

Directions:

1. Heat the smoker to 225°F and add hickory wood chips.

2. Combine sea salt, black pepper, onion powder, dry ginger, garlic powder, and garlic salt to taste and season both sides of the tenderloin.
3. Smoke unwrapped for 1(1/2) hours until internal temp reaches 150°F.
4. Remove from smoker and wrap with heavy-duty aluminum foil.
5. Place back into the smoker for an additional 30 minutes or until internal temp reaches 165°F.
6. When tenderloin is finished cooking, remove it from the smoker, then let rest for 15 minutes.
7. Slice tenderloin into slices and serve hot. Enjoy!

Nutrition:

Calories: 307.9 **Fat:** 12.3g **Protein:** 45.3g **Carbs:** 0.9g

485. Pork Butt Honey Mustard Smoked Bites

Preparation time: 30 minutes.

Cooking time: 5 hours.

Servings: 6

Ingredients:

- Mustard, to taste
- 4 pound Boston pork butt, no bones
- Butter, enough for the pork
- BBQ rub
- Brown sugar, to taste
- Honey, enough, for the pork
- BBQ sauce, to taste

Directions:

1. Prepare your Electric Smoker by heating it to a temperature of about 250°F.
2. Carefully trim and remove excess fat from the pork.
3. Place the meat in a disposable aluminum pan and properly coat it with BBQ rub and mustard.
4. Transfer the pan to your heated smoker and give about 2–3 hours of smoking. Or you can aim to reach an internal temperature of about 160°F.
5. Take out the meat after smoking and cut small cubes and transfer to a fresh aluminum pan.
6. Include some of the cooking juice and some butter.
7. Spread BBQ sauce and honey, according to your taste. Coat with brown sugar.
8. Cover the pan with a thick aluminum foil and transfer it to your smoker again.
9. Give another 80–90 minutes of smoking.

10. Take out and stir the meat bites and remove extra cooking juice.
11. Coat more sauce and transfer to your smoker again. You can increase the smoking temperature and reach up to 275°F.
12. Give about 25–32 minutes of smoking to get a caramelized texture.
13. Take out and allow to cool down for about 12–15 minutes.
14. Your dish is ready to be served. Enjoy!

Nutrition:

Calories: 257 **Fat:** 18g **Protein:** 15g **Carbs:** 8g

486. Taquito Rolls with Creamy Pulled Pork

Preparation time: 5 minutes.

Cooking time: 45 minutes.

Servings: 6

Ingredients:

- 4 ounces cream cheese
- 1/2 pound pulled pork
- 1/3 cup green salsa
- 1/2 cup cheddar cheese
- 1 pack small tortillas
- 1 lime

Directions:

1. Prepare your Electric Smoker by heating it to a temperature of about 250°F.
2. Transfer the cream cheese to an over-friendly bowl and give about 30 seconds in the oven to get a soft texture.
3. In a large enough bowl, make a mixture of cheddar cheese, pulled pork, cream cheese, ½ lime juice, and salsa.
4. Use this mixture to make taquito rolls and seal each one of them tightly.
5. Put the rolls on a large aluminum sheet and transfer it to your smoker.
6. Bake in the smoker for about 20–35 minutes. Make sure the edges start looking golden and crispy.
7. Take out and serve. Enjoy!

Nutrition:

Calories: 160 **Fat:** 7g **Protein:** 21g **Carbs:** 3g

SEAFOOD RECIPES

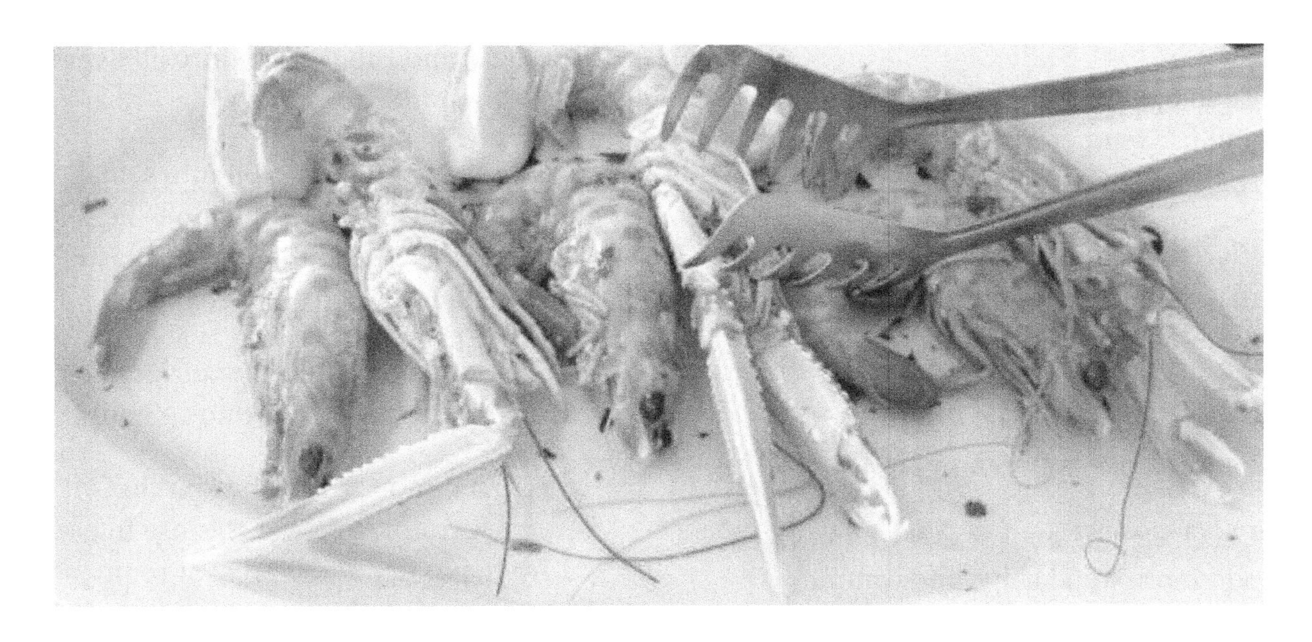

487. Brined Bass

Preparation time: 9 hours.
Cooking time: 4 hours.
Servings: 6

Ingredients:

- 2 pounds striped bass fillets, gutted and scaled
- 1/3 cup salt
- 1/4 cup brown sugar
- 1 tablespoon ground black pepper
- 2 dried bay leaves
- 2 slices of lemon
- 4 cups filtered water
- 1/2 cup dry white wine

Directions:

1. Place a pot at high heat, pour water, add salt and sugar and bring the mixture to a low boil or until salt and sugar are dissolved completely.
2. Then remove the pot from heat, cool brine at room temperature, and transfer it into a large container with a lid.
3. Add lemon slices along with black pepper, bay leaves and wine, stir until mixed, then add the bass, pour more water to cover fillets completely, and let soak for 4 to 8 hours in the refrigerator, covering the container.
4. Then remove the bass from brine, rinse well, pat dry with paper towels and let rest for 45 minutes at room temperature.
5. In the meantime, plug in the smoker, fill its tray with hickory wood chips

and water pan halfway through, and place the dripping pan above the water pan.

6. Then open the top vent, shut with lid, and use temperature settings to heat the smoker at 180°F.

7. Place bass fillets on the smoker rack, insert a meat thermometer, then shut with lid and set the timer to smoke for 2 to 4 hours or until meat thermometer registers an internal temperature between 145 to 160°F.

8. Check vent of smoker every hour and add more wood chips and water to maintain temperature and smoke.

9. Serve straight away.

Nutrition: Calories: 25 **Carbs:** 0g **Fat:** 2g **Protein:** 4g **Fiber:** 0g

488. Cured Salmon

Preparation time: 14 hours
Cooking time: 6 minutes
Servings: 3

Ingredients:

- 1(½) pound salmon filet, skinless and boneless
- 1 bunch of fresh dill, chopped
- 1/2 of lemon, thinly sliced
- 1/4 cup salt
- 1/4 cup brown sugar
- 2 tablespoons ground black pepper

Directions:

1. Stir together salt, black pepper, and sugar and rub this mixture all over the salmon filet.

2. Place seasoned salmon into a shallow baking dish, top with lemon slices and with dill, and wrap top with plastic wrap and then the whole dish.

3. Place this dish into the refrigerator to marinate salmon for 8 to 12 hours.

4. Then remove the dish from the refrigerator, uncover it, rinse fillet well, pat dry and let rest at room temperature for 2 hours.

5. When ready to cook, plug in the smoker, fill its tray with pecan wood chips and water pan halfway through, and place the dripping pan above the water pan.

6. Then open the top vent, shut with lid and use temperature settings to heat the smoker at 160°F.

7. Place salmon on smoker rack, insert a meat thermometer, then shut with lid and set the timer to smoke for 3 to 6 hours or until meat thermometer register an internal temperature of 130°F.

8. Check vent of the smoker every hour and add more wood chips and water to maintain temperature and smoke.

9. Serve straight away.

Nutrition: Calories: 210 **Carbs:** 0g **Fat:** 12.3g **Protein:** 22.5g **Fiber:** 0g

489. Shrimps

Preparation time: 15 minutes.

Cooking time: 30 minutes.

Servings: 6

Ingredients:

- 2 pounds shrimp, peeled, deveined and rinsed
- 2 tablespoons lemon juice
- 2 tablespoons chopped parsley
- 2 tablespoons onion powder
- 2 tablespoons garlic powder
- ¼ cup sea salt
- 3 tablespoons paprika
- 3 tablespoons ground black pepper
- 2 teaspoons cayenne pepper
- 2 tablespoons dried thyme
- 2 tablespoons olive oil

Directions:

1. Take a large foil pan, grease with oil and then place shrimps in it.
2. Stir the remaining ingredients except for lemon juice and sprinkle this mixture on all over shrimps until evenly coated.
3. Then plug in the smoker, fill its tray with hickory wood chips and water pan halfway through, and place the dripping pan above the water pan.
4. Then open the top vent, shut with lid and use temperature settings to heat the smoker at 250°F.
5. Drizzle 1 tablespoon of lemon juice over shrimps, then place the pan on smoker rack, then shut with lid and set the timer to smoke for 30 minutes or until shrimps are opaque, stirring in half.
6. When done, remove the pan from the smoker, drizzle remaining lemon juice over shrimps and serve.

Nutrition:

Calories: 60 Cal **Carbs:** 0 g **Fat:** 2 g, **Protein:** 10 g **Fiber:** 0 g.

490. Smoked Red Fish Fillets

Preparation time: 16 hours.

Cooking time: 1 hour.

Servings: 2

Ingredients:

- 2 fillets of redfish with skin, each about 12 ounces
- 1 teaspoon garlic powder
- 1/2 cup salt
- 1 teaspoon ground black pepper
- 1/2 cup brown sugar

- 1 teaspoon dried lemon zest
- 1 lemon, sliced

Directions:

1. Stir together garlic powder, salt, black pepper, sugar, and lemon zest until combined.
2. Take a glass baking dish, spread 1/3 of prepared spice mixture in the bottom, then later with one fillet, skin-side down and press lightly.
3. Sprinkle half of the remaining spice mixture over the fillet in the pan, then top with another filet, flesh-side down and then sprinkle the remaining spice mixture on top of it and around the side of the fish.
4. Close the dish with plastic wrap and then let marinate in the refrigerator for 8 to 12 hours.
5. Remove marinated fish from the dish, rinse well, and pat dry using paper towels.
6. Return fish into the refrigerator for 2 to 3 hours or until dried and then bring fish to room temperature for 45 minutes.
7. When ready to cook, plug in the smoker, fill its tray with hickory wood chips and water pan halfway through, and place the dripping pan above the water pan.
8. Then open the top vent, shut with lid and use temperature settings to heat the smoker at 120°F.
9. Place fish on smoker rack, insert a meat thermometer, then shut with lid and set the timer to smoke for 1 hour or more until meat thermometer registers an internal temperature of 140°F.
10. Check vent of smoker every hour and add more wood chips and water to maintain temperature and smoke.
11. Serve straight away.

Nutrition:

Calories: 27 **Carbs:** 0g **Fat:** 0.5g **Protein:** 5.3g **Fiber:** 0g

491. Lemon Pepper Tuna

Preparation time: 1 hour.

Cooking time: 4 hours, 10 minutes.

Servings: 6

Ingredients:

- 6 tuna steaks, each about 6 ounces
- 3 tablespoons salt
- 3 tablespoons brown sugar
- 1/4 cup olive oil
- ¼ cup lemon pepper seasoning
- 1 teaspoon minced garlic
- 12 slices of lemon

Directions:

1. Season tuna with salt and sugar until evenly coated on all sides, then place in a dish and cover with plastic wrap.
2. Place dish into the refrigerator for 4 hours or overnight, then rinse well and pat dry and coat well with garlic powder, lemon pepper seasoning, and oil.
3. Plug in the smoker, fill its tray with peach wood chips and water pan halfway through, and place the dripping pan above the water pan.
4. Then open the top vent, shut with lid and use temperature settings to heat the smoker at 120°F.
5. In the meantime, place seasoned tuna on smoker rack, insert a meat thermometer, then shut with lid and set the timer to smoke for 1 hour or more until meat thermometer registers an internal temperature of 140°F.
6. Check vent of smoker every hour and add more wood chips and water to maintain temperature and smoke.
7. When done, transfer tuna to a cutting board, let rest for 10 minutes and then serve with lemon slices.

Nutrition:

Calories: 275 Cal **Carbs:** 0.6g **Fat:** 23 g **Protein:** 17 g **Fiber:** 0 g.

492. Seasoned Shrimp Skewers

Preparation time: 10 minutes.

Cooking time: 35 minutes.

Servings: 4

Ingredients:

- 1(½) pound fresh large shrimp, peeled, deveined and rinsed
- 2 tablespoons minced basil
- 2 teaspoons minced garlic
- 1/2 teaspoon sea salt
- 1/2 teaspoon ground black pepper
- 1/3 cup olive oil
- 2 tablespoons lemon juice

Directions:

1. Place basil, garlic, salt, black pepper and oil in a large bowl, whisk until well combined, then add shrimps and toss until well coated.
2. Fill the smoker tray with hickory wood chips and water pan with water and white wine halfway through, and place the dripping pan above the water pan.
3. Then open the top vent, shut with lid and use temperature settings to heat the smoker at 225°F.
4. Thread shrimps on wooden skewers, six shrimps on each skewer.
5. Place shrimp skewers on smoker rack, then shut with lid and set the timer to

smoke for 35 minutes or shrimps are opaque.

6. When done, drizzle lemon juice over shrimps and serve.

Nutrition:

Calories: 168 **Carbs:** 2g **Fat:** 11g **Protein:** 14g **Fiber:** 0g

493. Marinated Trout

Preparation time: 7 hours.

Cooking time: 3 hours.

Servings: 8

Ingredients:

- 4 pounds trout fillets
- 1/2 cup salt
- 1/2 cup brown sugar
- 2 quarts water

Directions:

1. Pour water in a large container with lid, add salt and sugar and stir until salt and sugar are dissolved completely.
2. Add trout, pour more water to submerge trout in brine and refrigerate for 4 to 8 hours, covering the container.
3. Then remove trout from brine, rinse well and pat dry with paper towels.
4. Place trout on a cooling rack, skin side down, and cool in the refrigerator for 2 hours or until it is dried.
5. Then remove trout from the refrigerator and bring to room temperature.
6. In the meantime, plug in the smoker, fill its tray with maple wood chips and water pan halfway through, and place the dripping pan above the water pan.
7. Then open the top vent, shut with lid and use temperature settings to heat the smoker at 160°F.
8. In the meantime, place trout on smoker rack, insert a meat thermometer, then shut with lid and set the timer to smoke for 2(½) to 3 hours or more until meat thermometer registers an internal temperature of 145°F.
9. Check vent of smoker every hour and add more wood chips and water to maintain temperature and smoke.
10. Serve straight away.

Nutrition:

Calories: 49 **Carbs:** 0g **Fat:** 1.2g **Protein:** 8.8g **Fiber:** 0g

494. Smoked Salmon Recipe

Preparation time: 15 minutes.

Cooking time: 4 hours.

Servings: 5

Ingredients:

- 1 to 1(½) pound whole salmon filet, skin and bones removed
- 1 shot (jigger) unflavored vodka or tequila
- ¼ cup kosher salt
- ¼ cup brown or raw turbinado sugar
- 2 teaspoon cracked black pepper
- 1 bunch of fresh dill, chopped
- ½ lemon, thinly sliced
- Alder wood chips

Directions:

1. Spot the entire salmon filet in a shallow preparing dish, ideally glass or earthenware. Pour the liquor over the filet.
2. Blend salt, pepper, and sugar together and pat everywhere throughout the salmon. Top with the lemon cuts. Spot the dill in addition and tenderly press down.
3. Spread the highest point of the salmon firmly with cling wrap, tucking it down into the dish. Spot another layer of the fold around the dish to seal it firmly.
4. Spot the salmon in the icebox medium-term for roughly 8 to 12 hours.

Nutrition:

Calories: 223 **Carbs:** 4g **Fat:** 19g **Protein:** 9g

495. Smoked Cajun Spiced Shrimp

Preparation time: 10 minutes.

Cooking time: 30 minutes.

Servings: 6

Ingredients:

- 2 pounds large or jumbo cleaned shrimp
- ¼ cup sea salt
- 3 teaspoon paprika
- 3 teaspoon cracked black pepper
- 2 teaspoon garlic powder
- 2 teaspoon onion powder
- 2 teaspoon dried thyme
- 2 teaspoon cayenne pepper (adjust to your liking)
- Butter
- 1 lemon of Juice
- Chopped fresh parsley

Directions:

1. Whenever solidified, defrost the shrimp. Strip and devein the shrimp if no longer effectively arranged. Pat the shrimp dry.

2. Join the dry ingredients in a field with a top. Shake until every one of the fixings is mixed properly.

3. Coat the foil field, or skillet, with both margarine and olive oil. Spot the shrimp inside the box. Spoon as a splendid a part of the dry rub over the shrimp, as required, to coat it. Any extra dry rub will shop exceptional at the off danger which you don't make use of everything. Hurl the shrimp to coat the 2 aspects. Set the box apart at the same time as you set up the smoker.

4. Set up the smoker by using adding timber chips to the plate and water to the bowl. Heat the smoker to somewhere inside the range of 225°F and 250°F. Open the top vent.

5. While the smoker is up to temperature, press 1/2 of the lemon, squeeze over the shrimp and Set the dish in the smoker. Cook for 15minutes and give the shrimp a mix. Cook for a further 15 minutes until the shrimp is misty crimson.

6. Take the dish and crush the rest of the lemon squeeze at the shrimp. Include the parsley and serve a platter with cornmeal and a plate of combined vegetables.

7. Two or three first-rate side dishes for Cajun-spiced shrimp are gooey cornmeal and a harsh veggies plate of blended veggies. You could make these at the same time as the shrimp is smoking. Teach the shrimp a lesson over the cornmeal.

Nutrition:

Calories: 133 **Carbs:** 0g **Fat:** 6g **Protein:** 18g

496. Smoked Greek-Style Shrimp

Preparation time: 10 minutes.

Cooking time: 30 minutes.

Servings: 6

Ingredients:

- 2 pounds large shrimp, cleaned
- 3 teaspoons good extra virgin olive oil
- 3 tablespoons butter, melted
- ½ fresh lemon o Juice
- 5 garlic cloves, minced
- 1 tablespoon dried oregano
- 1/3 cup flat-leaf parsley, chopped
- 1 teaspoon sea or coarse kosher salt
- 1(½) cups crumbled feta cheese
- EVOO for serving
- Fresh lemon wedges for serving

Directions:

1. Set up the smoker by adding wood chips to the plate and water to the bowl. Heat the smoker to somewhere in the range of 225°F and 250°F. Open the top vent.

2. Whenever solidified, defrost the shrimp. Strip and devein the shrimp if not officially arranged. Pat the shrimp dry.

3. In an enormous bowl, combine the EVOO, spread, and lemon juice. Permit cooling. Blend in the garlic, oregano, parsley, and salt. Add the shrimp and hurl to coat equitably. Spot the shrimp in the container or dish with the majority of the sauce.

4. At the point when the smoker is up to temperature, place the dish in the smoker. Cook for 15 minutes. Give the shrimp a mix. Include the feta cheddar top and smoke for an additional 15 minutes until the shrimp is an obscure pink and the cheddar is somewhat dissolved.

5. Take the dish out from the smoker and spot the shrimp and all the skillet squeezes on a serving platter or bowl. Shower with EVO and present with lemon wedges and side dishes.

Nutrition:

Calories: 379 **Carbs:** 43g **Fat:** 9g **Protein:** 32g

497. Smoked Ahi Tuna Steaks

Preparation time: 4 hours.

Cooking time: 2 hours, 15 minutes.

Servings: 5

Ingredients:

- 6 Ahi tuna steaks, approximately 1-inch thick and weighing approximately 4 to 6-ounces each
- 3 tablespoons kosher salt
- 3 tablespoons light brown sugar or raw turbinado sugar
- ¼ cup extra virgin olive oil (EVOO)
- Lemon pepper seasoning in a shaker jar
- 1 teaspoon ground garlic
- 12 thin slices of fresh lemon

Directions:

1. Coat the fish steaks with salt and sugar on all sides. Spot these in a fixed holder or baggie and refrigerate for 4 hours or medium-term.

2. Add water to the skillet in your smoker. Spot wood contributes to the plate. Heat the smoker to 190°F.

3. Take the fish steaks to a clean surface and wipe up a huge portion of the dry saline solution. Coat the two sides with EVOO, lemon pepper flavoring, and garlic powder.

4. Set the fish steaks legitimately on the smoker rack and put 2 cuts of lemon over every steak. Spot the rack back in the smoker. Smoke for 60 minutes.

5. At 60 minutes, check the interior temperature with a computerized thermometer. You are searching for 140 to 145°F. Keep on smoking until you accomplish that temperature. This will take roughly 60 to 105 minutes.

6. Take to a slicing board to rest for only a couple of minutes. Present with new lemon wedges, corn salsa, and cuts of ready avocado.

Nutrition:

Calories: 140 **Carbs:** 0g **Fat:** 2g **Protein:** 29g

498. Smoked Seasoned Shrimp Skewers

Preparation time: 15 minutes.

Cooking time: 35 minutes.

Servings: 4

Ingredients:

- 1(½) pounds large shrimp, approximately 25 to 30 count fresh or frozen (thawed)
- 1/3 cup good extra virgin olive oil (EVOO)
- 4 garlic cloves, minced
- 2 tablespoons minced fresh basil leaves
- ½ teaspoon sea salt
- ½ teaspoon cracked black pepper
- Dry white wine for smoking
- Lemon for serving

Directions:

1. Spot the EVOO, garlic, basil, salt, and pepper in a big bowl. Whisk collectively.
2. Clean the shrimp by using evacuating the shells and eliminating the intestinal tract from of the shrimp.
3. Add the shrimp to the bowl and hurl to coat with the dressing. Put in a safe spot at the same time as you drench your sticks and install your smoker.
4. Take a rack out from the smoker to place the sticks on. Set up your smoker by including timber chips to the plate and half of water + half white wine to the bowl.
5. Heat the smoker to 225°F. Open the pinnacle vent.
6. Set up to six shrimp on every stick via penetrating the top stop and the final component. They need to resemble the letter C mendacity degree.
7. Spot the shrimp sticks on the rack and placed the rack inside the smoker. Smoke for 35 minutes, or till the shrimp are surely murky and a crimson/white tone.

8. Crush some lemon over the shrimp before serving.

Nutrition:

Calories: 260 **Carbs:** 2g **Fat:** 17g **Protein:** 25g

499. Smoked Whole Snapper With Chimichurri Recipe

Preparation time: 15 minutes.

Cooking time: 3 hours.

Servings: 4

Ingredients:

- 4 to 5 pounds whole snapper, scaled and gutted
- 2 tablespoons butter at room temperature
- 2 tablespoons extra virgin olive oil (EVOO)
- ½ fennel bulb + fronds, woody stalks and core discarded, sliced thinly
- ½ small white or yellow onion, sliced thinly
- 1 whole lemon sliced thinly into rounds
- 1(½) teaspoon sea salt
- 1 teaspoon cracked black pepper

Directions:

1. In a medium container, integrate the margarine and EVOO. Include salt and pepper. Cut the fennel, onion, and lemon. Save a portion of the verdant fennel fronds.
2. Set up your smoker using adding timber chips to the plate and water to the bowl. Heat the smoker to 225°F. Open the pinnacle vent.
3. Liberally coat within and outside of the fish with the margarine and EVOO + salt and pepper. Stuff the hole with the fennel cuts, onion, fennel fronds, and 1/2 of the lemon cuts. Spot the fish on 2 sheets of aluminum foil. Turn up the rims of the foil to maintain the dampness.
4. Set the fish at the foil pontoon within the smoker on the center rack. Cook for three to 4 hours until the internal temperature arrives at 145°F to a 150°F. This is roughly 45 minutes for every pound of fish.
5. Serve the fish complete or spot it on a reducing board and filet it, taking the backbone and any stick bones. Serve it on a platter with chimichurri sauce and pureed potatoes

Nutrition:

Calories: 127 **Carbs:** 0g **Fat:** 3g **Protein:** 25g

500. Brined & Smoked Trout Fillets

Preparation time: 7 hours.

Cooking time: 3 hours.

Servings: 8

Ingredients:

- 2 quarts filtered water
- ½ cup of kosher salt
- ½ cup brown sugar

Directions:

1. Take the stick bones from the trout fillets.
2. Add the salt, sugar, and water in a huge compartment, kind of 1 to 2-gallon length. Blend nicely until the salt and sugar have disintegrated. Submerge the trout fillets in the saline solution and unfold the holder. Refrigerate for three and as long as 8 hours.
3. Take the trout from the saltwater and wash below the water. Expel the overabundance dampness from the filets with smooth paper towels. Spot the trout filets, skin aspect down, on a cooling rack this is fitted inside a sheet dish. Spot the skillet inside the fridge for 2 hours and as much as medium-term.
4. Take 2 racks from the smoker and see the filets at the racks pores and skin facet down. Enable the fish to relax while you set up the smoker. Fill the water bowl half manner. Spot wood contributes to the plate. Birch, maple, or very well feature admirably. Open the pinnacle vent. Turn the smoker on and heat to one 160°F.
5. Set the racks with fish within the smoker. Cook the trout for 2(1/2) to 3 hours or until the fish arrives at an inner temperature of one hundred 45°F. Renew the wood chips and water as required, probable like clockwork.
6. Serve the smoked fish with toasted loaf cuts and an invigorating, inexperienced plate of blended vegetables. You can acquire canapé crostinis by means of placing a bit serving of blended veggies at the toasts and setting a few cuts of fish on the pinnacle.

Nutrition:

Calories: 150 **Carbs:** 0g **Fat:** 6g **Protein:** 22g

501. Tender Salmon Fillet

Preparation time: 10 minutes.

Cooking time: 60 minutes.

Servings: 4

Ingredients:

- 1(½) pounds salmon fillet
- 1 teaspoon black pepper
- 1 tablespoon Dijon mustard
- 1 teaspoon kosher salt

Directions:

1. Heat the smoker to 225°F/107°C using the cherry wood chips.
2. Season salmon with mustard, pepper, and salt.
3. Place salmon in the smoker and cook until the internal temperature reaches to 145°F.
4. Serve and enjoy

Nutrition:

Calories: 229 **Total fat:** 10.7g **Saturated fat:** 1.5g **Protein:** 33.2g **Carbs:** 0.6g **Fiber:** 0.3g **Sugar:** 0g

502. Healthy & Delicious Smoke Salmon

Preparation time: 10 minutes.

Cooking time: 2 hours.

Servings: 6

Ingredients:

- 1 salmon fillet
- 1 teaspoon chili powder
- 1 tablespoon garlic powder
- ¼ cup sugar
- ½ cup brown sugar
- ½ teaspoon paprika
- 2 tablespoon salt

Directions:

1. Mix together chili powder, garlic powder, sugar, brown sugar, paprika, and salt.
2. Rub chili powder mixture over the salmon fillet.
3. Heat the smoker to 225°F/107°C using the cherry wood chips.
4. Place salmon in the smoker and cook for 2 hours.
5. Serve and enjoy

Nutrition: Calories: 123 **Total fat:** 1.9g **Saturated fat:** 0.3g **Protein:** 6.1g **Carbs:** 21.5g **Fiber:** 0.4g **Sugar:** 20.4g

503. Crab Legs

Preparation time: 10 minutes.

Cooking time: 30 minutes.

Servings: 4

Ingredients:

- 6 pounds crab legs
- 2 tablespoon sweet rub
- 2 lemon juice
- 6 garlic cloves, minced
- ¼ cup fresh basil, chopped
- 1 cup butter, melted

Directions:

1. Heat the smoker to 225°F/107°C using the cherry wood chips.
2. Place crab legs in a foil pan.
3. Mix together butter, basil, garlic, lemon juice, and sweet rub and pour over crab legs.
4. Place in smoker and cook for 30 minutes.
5. Serve and enjoy

Nutrition:

Calories: 596 **Total fat:** 11g **Saturated fat:** 28g **Protein:** 32g **Carbs:** 11g **Fiber:** 2g **Sugar:** 1g

504. Flavorful Smoked Trout

Preparation time: 15 minutes.

Cooking time: 2 hours.

Servings: 4

Ingredients:

- 1 pound trout
- 1 teaspoon black pepper
- 1 teaspoon fennel seeds
- 1 teaspoon mustard seeds
- 1 tablespoon olive oil
- 1 teaspoon kosher salt

Directions:

1. Heat the smoker to 225°F/107°C using the applewood chips.
2. Coarsely grind the spices.
3. Brush fish with oil and rub with ground spices.
4. Place fish into the smoker and cook for 1(1/2) to 2 hours or until internal temperature reaches 125°F.
5. Serve and enjoy.

Nutrition:

Calories: 252 **Total fat:** 13.4g **Protein:** 30.6g **Carbs:** 0.9g **Saturated fat:** 2.2g **Fiber:** 0.5g **Sugar:** 0.1g

505. Garlic Herb Shrimp

Preparation time: 10 minutes.

Cooking time: 30 minutes.

Servings: 4

Ingredients:

- 1 pound fresh shrimp, peeled and deveined
- 1 teaspoon garlic powder
- 1 tablespoon dried basil
- 1 tablespoon dried oregano
- 2 tablespoon olive oil
- 1 teaspoon salt

Directions:

1. In a bowl, toss shrimp with oil, garlic powder, basil, oregano, and salt.
2. Transfer shrimp in foil pan.
3. Heat the smoker to 225°F/107°C using the alder wood chips.
4. Place shrimp into the smoker and cook for 30 minutes.
5. Serve and enjoy

Nutrition:

Calories: 201 **Total Fat:** 9.1g
Saturated Fat: 1.6g **Protein:** 26.1g
Carbs: 3g **Fiber:** 0.6g **Sugar:** 0.2g

506. Smoked Scallops

Preparation time: 10 minutes.

Cooking time: 20 minutes.

Servings: 6

Ingredients:

- 2 pounds scallops
- Pepper
- Salt

Directions:

1. Heat the smoker to 225°F/107°C using the cherry wood chips.
2. Season scallops using the pepper and salt and place them into the smoker and cook for 20 minutes.
3. Serve and enjoy.

Nutrition:

Calories: 133 **Total Fat:** 1.2g
Saturated Fat: 0.1g **Protein:** 25.4g
Carbs: 3.6g **Fiber:** 0g **Sugar:** 0g

507. Garlic Prawns

Preparation time: 10 minutes.

Cooking time: 30 minutes.

Servings: 4

Ingredients:

- 24 large prawns
- 4 garlic cloves, minced

For rub:

- 1 teaspoon onion powder
- 1 teaspoon garlic powder
- 2 teaspoon paprika
- ¼ cup brown sugar
- 1 tablespoon kosher salt

Directions

1. Mix all the rub ingredients.
2. Add garlic and prawns into the bowl and toss well to coat. Place in the refrigerator for 1-2 hours.
3. Heat the smoker to 200°F/93°C using the hickory wood chips.
4. Thread marinated prawns onto the soaked wooden skewers, then place into the smoker.
5. Cook for 30 minutes and serve.

Nutrition: Calories: 86 **Total Fat:** 0.7g **Saturated Fat:** 0.2g **Protein:** 8.1g **Carbs:** 12g **Fiber:** 0.6g **Sugar:** 9.3g

508. Herb Marinade Fish Fillets

Preparation time: 10 minutes.

Cooking time: 2 hours, 30 minutes.

Servings: 4

Ingredients:

- 4 catfish fillets

For marinade:

- 3 tablespoon sugar
- 1 teaspoon cayenne pepper
- 1 teaspoon black pepper
- 1 tablespoon basil
- 1 tablespoon thyme
- 2 tablespoons oregano
- 2 garlic cloves, minced
- 1 lemon juice
- ½ cup red wine vinegar
- 1 cup olive oil
- 1 tablespoon salt

Directions:

1. Mix all the marinade ingredients.
2. Place fish fillets in marinade and coat well. Place in the refrigerator for 1 hour.
3. Heat the smoker to 225°F using the alder wood chips.

4. Place marinated fish fillets in the smoker and cook for 2 hours 30 minutes.
5. Serve and enjoy.

Nutrition:

Calories: 702 **Total Fat:** 62.9g **Saturated Fat:** 9.5g **Protein:** 25.4g **Carbs:** 12.3g **Fiber:** 1.5g; **Sugar:** 9.3g

509. Smoked Tilapia

Preparation time: 10 minutes.

Cooking time: 2 hours.

Servings: 4

Ingredients:

- 4 tilapia fillets
- 1 cup brown sugar
- 2 tablespoon fish sauce
- ¼ cup olive oil
- 3 garlic cloves
- 1 tablespoon ginger root, peeled and chopped
- 2 grapefruits, peeled and quartered
- ½ teaspoon salt

Directions:

1. Add grapefruit, brown sugar, fish sauce, garlic, ginger, and salt into the blender and blend until smooth.
2. In a large bowl, place tilapia fillets. Pour blended mixture over tilapia fillets and coat well. Put in the refrigerator for 2 hours.
3. Heat the smoker to 275°F/135°C using the alder wood chips.
4. Place tilapia fillets in the smoker and cook for 2 hours.
5. Serve and enjoy.

Nutrition:

Calories: 414 **Total fat:** 14.7g **Saturated fat:** 2.8g **Protein:** 33.1g **Carbs:** 42g **Fiber:** 0.8g **Sugar:** 40g

510. Teriyaki Fish Fillets

Preparation time: 10 minutes.

Cooking time: 2 hours.

Servings: 4

Ingredients:

- 4 tilapia fillets
- 1 tablespoon sriracha sauce
- 2/3 cup honey
- 1 cup teriyaki sauce

Directions:

1. In a shallow dish, mix together teriyaki sauce, sriracha, and honey until well blended.
2. Add fish fillets to the dish and coat well with marinade. Place in the refrigerator for 2 hours.
3. Heat the smoker to 275°F/135°C using the alder wood chips.
4. Place marinated fish fillets in the smoker and cook for 2 hours.

Nutrition:

Calories: 401 **Total fat:** 4.5g
Saturated fat: 1.4g **Protein:** 36.4g
Carbs: 58g **Fiber:** 0.2g; **Sugar:** 56.8g

511. Flavorful Smoked Tilapia

Preparation time: 10 minutes.

Cooking time: 2 hours.

Servings: 6

Ingredients:

- 6 tilapia fillets
- ½ teaspoon lemon pepper
- ½ teaspoon garlic powder
- 2 tablespoon fresh lemon juice
- 3 tablespoon olive oil
- 1 teaspoon kosher salt

Directions:

1. Heat the smoker to 225°F/107°C using maple wood chips.
2. Mix the oil, lemon pepper, garlic powder, lemon juice, and salt.
3. Brush oil mixture onto both sides of the fish fillets.
4. Place fish fillets in the smoker and cook for 1(½) to 2 hours.
5. Serve and enjoy.

Nutrition:

Calories: 202 **Total fat:** 9.1g
Saturated fat: 2g **Protein:** 32.1g
Carbs: 0.4g **Fiber:** 0.1g **Sugar:** 32.1g

512. Tasty Buttery Shrimp

Preparation time: 10 minutes.

Cooking time: 25 minutes.

Servings: 4

Ingredients:

- 15 large shrimp
- 2 tablespoon fresh lemon juice
- 1 tablespoon Italian seasoning
- 2 rosemary spring
- 2 garlic cloves, minced
- ½ cup butter, melted

Directions:

1. Heat the smoker to 275°F/135°C using the pecan wood chips.
2. Add all the ingredients into the large bowl and toss well.
3. Transfer shrimp mixture into the foil pan. Place in smoker and cook for 20–25 minutes.
4. Serve and enjoy.

Nutrition:

Calories: 243 **Total fat:** 24.5g **Saturated fat:** 14.9g **Protein:** 5.1g **Carbs:** 1.4g **Fiber:** 0.1g **Sugar:** 0.5g

513. Chipotle Shrimp

Preparation time: 10 minutes.
Cooking time: 30 minutes.
Servings: 4

Ingredients:

- 1(½) pounds jumbo shrimp, peeled and deveined
- 4 tablespoon butter, melted
- 2 tablespoon fresh lime juice
- 2 tablespoon olive oil
- 2 teaspoon chipotle chili, chopped
- 2 garlic cloves, minced
- 2 green onion, minced
- 3 tablespoon fresh parsley, chopped
- Pepper
- Salt

Directions:

1. Heat the smoker to 250°F/135°C using pecan wood chips.
2. Add all the ingredients into the large bowl and toss well. Transfer shrimp mixture into the foil pan and place into the smoker.
3. Cook shrimp for 30 minutes.
4. Serve and enjoy

Nutrition: Calories: 289 **Total fat:** 18.6g **Saturated fat:** 8.3g **Protein:** 30.8g **Carbs:** 1.3g **Fiber:** 0.3g **Sugar:** 3.3g

514. Smoked Halibut

Preparation time: 10 minutes.

Cooking time: 45 minutes.

Servings: 6

Ingredients:

- 4 pounds fresh halibut fillets
- 2 garlic cloves, minced
- 6 tablespoon butter, melted
- Pepper
- Salt

Directions:

1. In a shallow dish, mix together butter, garlic, pepper, and salt.
2. Place fish fillets in the dish and coat well. Place in the refrigerator for 1 hour.
3. Heat the smoker to 225°F/107°C using alder wood chips.
4. Place marinated fish fillets in the smoker and cook for 45 minutes or until the internal temperature reaches 140°F.
5. Serve and enjoy.

Nutrition:

Calories: 219 **Total fat:** 13.3g **Saturated fat:** 7.3g **Protein:** 20.9g **Carbs:** 0.9g **Fiber:** 0g **Sugar:** 0.2g

515. Smoked Tuna

Preparation time: 10 minutes.

Cooking time: 7 hours.

Servings: 4

Ingredients:

- 4 tuna steaks
- 1-gallon water
- 1 cup honey
- ¼ teaspoon garlic, chopped
- 1(1/8) cup sugar
- 1 teaspoon pepper
- 3/8 cup salt

Directions:

1. Add all the ingredients except tuna steaks into the pot and stir well.
2. Add tuna steaks. Cover and place in the refrigerator overnight.
3. Heat the smoker to 140°F/60°C using the applewood chips.
4. Place marinated tuna steaks in the smoker and cook for 7 hours.
5. Serve and enjoy.

Nutrition:

Calories: 700 **Total Fat:** 19.3g **Saturated Fat:** 4.5g **Protein:** 8.1g **Carbs:** 133.2g **Fiber:** 0.9g **Sugar:** 125.9g

516. Smoked Snapper Fillet

Preparation time: 10 minutes.

Cooking time: 60 minutes.

Servings: 6

Ingredients:

- 1(½) pounds red snapper fillets
- 1 tablespoon garlic, granulated
- 3 tablespoon brown sugar
- 2 quarts water
- 1 tablespoon maple syrup
- 1 tablespoon black pepper
- 2 tablespoon olive oil
- Kosher salt

Directions:

1. **For the brine:** Add water, salt, garlic, and 2 tablespoon brown sugar in a pot and stir well.
2. Add fish fillets in brine and set aside for 2 hours.
3. Mix together olive oil, 1 tablespoon brown sugar, and pepper and rub over fish fillets.
4. Heat the smoker to 225°F/107°C using the applewood chips.
5. Place fish fillets in the smoker and cook for 60 minutes.
6. Brush fish fillets with maple syrup and serve.

Nutrition:

Calories: 216 **Total fat:** 6.7g **Saturated fat:** 1.1g **Protein:** 30g **Carbs:** 7.8g **Fiber:** 0.3g **Sugar:** 6.4g

517. Delicious Trout Fillets

Preparation time: 10 minutes.

Cooking time: 3 hours.

Servings: 4

Ingredients:

- 4 trout fillets
- 1 teaspoon lemon pepper
- ¼ cup teriyaki sauce
- ¼ cup soy sauce
- 2 cups of water
- ½ tablespoon salt

Directions:

1. In a bowl, mix together water, soy sauce, teriyaki sauce, and salt.
2. Place fish fillets in a bowl. Cover and place in the refrigerator. Store for overnight.
3. Heat the smoker to 225°F/107°C using the alder wood chips.
4. Place marinated fish fillets in the smoker and cook for 3 hours.
5. Serve and enjoy.

RUBS, SAUCES, MARINADES, AND GLAZES

518. BBQ Sauce for Chicken

Preparation time: 5 minutes.

Cooking time: 5–8 minutes.

Servings: 1(½)–2 cups

Ingredients:

- 1 cup ketchup
- ¼ cup brown sugar
- ¼ cup apple cider vinegar
- 2 tablespoons smoked paprika
- 1 tablespoon extra-virgin olive oil
- 1 tablespoon chili powder
- 2 teaspoons garlic powder
- ½ teaspoon salt

Directions:

1. Combine the ingredients in a saucepan.
2. Place on the stove and heat on medium.
3. Light boil and simmer for 5–8 minutes to blend the flavors.
4. Cool before storing in the fridge, covered, for no longer than a week.

Nutrition:

Total calories: 126 **Protein:** 1.1g **Carbs:** 24.1g **Fat:** 3.7g **Fiber:** 0.2g

519. BBQ Sauce for Beef

Preparation time: 5 minutes.

Cooking time: 35 minutes.

Servings: 2(½) cups.

Ingredients:

- 3 tablespoons extra-virgin olive oil
- 1 cup diced onion
- 3 minced garlic cloves
- 1(½) cups tomato sauce
- ¾ cup pure maple syrup
- ½ cup beef stock
- ½ cup apple cider vinegar
- 3 tablespoons chipotle chili powder
- 2 tablespoons Worcestershire sauce
- ½ teaspoon salt

Directions:

1. Heat a saucepan and add oil.
2. When hot, cook onion and garlic until softened.
3. Add the rest of the ingredients and bring to a boil.
4. Reduce heat and simmer for 30 minutes.
5. Cool before using, and store extra in a covered container in the fridge for up to a week.

Nutrition:

Total calories: 235 **Protein:** 1.5g **Carbs:** 39.2g **Fat:** 8.7g **Fiber:** 1.6g

520. Mustard Sauce for Pork

Preparation time: 5 minutes.

Cooking time: 27 minutes.

Servings: 4 cups

Ingredients:

- 1 cup yellow mustard
- ½ cup balsamic vinegar
- 1/3 cup brown sugar
- 2 tablespoons butter
- 1 tablespoon fresh lemon juice
- 2 teaspoons Worcestershire sauce
- ½ teaspoon chili powder

Directions:

1. Put all the ingredients in a saucepan.
2. Heat until simmering, stirring.
3. Reduce the heat to low after 1–2 minutes of a light boil, then simmer for 25 minutes.
4. Cool to room temperature before storing for 3–4 days in the fridge, or use right away!

Nutrition:

Total calories: 148 **Protein:** 2.8g **Carbs:** 16g **Fat:** 8.3g **Fiber:** 2.1g

521. Spicy-Citrus Cocktail Sauce

Preparation time: 5 minutes.

Cooking time: 6 minutes.

Servings: 1 ¼ cups.

Ingredients:

- 1 cup ketchup
- ¼ cup orange juice
- 1 minced chipotle chile, jarred in adobo
- 2 tablespoons diced onion
- 1 tablespoon Worcestershire sauce
- 2 teaspoons adobo sauce
- 2 teaspoons dry cilantro
- 1 teaspoon orange zest
- Pinch of red pepper flakes, or more

Directions:

1. Prepare the ingredients in a saucepan, then heat on medium.
2. When simmering, reduce heat, so the sauce is barely bubbling.
3. Cook for 5–6 minutes to blend flavors.
4. Let it rest at room temperature before storing in the fridge.
5. You can serve cocktail sauce cold or warm.

Nutrition: Total calories: 58 **Protein:** 0.9g **Carbs:** 14.6g **Fat:** 0.2g **Fiber:** 0.2g

522. Brisket Dry Rub

Preparation time: 5 minutes.

Cooking time: Overnight.

Servings: 1

Ingredients:

- ½ cup onion powder
- ¼ cup salt
- ¼ cup sweet paprika
- ¼ cup brown sugar
- ½ cup black pepper
- ½ cup garlic powder

Directions:

1. Mix all the seasonings together.
2. Rub on your brisket!
3. For best results, let the seasoned brisket sit in the fridge, covered, overnight.

Nutrition:

Total calories: 23 **Protein:** 0g **Carbs:** 5.9g **Fat:** 0g **Fiber:** 0g

523. Pork Dry Rub

Preparation time: 5 minutes.

Cooking time: Overnight.

Servings: 3–4

Ingredients:

- 1 cup brown sugar
- 3 tablespoons smoked paprika
- 1 tablespoon chili powder
- 2 teaspoons garlic powder
- 2 teaspoons black pepper
- 2 teaspoons salt
- 2 teaspoons Italian seasoning
- ½ teaspoon onion powder
- ½ teaspoon ground mustard

Directions:

1. Mix all the ingredients together.
2. Rub on your pork.
3. Store in the fridge overnight before cooking ribs.

Nutrition:

Total calories: 69 **Protein:** 0g **Carbs:** 17.8g **Fat:** 0g **Fiber:** 0g

524. Dry Rub for Salmon

Preparation time: 5 minutes.

Cooking time: 24 hours.

Servings: 3(¼) cups.

Ingredients:

- 2 cups brown sugar
- 1 cup kosher salt
- 1 tablespoon black pepper
- 1 tablespoon celery salt
- 1 tablespoon onion powder
- 1 tablespoon garlic powder

Directions:

1. Mix all the ingredients together.
2. To use, rub all over your salmon
3. Cover in plastic wrap, then store in the fridge for 24 hours.
4. Rinse before smoking.

Nutrition:

Energy: 583kcal **Carbohydrate:** 149.89g **Calcium**: 171mg **Magnesium:** 24mg **Phosphorus:** 33mg **Iron:** 1.9mg **Potassium:** 322mg

525. Fruit-Spiced Brine for Pork

Preparation time: 5 minutes.

Cooking time: 15 minutes.

Servings: 2

Ingredients:

- 2 quarts apple juice
- 2 quarts orange juice
- 2 cups salt
- ½ cup brown sugar
- 10 whole cloves
- 1 tablespoon ground nutmeg
- 1-gallon cold water

Directions:

1. Pour apple cider and orange juice into a big pot.
2. Heat on medium-high, adding salt, brown sugar, cloves, and nutmeg.
3. Simmer for 15 minutes, until the sugar and salt have dissolved.
4. Take the pot off the burner and cool for at least 40–60 minutes.
5. Pour the cold water.
6. To brine, pour over pork and store in the fridge, covered, for 1 hour per pound of meat.
7. Rinse meat before cooking it.

526. 6-Ingredient Turkey Brine

Preparation time: 5 minutes.

Cooking time: 0 minutes.

Servings: 12

Ingredients:

- 2 gallons water
- 1 ½ cups canning salt
- 1/3 cup brown sugar
- ¼ cup Worcestershire sauce
- 3 tablespoons minced garlic
- 1 tablespoon black pepper

Directions:

1. Mix all the ingredients in a container big enough for your turkey.
2. Soak the turkey in the brine, covered, in the fridge for 1–2 days.
3. Rinse meat before smoking.

Nutrition:

Total fat: 0g **Saturated fat:** 0g **Cholesterol:** 0mg **Sodium:** 11596mg **Total carbohydrate:** 6g **Dietary fiber:** 0.2g **Total sugars:** 4.9g **Protein:** 0.2g

527. Brine for Fish

Preparation time: 5 minutes.

Cooking time: 0 minutes.

Servings: 8–10

Ingredients:

- 8 cups water
- 2 cups soy sauce
- 1(½) cups brown sugar
- ½ cup kosher salt
- 1(½) tablespoons ground garlic

Directions:

1. Mix all the ingredients.
2. Pour into a bag with the fish, so it's all covered.
3. Marinate for at least 8 hours.
4. Pat fish dry before smoking.

Nutrition:

Total calories: 110 **Protein:** 3.2g **Carbs:** 25.2g **Fat:** 0g **Fiber:** 0.4g

528. Ranch Mix

Preparation time: 5 minutes.

Cooking time: 0 minutes.

Servings: 16

Ingredients:

- 2 teaspoons pepper
- 0.5 cup powdered milk
- 2 teaspoons salt
- ½ teaspoon paprika
- 1 tablespoon parsley flakes, dried-out
- 1 tablespoon onion powder
- 2 tablespoon onion, dried and minced
- 2 teaspoons garlic powder

Directions:

1. Mix the ingredients well, then store in a dry and cool place for a maximum of 6 months.

Nutrition:

Protein: 1g **Sugar:** 1.2g **Fiber:** 0.3g **Carbohydrates:** 2.5g **Sodium:** 303.9mg **Cholesterol:** 0.4mg **Total fat:** 0.1g **Total calories:** 14

529. Easy and Tasty Blend

Preparation time: 5 minutes

Cooking time: 0 minutes

Servings: 84 teaspoons

Ingredients:

- ¼ cup black pepper, grounded
- 1 cup salt, kosher
- ¼ cup garlic powder
- ¼ cup onion powder

Directions:

1. Mix all the ingredients very well. You can keep it inside any sealed container located in any place that's dry as well as cool for a maximum of 6 months.

Nutrition:

Protein: 0.1g **Fiber:** 0.2g **Carbohydrates:** 0.8g **Sodium:** 1347.9mg

530. Mopping Sauce, Carolina Style

Preparation time: 5 minutes.

Cooking time: 0 minutes.

Servings: 8

Ingredients:

- 0.25 teaspoon black pepper, ground
- 0.5 teaspoon salt
- 1 teaspoon mustard, dry
- 2 tablespoons brown sugar, packed
- 1 teaspoon onion powder
- 1 tablespoon hot sauce
- 1 tablespoon red pepper flakes
- 1 cup cider vinegar
- 1 teaspoon garlic powder
- 1 cup white vinegar, distilled

Directions:

1. Mix all the ingredients altogether. Keep them inside an airtight container for a maximum of one month.

Nutrition:

Protein: 0.3g **Sugar:** 2.7g **Fiber:** 0.4g **Carbohydrates:** 3.9g **Sodium:** 184.6mg **Monounsaturated Fats:** 0.1g **Polyunsaturated Fats:** 0.1g **Total fats:** 0.2g **Total calories:** 28

531. Simple Barbecue Sauce

Preparation time: 5 minutes.
Cooking time: 0 minutes.
Servings: 8

Ingredients:

- 0.25 cup cider vinegar
- 15-ounce can tomato sauce, organic
- 1/2 teaspoon black pepper, ground
- 1/2 teaspoon salt
- 1 tablespoon garlic powder
- 1 teaspoon hot sauce
- 1 tablespoon mustard, dry
- 1 tablespoon onion powder
- 1 teaspoon paprika, smoked
- 1 tablespoon paprika, sweet
- 2 tablespoon lemon juice, fresh

Directions:

1. In a small pan, put the tomato sauce in and then add all the rest of the ingredients. Whisk them all well to coming.
2. Simmer the sauce for 15 minutes until it is thickened.

Nutrition: Protein: 2.6g **Sugar:** 36.2g **Fiber:** 3.9g **Carbohydrates:** 126.6g **Sodium:** 7365mg **Polyunsaturated fats:** 0.5g **Monounsaturated fats:** 0.1g **Saturated fats:** 0.2g **Total fats:** 0.9g **Total calories:** 116g

532. Barbecue Sauce, Jamaican Style

Preparation time: 5 minutes.

Cooking time: 0 minutes.

Servings: 3

Ingredients:

- ¼ teaspoon allspice
- 2 tablespoon Cider vinegar
- ¼ teaspoon nutmeg, ground
- ¼ teaspoon cinnamon, ground
- 1 teaspoon salt
- 1 teaspoon dry mustard, ground
- 2 tablespoon Worcestershire sauce
- 2 tablespoon sugar, dark brown
- 0.5 cup water
- 2 tablespoon lime juice
- 1 cup ketchup
- 1 Scotch Bonnet pepper, whole
- 0.5 cup onion, chopped
- 1/2 teaspoon pepper
- 1 tablespoon olive oil

Directions:

1. Grease your saucepan by using oil with the temperature in the meter, until it's hot.
2. Then, the sauce your pepper as well as your onion until they are soft.
3. Then, add in your remaining raw materials.
4. Heat your mixture to its boiling point, and let it settle down until you get the thickness you want. Take out the pepper and throw it away.
5. Take the sauce off the heat and let it cool.
6. Place the sauce in the refrigerator in a mason jar for a maximum of 2 weeks.

Nutrition:

Protein: 0.2g **Sugar:** 16.1g **Fiber:** 0.3g **Carbohydrates:** 20.6g **Sodium:** 872.7mg **Polyunsaturated fats:** 0.3g **Monounsaturated fats:** 1.7g **Saturated fats:** 0.4g **Total fats:** 2.4g **Total calories:** 103

533. Best Smoker Sauce

Preparation time: 5 minutes.

Cooking time: 0 minutes.

Servings: 32

Ingredients:

- ½ cup soy sauce
- 1 tablespoon ginger, ground
- Salt, to taste
- 8-ounces of your favorite salsa
- 18-ounce of your favorite BBQ sauce
- 1 tablespoon crushed garlic
- 3/4 cup brown sugar, packed
- 1 tablespoon black pepper, grounded

Directions:

1. Mix all the ingredients together. Store in the fried until you're ready to use.

Nutrition:

Protein: 0.4g **Carbohydrates:** 11.5g **Sodium:** 424mg **Fat:** 0.1g **Total calories:** 47

534. Garlic and Honey Sauce

Preparation time: 10 minutes.

Cooking time: 15 minutes.

Servings: 16

Ingredients:

- 0.25 cup water
- 1 teaspoon ginger root, fresh
- 0.25 cup soy sauce
- 0.5 cup honey
- 0.5 cup brown sugar
- 2 cup chicken broth
- 6 garlic cloves, crushed

Directions:

1. Mix all the ingredient and pour into a pot. Bring it to a boil and let it simmer until all the ingredients are combined for about 5 minutes.

Nutrition:

Protein: 0.5g **Carbohydrates:** 16.8g **Sodium:** 348mg **Fat:** 0.1g **Total calories:** 66

535. Carne Asada Marinade

Preparation time: 15 minutes.

Cooking time: 2 hours.

Servings: 5

Ingredients:

- 1 cloves garlic, chopped
- 1 teaspoon lemon juice
- 1/2 cup extra virgin olive oil
- 1/2 teaspoon salt
- 1/2 teaspoon pepper

Directions:

1. Mix all the ingredients in a bowl.
2. Pour the beef into the bowl and allow to marinate for 2–3hours before grilling.

Nutrition:

Calories: 465kcal **Carbs:** 26g **Fat:** 15g **Protein:** 28g

536. Grapefruit Juice Marinade

Preparation time: 10 minutes.

Cooking time: 1 hour.

Servings: 3

Ingredients:

- 1/2 reduced-sodium soy sauce
- 1 cups grapefruit juice, unsweetened
- 1–1/2 pound chicken, bone and skin removed
- 1/4 brown sugar

Directions:

1. Thoroughly mix all the ingredients in a large bowl.
2. Add the chicken and allow it to marinate for 2–3 hours before grilling.

Nutrition:

Calories: 489kcal **Carbs:** 21.3 **Fat:** 12g **Protein:** 24g

537. Steak Marinade

Preparation time: 5 minutes.

Cooking time: 10 minutes.

Servings: 2

Ingredients:

- 1 tablespoon Worcestershire sauce
- 1 tablespoon red wine vinegar
- 1/2 cup barbeque sauce
- 1 tablespoon soy sauce
- 1/4 cup steak sauce
- 1 clove garlic, minced
- 1 teaspoon mustard
- Pepper and salt to taste

Directions:

1. Mix all the ingredients thoroughly.
2. Use immediately or keep refrigerated.

Nutrition:

Calories: 303kcal **Carbs:** 42g **Fat:** 10g
Protein: 2.4g

RECIPE INDEX

Made in the USA
Coppell, TX
07 November 2021